X-RAY AND NEUTRON STRUCTURE ANALYSIS IN MATERIALS SCIENCE

X-RAY AND NEUTRON STRUCTURE ANALYSIS IN MATERIALS SCIENCE

Edited by

J. Hašek

Institute of Macromolecular Chemistry
Czechoslovak Academy of Sciences
Prague, Czechoslovakia

Plenum Press • New York and London

Library of Congress Cataloging in Publication Data

International Conference on Advanced Methods in X-Ray and Neutron Structure Analysis
of Materials (1987: Karlovy Vary, Czechoslovakia)
    X-Ray and neutron structure analysis in materials science / edited by J. Hašek.
       p.   cm.
    "Proceedings of an International Conference on Advanced Methods in X-Ray and
Neutron Structure Analysis of Materials, held in Karlovy Vary, Czechoslovakia, October
5–9, 1987" — T.p. verso.
    Includes bibliographies and index.
    ISBN-13:978-1-4612-8072-9   e-ISBN-13:978-1-4613-0767-9
    DOI: 10.1007/978-1-4613-0767-9

    1. X-ray crystallography — Congresses. 2. X-ray — Scattering — Congresses. 3. Neutrons —
Scattering — Congresses. I. Hašek, J. (Jindřich) II. Title.
QD945.I52 1987   88-33365
620.1'1299 — dc19   CIP

Proceedings of an International Conference on Advanced Methods in
X-Ray and Neutron Structure Analysis of Materials,
held in Karlovy Vary, Czechoslovakia, October 5–9, 1987

© 1989 Plenum Press, New York
Softcover reprint of the hardcover 1st edition 1989

A Division of Plenum Publishing Corporation
233 Spring Street, New York, N.Y. 10013

During the last few decades, crystallography has become a wide and economically important field of science with many interesting applications in materials research, in different branches of physics, chemistry, geology, pharmacology, biochemistry, electronics, in many technological processes, machinery, heavy industry, etc. Twenty Nobel prizes awarded for achievements belonging to this field only underline its distinction. Crystallography has become a commonly used term, but - like a whale - it is much easier to recognize than to describe because of an extreme diversity of subjects involved which range from highly sophisticated theories to the development of routine technological processes or testing of materials in production. It is apparent that only some aspects of selected topics could be included on a single occasion.

The conference "ADVANCED METHODS IN X-RAY AND NEUTRON STRUCTURE ANALYSIS OF MATERIALS" held in Karlovy Vary (Czechoslovakia) on October 5-9, 1987, was intended to cover the most important crystallographic aspects of materials science. The conference was attended by 250 people from 16 countries (Belgium, Bulgaria, China, Czechoslovakia, Finland, France, FRG, GDR, Hungary, Italy, The Netherlands, Poland, Sweden, USA, USSR and Yugoslavia). To advance the cooperation of people from different fields of structure analysis, no strict requirements were laid on the topics presented in poster communications, however the choice of invited lectures and oral contributions was aimed to promote the following topics:
- phase analysis, identification and structure analysis of powder materials
- structure analysis of amorphous materials and polymers
- real structure of crystalline materials
- thin layers and surface coatings (single and polycrystalline)
- difficult problems in crystal structure determination (modulated and polytype structures)
- data acquisition and processing.

This volume consists of 53 papers selected from 160 active contributions presented at the conference.

The principal subject of the first section is the identification of materials and quantitative phase analysis of powder materials. Four papers (by J. Fiala, B. L. Davis, A. Griger, S. Popovič and B. Gržeta) provide a sufficient theoretical background and a comprehensive review of the newest development in the methodology of quantitative analysis of a composition of mixtures. The following nine contributions add a number of significant applications to metals, their corrosion products, alloys, garnets, industrially important minerals and minerals deposited in the human body. The section is closed by papers dealing with the optical simulation of diffraction patterns.

The second section is devoted to the structure determination of amorphous materials and polymers. Highly interesting are the structure models of amorphous alloys given by L. Rang-Su. The neutron structure analysis of

v

amorphous $Fe_{75}B_{25}$ and tellurite glasses are undoubtedly of great interest and practical impact. X-ray analyses of styrene-methacrylate copolymers and polypropylene fibres close this section.

An introduction to the investigation of the <u>real structure of powder materials</u> in the third section is given by P. Klimanek. Six papers in this section offer examples of studies of different structure defects, texture and interdiffusion in solids using a wide spectrum of systems (rolled FeSi, materials for microelectronics and minerals).

The next section gives a comprehensive view of <u>polycrystalline thin films and surface coatings</u>. The first paper by R. I. Barabash and M. A. Krivoglaz summarizes the theoretical aspects of scattering by thin surface layers influenced inevitably by different types of defects and stresses. An extensive methodological treatment of a number of specific systems is given in papers by H. Oettel, V. Valvoda, E. L. Haase, R. Delhez, J. Pielaszek and I. Tomov. From the material point of view, one can find interesting structure analyses of laser modified surface layers of alloys, analyses of ceramic coatings, nitrided steels, TiC, SnO_x, and others, with applications in microelectronics as resistors or potentiometers in hybrid circuits, or as diffusion barriers in integrated circuits. They can be used further as selectively transmitting coatings on architectural glass, in jewelry production, as thermal resistant and wear resistant coatings on cutting tools and as anticorrosive and antiabrasive coatings with interesting applications.

The fifth section dealing with <u>single crystal thin layers</u> is introduced by a comprehensive treatment of garnet films by S. Lagomarsino with applications in lasers, microwave devices, new types of magnetic memories and optoelectronics. A group of crystallographers from the Institute of Crystallography in Moscow describe new methods for the study of structure defects in epitaxial layers based on a standing wave technique (silicone films doped by B and Ge).

The sixth chapter is devoted to <u>special problems in single crystal structure determination</u>. The first paper formulates the conditions for optimal solution of the phase problem and reviews new methods of the crystal structure determination based on the principle of distribution fitting of seminvariants. These methods are suitable for well-ordered single crystals.

The paper written by P. M. de Wolff which offers the reader a new view of the use of <u>superspace groups</u> for incommensurate crystal structures is followed by an informative survey of <u>modulated crystal structures</u> determined at the State University at Buffalo (F. Coppens, V. Petříček), explaining a pressure dependent low-temperature organic superconductivity, transition to the ferroelectric phase and metal-insulator transitions of the organic materials under study.

The paper by Yu. M. Tairov and V. F. Tsvetkov introduces the part devoted to <u>polytype structures</u>. The growing interest in polytypism during the last two decades is mainly due to the fact that electrophysical properties of polytypic substances (e.g., ZnS, SiC, ferrites, chalcogenites) which depend on the actual stacking of layers can be "tuned" simply by their controlled growth. An introduction to the OD symmetry of these structures is given by K. Fichtner. A proper understanding of their symmetry plays a significant role in explaining the physico-chemical properties of polytypes and helps in interpreting the diffraction pattern. Methods of the structure solution of polytypes are discussed by B. B. Zvyagin and B. I. Nikolin. The section is closed by an analysis of statistical parameters describing the polytype stacking.

The last section is devoted to data acquisition and to special instrumental problems - an evaluation of different approximations of diffraction peak profiles, a geometry for multiple scattering, area detectors for powder diffractometers, and X-ray diffractometer for the lattice mismatch measurement in heterostructures.

It must be emphasized that the volume is not meant to provide a full coverage of all latest developments. However, it gives an informative survey of the contemporary state of the art and of new trends in research at universities, research and industrial laboratories. The final manuscripts were subjected to only a brief scrutiny for scientific and language correctness. Therefore, the views expressed in this volume do not necessarily correspond to those of the editors. The full responsibility for exactness, originality and language rests with the individual authors.

Finally, we express our gratitude to all contributors who willingly cooperated in the preparation of this volume, and to all who contributed significantly to the success of the conference in Karlovy Vary, namely, to all members of the organizing committee (J. Fiala, Z. Weiss, I. Vostřáková, A. Stopková, Z. Čížek, V. Petříček, V. Šubrtová, K. Malý and J. Pilná) for their efforts ensuring the smooth course of the conference. The organizers are also indebted to J. Tříska, Director of the Physical Institute of the Czechoslovak Academy of Sciences, Vl. Kubánek, Director of the Institute of Macromolecular Chemistry of the Czechoslovak Academy of Sciences, Z. Kletečka, Director of the Central Research Institute of ŠKODA works, and J. Garaj, Chairman of the Czechoslovak National Committee of IUCr for their great interest in and support of this conference. Lastly a special note of appreciation is due to Mrs E. Biskupová for the typing and artwork involved in producing this book.

Jindřich Hašek
Jaroslav Fiala
Václav Valvoda
Václav Petříček
Slavomil Ďurovič

CONTENTS

POWDER DIFFRACTION ANALYSIS

CRYSTAL STRUCTURE DETERMINATION

DATA ACQUISITION

I. POWDER DIFFRACTION ANALYSIS

IDENTIFICATION OF X-RAY DIFFRACTION PATTERNS

OF MULTICOMPONENT MIXTURES

Jaroslav Fiala

Central Research Institute ŠKODA

316 00 Plzeň, Czechoslovakia

INTRODUCTION

A single phase diffraction pattern can usually be readily identified by searching a data base of reference powder patterns, provided of course that the substance in question is present in the data base used. The real problems of phase analysis occur when the measured pattern results from a mixture of substances. Usually, no more than five constituents in a mixture may reliably be resolved[1,2] in case that there is no additional information available by means of which the number of substances that have to be taken into consideration can sufficiently be reduced. Physical reasons of this limitation are quantitatively examined and the ways how to overcome it are elucidated in the present paper.

INFORMATION CONTENT

It is easy to understand why unraveling mixture patterns is much more difficult than identifying single-phase substances, when we take into consideration the fact that the number of qualitatively different mixtures of the 46 000 substances from the Powder Diffraction File of the International Centre for Diffraction Data amounts to $2^{46000} = 10^{14000}$. For an unequivocal identification of these mixtures there would be necessary an information capacity of $\log_2(2^{46000}) = 46000$ bits[3], while for identification of their 46000 constituents (as single-phase substances) it is sufficient $\log_2(46000) \doteq 15.5$ bits. It is true that a mixture of more than 100 constituents gives no measurable diffraction pattern at all, so that we can limit out reasonings to mixtures of no more than 100 components. But the number of such mixtures is still very large – $46000^{100} \doteq 10^{470}$ so that their identification would require an information capacity of $\log_2(10^{470}) \doteq 1600$ bits which is about 100 times more than we need for identification of pure constituents. By the way, this number is comparable with the information content of an X-ray powder diffraction pattern, i.e. 1800 bits = $\log_2$ (number of different distinguishable patterns) = $\log_2(2^{1800})$. (Uncertainty regarding the location of a diffraction peak as given by real structure effects is estimated at $0.1°$ 2ϑ so that 1800 diffraction peaks can be distinguished in the whole $180°2\vartheta$-range.)

STRIPPING STRATEGY

Therefore, a number of procedures has been developed that allow to identify the different groups of diffraction lines corresponding to the individual constituents. The complex diffraction patterns are differentiated using conditions that increase or diminish the contribution of certain phases in mixture (by froth flotation, sieving, differential settling, heavy liquid separation, magnetic separation, hand picking, selective dissolution, heavy ion bombardment or thermal treatment[4-10]) or using different wavelengths to exploit the anomalous dispersion effects and microabsorption[4,11] that influence the contributions of the particular components in mixture pattern. The pairs of differentiated mixture patterns are then scaled (multiplying by a constant so that peaks common to both patterns have the same height) and subtracted. The difference pattern is usually simpler than the parent patterns. Sometimes it is even made of diffraction peaks of only one phase so that it can easily be identified.

SIMULTANEOUS SOLUTION

Extension of the methods described in the preceding chapter results in a number of procedures in which all patterns taken under different experimental conditions are compared simultaneously.

Denoting by $\vec{x}_1(2\vartheta)$, $\vec{x}_2(2\vartheta)$, ..., $\vec{x}_p(2\vartheta)$ the diffraction patterns of p mixtures of k unknown constituents, we have

$$\vec{x}_i(2\vartheta) = \sum_{j=1}^{k} c_{ij}\vec{y}_j(2\vartheta); \quad i = 1, 2, 3, \ldots, p \tag{1}$$

where c_{ij} represents the concentration of the j-th component in the i-th mixture. The diffraction patterns $\vec{y}_1(2\vartheta)$, $\vec{y}_2(2\vartheta)$, ...,$\vec{y}_k(2\vartheta)$ of the constituents as well as the mixture patterns $\vec{x}_i(2\vartheta)$ can be viewed as continuous functions of a real variable 2ϑ expressing directional distribution of the diffracted intensity, i.e. as vectors (pattern vectors) of an infinite-dimensional Hilbert space[13]. Suppose that Ω_j denotes the 2ϑ-range for which $\vec{y}_j(2\vartheta) \neq 0$; $\vec{y}_i(2\vartheta) = 0$; $i \neq j$; $j = 1,\ldots,k$. Then for $2\vartheta \in \Omega_j$ it holds

$$\frac{\vec{x}_i(2\vartheta)}{\vec{x}_1(2\vartheta)} = \frac{c_{ij}\vec{y}_j(2\vartheta)}{c_{1j}\vec{y}_j(2\vartheta)} = \frac{c_{ij}}{c_{1j}} = \gamma_{ij} = \text{const}; \quad j=1,2,\ldots,k \tag{2}$$

The ratio pattern consists of a number of flat regions linked by curves and slopes (see Fig. 1). A flat region occurs wherever a single constituent dominates the pattern, whether it has a peak there or not. The number of flat regions of differing height will give the number of constituents, and allow assembling a set of γ_{ij}'s on the basis of which the concentrations c_{ij} can be calculated. Using the conditions

$$\sum_{j=1}^{k} c_{ij} = 1; \quad i = 2, 3, \ldots, p$$

we obtain the equations

$$\sum_{j=1}^{k} \frac{c_{ij}}{c_{1j}} c_{1j} = \sum_{j=1}^{k} \gamma_{ij} c_{1j} = 1; \quad i = 2, 3, \ldots, p,$$

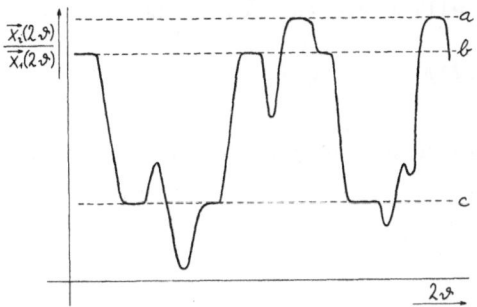

Fig. 1. Ratio of diffraction patterns of two mixtures; flat regions of three different heights (a, b, c) are apparent in the ratio pattern, indicating that the number of constituents in the mixtures is three.

from which c_{11}, c_{12}, ..., c_{1k} can be determined and, making use of (2), we calculate further

$$c_{i1}, c_{i2}, ..., c_{ik} \text{ for } i = 2, 3, ..., p.$$

If we now pass from the Hilbertian representation to the Euclidean coding, i.e., if we digitize diffraction patterns $\vec{x}_i(2\vartheta)$

$$x_i(2\vartheta_1), x_i(2\vartheta_2), ..., x_i(2\vartheta_n) = x_{i1}, x_{i2}, ..., x_{in} ; \quad i = 1, 2, ..., p$$

as well as $\vec{y}_j(2\vartheta)$

$$y_j(2\vartheta_1), y_j(2\vartheta_2), ..., y_j(2\vartheta_n) = y_{j1}, y_{j2}, ..., y_{jn} ; \quad j = 1, 2, ..., k,$$

then the expressions (1) take the following form

$$\hat{X} = \hat{C} \hat{Y}, \tag{3}$$

where

$$\hat{X} = \begin{bmatrix} x_{11}, x_{12}, ..., x_{1n} \\ x_{21}, x_{22}, ..., x_{2n} \\ \cdots \cdots \cdots \cdots \\ x_{o1}, x_{p2}, ..., x_{pn} \end{bmatrix} \qquad \hat{C} = \begin{bmatrix} c_{11}, c_{12}, ..., c_{1k} \\ c_{21}, c_{22}, ..., c_{2k} \\ \cdots \cdots \cdots \cdots \\ c_{p1}, c_{p2}, ..., c_{pk} \end{bmatrix}$$

$$\hat{Y} = \begin{bmatrix} y_{11}, y_{12}, ..., y_{1n} \\ y_{21}, y_{22}, ..., y_{2n} \\ \cdots \cdots \cdots \cdots \\ y_{k1}, y_{k2}, ..., y_{kn} \end{bmatrix}$$

From this equation we obtain finally the diffraction patterns of the unknown constituents

$$\hat{Y} = (\hat{C}'\hat{C})^{-1}\hat{C}'\hat{X}.$$

The solution of (1) will yield the following equations

$$\vec{y}_j(2\vartheta) = \sum_{i=1}^{p} \Gamma_{ji} \, \vec{x}_i(2\vartheta); \quad j = 1, 2, \ldots, k \tag{5}$$

the coefficients Γ_{ji} of which are subject to the constraints

$$\sum_{i=1}^{p} \Gamma_{ji} = 1; \quad j = 1, 2, \ldots, k \tag{6}$$

and

$$\sum_{i=1}^{p} \Gamma_{ji} \, \vec{x}_i(2\vartheta) \geq 0 \tag{7}$$

(non-negativity of intensities). For $k = p = 3$ this set of constraints

$$\Gamma_{11} + \Gamma_{12}Q_1(2\vartheta) + \Gamma_{13}Q_2(2\vartheta) \geq 0 \tag{8}$$

$$\Gamma_{21} + \Gamma_{22}Q_1(2\vartheta) + \Gamma_{23}Q_2(2\vartheta) \geq 0 \tag{9}$$

$$\Gamma_{31} + \Gamma_{32}Q_1(2\vartheta) + \Gamma_{33}Q_2(2\vartheta) \geq 0 \tag{10}$$

delimits a triangle in the plane spanned by two coordinate vectors (variables)

$$Q_1(2\vartheta) = \frac{\vec{x}_2(2\vartheta)}{\vec{x}_1(2\vartheta)} ; \qquad Q_2(2\vartheta) = \frac{\vec{x}_3(2\vartheta)}{\vec{x}_1(2\vartheta)} \tag{11}$$

(ratio patterns). This is the basis of the Bezjak's graphical method[14,15] of X-ray diffraction phase analysis. The points corresponding to the individual 2ϑ-values of the analysed mixture patterns $\vec{x}_1(2\vartheta)$, $\vec{x}_2(2\vartheta)$, $\vec{x}_3(2\vartheta)$ fill up the inside of the triangle (Fig. 2) the contours of which determine the values of coefficients Γ_{ji} (see equations (8)-(10) and (6)). The desired diffraction patterns of the unknown constituents $\vec{y}_1(2\vartheta)$, $\vec{y}_2(2\vartheta)$, and $\vec{y}_3(2\vartheta)$ are then calculated from the equations (5).

Pure component patterns can easily be estimated from mixture patterns in case that we know (at least) one isolated diffraction line of each of the components. In the initial stage of analysis we confine ourselves to only those unoverlapped lines rearranging the constituents for the sake of simplicity of notation so that the j-th phase corresponds to the j-th unoverlapped line, the intensity value of which being y_j. The equation (3) is then simplified to

$$\begin{bmatrix} x_{11}, x_{12}, \ldots, x_{1k} \\ x_{21}, x_{22}, \ldots, x_{2k} \\ \cdot \quad \cdot \quad \cdot \quad \cdot \quad \cdot \\ x_{p1}, x_{p2}, \ldots, x_{pk} \end{bmatrix} = \begin{bmatrix} c_{11}, c_{12}, \ldots, c_{1k} \\ c_{21}, c_{22}, \ldots, c_{2k} \\ \cdot \quad \cdot \quad \cdot \quad \cdot \quad \cdot \\ c_{p1}, c_{p2}, \ldots, c_{pk} \end{bmatrix} \cdot \begin{bmatrix} y_1, 0, \ldots, 0 \\ 0, y_2, \ldots, 0 \\ \cdot \quad \cdot \quad \cdot \quad \cdot \\ 0, 0, \ldots, y_k \end{bmatrix} \tag{12}$$

i.e.

6

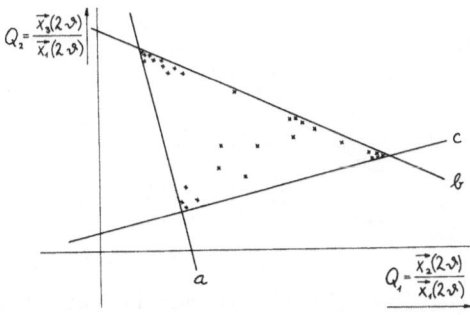

Fig. 2. Graphical method for determining of Γ_{ij} ; these values are required for the synthesis

$$\vec{y}_j(2\vartheta) = \sum_{i=1}^{p} \Gamma_{ji} \, \vec{x}_i(2\vartheta)$$

of the desired patterns $\vec{y}_j(2\vartheta)$ of constituents based on diffraction data $\vec{x}_i(2\vartheta)$ of their different mixtures. Crosses (x) correspond to the individual 2ϑ-values of the analysed mixture patterns. Straight lines a, b, and c correspond to the expressions (8), (9), and (10).

$$x_{i\ell} = c_{i\ell} y_\ell; \quad i = 1, 2, \ldots, p; \; \ell = 1, 2, \ldots, k. \tag{13}$$

Inserting the value of $c_{i\ell}$ found in equations (13) into

$$\sum_{\ell=1}^{k} c_{i\ell} = 1; \quad i = 1, 2, \ldots, p \tag{14}$$

we obtain the following set of equations

$$\sum_{\ell=1}^{k} \frac{x_{i\ell}}{y_\ell} = 1; \quad i = 1, 2, \ldots, p$$

from which the unknown values of $y_1, y_2, \ldots, y_k$ can be determined. Substitution of these quantities into (13) yields concentrations c_{ij} and, finally, with the help of equation (4), the intensity values y_{ij}, i.e. the desired diffraction patterns of unknown components[16-19]. Unoverlapped lines of a particular component can be recognized making use of the fact that the values of their intensities measured on different mixture patterns are strongly correlated[20-24] (see Fig. 3).

FACTOR ANALYSIS

Diffraction pattern $\vec{x}$ of the mixture under analysis is a linear combination

$$\vec{x} = \sum_{j=1}^{k} c_j \vec{y}_j$$

7

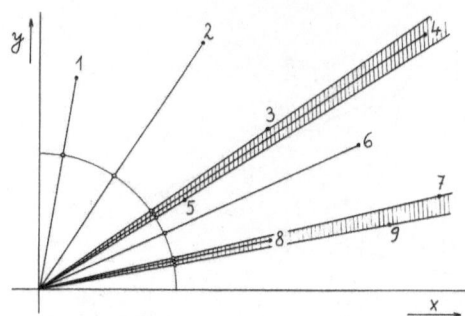

Fig. 3. Recognizing unique ("pure") lines: bunches of almost identically oriented radius vectors - clusters of their intersections (o) with the circle $x^2 + y^2 = c$. The points with numbers (•) correspond to the individual diffraction lines, their x- and y-coordinates expressing intensities of the given line on the 1st and 2nd mixture pattern respectively.

of the diffraction patterns $\vec{y}_1$, $\vec{y}_2$, ..., $\vec{y}_k$ of its (unknown) components. Diffraction patterns $\vec{x}_i$ of different samples of the analysed substance or diffraction patterns of this substance acquired under different experimental conditions

$$\vec{x}_i = \sum_{j=1}^{k} c_{ij} \vec{y}_j; \quad i = 1, 2, \ldots, p \tag{15}$$

have similar features; this is given by the common basis of the constituent patterns, $\vec{y}_1$, $\vec{y}_2$, ..., $\vec{y}_k$, which are not accessible to direct measurement. Thus, the deciphering of mixture patterns $\vec{x}_1$, $\vec{x}_2$, ..., $\vec{x}_p$ can be treated quite generally as the task to find their common factors $\vec{y}_j$; j = 1,2,...,k.

Such a problem, to ascertain latent roots of various effects or properties that occur simultaneously, explaining thus the mutual relationship between observed phenomena, is quite common in both natural and social sciences [25-28]. A general mathematical technique for the solution of this type of problems is factor analysis [29-32], developed by psychologists and sociologists at the beginning of this century. In the sixties, factor·analysis began to be used to interpret chemical data sets [33-36]. It has been proven to be useful in resolving multicomponent mixtures.

The mathematical basis is an algebraic rule according to which the matrix

$$\hat{X} = \begin{bmatrix} \vec{x}_1 \\ \vec{x}_2 \\ \cdots \\ \vec{x}_p \end{bmatrix} = \begin{bmatrix} x_{11}, x_{12}, \ldots, x_{1n} \\ x_{21}, x_{22}, \ldots, x_{2n} \\ \cdots \cdots \cdots \cdots \\ x_{p1}, x_{p2}, \ldots, x_{pn} \end{bmatrix} = \begin{bmatrix} \vec{d}_1, \vec{d}_2, \ldots, \vec{d}_n \end{bmatrix} \tag{16}$$

consisting of p rows - mixture patterns

$$\vec{x}_i = \begin{bmatrix} x_{i1}, x_{i2}, \ldots, x_{in} \end{bmatrix}; \quad i = 1, 2, \ldots, p$$

can be decomposed into two matrices

$$
\hat{X} \propto
\begin{bmatrix}
\sqrt{\lambda}_1, & \sqrt{\lambda}_2, & \ldots, & \sqrt{\lambda}_n \\
\sqrt{\lambda}_1, & \sqrt{\lambda}_2, & \ldots, & \sqrt{\lambda}_n \\
\cdot & \cdot & \cdots & \cdot \\
\sqrt{\lambda}_1, & \sqrt{\lambda}_2, & \ldots, & \sqrt{\lambda}_n
\end{bmatrix}
\cdot
\begin{bmatrix}
\vec{q}_1' \\
\vec{q}_2' \\
\\
\vec{q}_n'
\end{bmatrix}
\tag{17}
$$

where λ_i and $\vec{q}_i$,

$$
\vec{q}_i =
\begin{bmatrix}
q_{1i} \\
q_{2i} \\
\ldots \\
q_{ni}
\end{bmatrix}
; \quad i = 1, 2, \ldots, n
$$

are eigenvalues and eigenvectors of the Gramian

$$
\hat{G} =
\begin{bmatrix}
(\vec{d}_1, \vec{d}_1), (\vec{d}_1, \vec{d}_2), \ldots, (\vec{d}_1, \vec{d}_n) \\
(\vec{d}_2, \vec{d}_1), (\vec{d}_2, \vec{d}_2), \ldots, (\vec{d}_2, \vec{d}_n) \\
\cdot \quad \cdot \quad \cdots \quad \cdot \\
(\vec{d}_n, \vec{d}_1), (\vec{d}_n, \vec{d}_2), \ldots, (\vec{d}_n, \vec{d}_n)
\end{bmatrix}
,
\tag{18}
$$

respectively:

$$
[\hat{G}\vec{q}_1, \hat{G}\vec{q}_2, \ldots, \hat{G}\vec{q}_n] = [\lambda_1\vec{q}_1, \lambda_2\vec{q}_2, \ldots, \lambda_n\vec{q}_n] .
$$

Hence, according to (17), the magnitude of the eigenvalue is a gauge of the importance of the corresponding eigenvector. Eigenvectors associated with the largest eigenvalues are most important and eigenvectors associated with the smallest eigenvalues are least important. In practice, the least important eigenvalues are negligible and are dropped from the analysis[20]. In fact, their retention would simply regenerate the experimental error. The minimum number of eigenvalues (k) required to reproduce the data within experimental error represents the number of factors (latent variables) involved. So this number tells us how many constituents our mixtures consist of. The corresponding eigenvectors $\vec{q}_j' = [q_{1j}, q_{2j}, \ldots, q_{nj}]$, $j = 1, 2, \ldots, k$; $k \leq n$ (the eigenvectors corresponding to k largest eigenvalues $\lambda_1, \lambda_2, \ldots, \lambda_k$) represent the k (abstract) factors satisfactorily reproducing the measured data

$$
\vec{x}_i = \sum_{j=1}^{k} v_{ij} \vec{q}_j' ; \quad i = 1, 2, \ldots, p;
$$

in matrix notation

$$\hat{X} \doteq \begin{bmatrix} v_{11}, v_{12}, \ldots, v_{1k} \\ v_{21}, v_{22}, \ldots, v_{2k} \\ \cdot \cdot \cdot \cdot \cdot \cdot \cdot \cdot \\ v_{p1}, v_{p2}, \ldots, v_{pk} \end{bmatrix} \cdot \begin{bmatrix} \vec{q}_1' \\ \vec{q}_2' \\ \cdot \cdot \\ \vec{q}_k' \end{bmatrix} = \hat{V} \begin{bmatrix} q_{11}, q_{21}, \ldots, q_{n1} \\ q_{12}, q_{22}, \ldots, q_{n2} \\ \cdot \cdot \cdot \cdot \cdot \cdot \cdot \cdot \\ q_{1k}, q_{2k}, \ldots, q_{nk} \end{bmatrix} = \hat{V} \hat{W}$$

where

$$\hat{V} = \hat{X}\hat{W}' (\hat{W} \hat{W}')^{-1} \tag{19}$$

The sought constituent patterns $\vec{y}_1, \vec{y}_2, \ldots, \vec{y}_k$

$$\hat{Y} = \begin{bmatrix} \vec{y}_1 \\ \vec{y}_2 \\ \cdot \cdot \\ \vec{y}_k \end{bmatrix} \cdot \begin{bmatrix} y_{11}, y_{12}, \ldots, y_{1n} \\ y_{21}, y_{22}, \ldots, y_{2n} \\ \cdot \cdot \cdot \cdot \cdot \cdot \cdot \cdot \\ y_{k1}, y_{k2}, \ldots, y_{kn} \end{bmatrix}. \tag{20}$$

in (15) or (3) must also be linear superpositions of the significant eigen-vectors $\vec{q}_1', \vec{q}_2', \ldots, \vec{q}_k'$ of the Gramian of the data matrix $\hat{X}$

$$\vec{y}_j = \sum_{m=1}^{k} T_{jm} \vec{q}_m' ; \quad j = 1, 2, \ldots, k;$$

in matrix notation

$$\hat{Y} = \hat{T} \hat{W} . \tag{21}$$

So we have found the subspace in which the constituent diffraction patterns must lie. And we are left with the problem to specify where in this sub-space the constituent patterns lie, to specify the T's.

The transformation matrix

$$\hat{T} = \begin{bmatrix} T_{11}, T_{12}, \ldots, T_{1k} \\ T_{21}, T_{22}, \ldots, T_{2k} \\ \cdot \cdot \cdot \cdot \cdot \cdot \cdot \cdot \\ T_{k1}, T_{k2}, \ldots, T_{kk} \end{bmatrix} \tag{22}$$

can be determined from inequality constraints

$$\hat{Y} = \hat{T} \hat{W} \geq \hat{0} \tag{23}$$

and

$$\hat{C} = \hat{X} \hat{W}' \hat{T}' (\hat{T} \hat{W} \hat{W}' \hat{T}')^{-1} \geq \hat{0} \tag{24}$$

10

expressing conditions of non-negativity of intensities (23) and of concentrations (24), respectively (confirmatory factor analysis [37]). The uniqueness of the solution depends on the actual form of the simplex (23), which can be investigated only in the points on the sphere

$$\sum_{j=1}^{k} \sum_{m=1}^{k} T_{jm}^2 = 1 \; . \tag{25}$$

ERROR PROPAGATION

In case that the transformation $\hat{T}$ is specified by the conditions (23), (24) and (25) uniquely, the a posteriori estimation of errors in the determined constituent patterns $\hat{Y}$ is given as

$$\frac{\| \Delta \hat{Y} \|}{\| \hat{Y} \|} \leq \text{cond } \hat{T} \cdot \text{cond } \hat{V} \cdot \frac{\| \Delta \hat{X} \|}{\| \hat{X} \|}$$

Here $\hat{V}$ is matrix defined by equation (19),

$$\| \hat{A} \| = \max \| \hat{A}\vec{x} \|$$

$$\| \vec{x} \| = 1$$

is Euclidean norm of matrix $\hat{A}$ ($\| \vec{x} \|$ = square root of sum of squares of vector elements) and

$$\text{cond } \hat{A} = \| \hat{A} \| \cdot \| \hat{A}^+ \|$$

is condition number of the matrix $\hat{A}$[38,39]; $\hat{A}^+$ is pseudoinverse of matrix $\hat{A}$ ($\hat{A} \, \hat{A}^+ \, \hat{A} = \hat{A}$)[40].

An a priori estimate of the error $\Delta \hat{Y}$ is

$$\frac{\| \Delta \hat{Y} \|}{\| \hat{Y} \|} \leq \text{cond } \hat{C} \cdot \frac{\| \Delta \hat{X} \|}{\| \hat{X} \|} \quad ,$$

where $\Delta \hat{X}$ represents experimental errors and real structure effects; $\hat{C}$ is concentration matrix of the analysed mixtures.

CONCLUSION

Factor analytical approach provides an efficient and universal method for the resolution of complex mixtures by X-ray powder diffraction.

REFERENCES

1. J. Fiala, Spectral data bases for chemical compounds identification, Computer Phys.Comm. 33:85 (1984).
2. E. K. Vasiliev and M. S. Nakhmanson, "Kachestvennyi Rentgenofazovyi Analiz (Qualitative X-ray Diffraction Phase Analysis)", in Russian, Nauka, Novosibirsk (1986).

3. L. Brillouin, "Science and Information Theory", Academic Press, New York (1956).

4. M. C. Nichols, D. K. Smith, and Q. Johnson, Differential X-ray diffraction: a theoretical basis for a technique based on wavelength variation, J.Appl.Cryst. 18:8 (1985).

5. D. G. Schulze, Identification of soil iron oxide minerals by differential X-ray diffraction, Soil Sci.Soc.Amer.J. 45:437 (1981).

6. R. B. Bryant, N. Curi, C. B. Roth, and D. P. Franzmeier, Use of an internal standard with differential X-ray diffraction analysis for iron oxides, Soil Sci.Soc.Amer.J. 47:168 (1983).

7. G. Brown and I. G. Wood, Estimation of iron oxides in soil clays by profile refinement combined with differential X-ray diffraction, Clay Miner. 20:15 (1985).

8. D. G. Schulze, Correction of mismatches in 2ϑ scales during differential X-ray diffraction, Clays Clay Miner. 34:681 (1986).

9. L. Cartz, F. G. Karioris, and M. S. Wong, Heavy ion bombardment and X-ray powder patterns of mixtures, Rad.Effects Lett. 85:273 (1985).

10. L. Cartz, F. G. Karioris, and M. S. Wong, Analysis of mineral powder mixtures by heavy ion bombardment and X-ray diffraction, Rad.Effects Lett. 97:235 (1986).

11. I. G. Wood, L. Nichols, and G. Brown, X-ray anomalous scattering difference patterns in qualitative and quantitative powder diffraction analysis, J.Appl.Cryst. 19:364 (1986).

12. T. Hirschfeld, Computer resolution of infrared spectra of unknown mixtures, Anal.Chem. 48:721 (1976).

13. J. Fiala, Optimization of powder diffraction identification, J.Appl.Cryst. 9:429 (1976).

14. A. Bezjak, I. Šmit, and V. Alujević, Determination of the X-ray diffraction curve of amorphous phase, Croatica Chem.Acta 54:61 (1981).

15. A. Bezjak, V. Runje, Quantitative phase analysis of the hydrothermally treated system γ-Ca_2SiO_4-CaO, J.Amer.Ceram.Soc. 64:129 (1981).

16. F. H. Chung, Quantitative interpretation of X-ray diffraction patterns of mixtures, I. Matrix-flushing method for quantitative multicomponent analysis, J.Appl.Cryst. 7:519 (1974).

17. F. H. Chung, Quantitative interpretation of X-ray diffraction patterns of mixtures. II. Adiabatic principle of X-ray diffraction analysis of mixtures, J.Appl.Cryst. 7:526 (1974).

18. F. H. Chung, Quantitative interpretation of X-ray diffraction patterns of mixtures. III. Simultaneous determination of a set of reference intensities, J.Appl.Cryst. 8:17 (1975).

19. L. S. Zevin, A method of quantitative phase analysis without standards, J.Appl.Cryst. 10:147 (1977).

20. J. Fiala, Powder diffraction analysis of a three component sample, Anal.Chem. 52:1300 (1980).

21. J. Fiala, A new method for powder diffraction phase analysis, Cryst.Res.Technol. 17:643 (1982).

22. M. Čerňanský, Matrix formulation of the iterative method of phase analysis, J.Appl.Cryst. 20:260 (1987).

23. W. Windig, J. Haverkamp, and P. G. Kistemaker, Interpretation of sets of pyrolysis mass spectra by discriminant analysis and graphical rotation, Anal.Chem. 55:81 (1983).

24. W. Windig, E. Jakab, J. M. Richards, and H. L. C. Meuzelaar, Self-modeling curve resolution by factor analysis of a continuous series of pyrolysis mass spectra, Anal.Chem. 59:317 (1987).

25. K. G. Jöreskog, J. E. Klovan, and R. A. Reyment, "Geological Factor Analysis", Elsevier, Amsterdam (1976).

26. D. J. Aigner and A. S. Goldberger, "Latent Variables in Socio-Economic Models", North Holland, Amsterdam (1977).

27. Y. Ahmavaara, "On the Unified Factor Theory of Mind", S. Tiedeakatemia, Helsinki (1957).

28. P. Murray-Rust and R. Bland, Computer retrieval and analysis of molecular geometry. II. Variance and its interpretation, _Acta Cryst_. B34: 2527 (1978).

29. H. H. Harman, "Modern Factor Analysis", The University of Chicago Press, Chicago (1967).

30. P. F. Lazarsfeld and N. W. Henry, "Latent Structure Analysis", Houghton Mifflin, New York (1968).

31. R. J. Rummel, "Applied Factor Analysis", Northwestern University Press, Evanston (1970).

32. E. R. Malinowski and D. G. Howery, "Factor Analysis in Chemistry", John Wiley and Sons, New York (1980).

33. D. Macnaughtan L. B. Rogers, and G. Wernimont, Principle-component analysis applied to chromatographic data, _Anal.Chem_. 44:1421 (1972).

34. A. Meister, Estimation of component spectra by the principal components method, _Anal.Chim.Acta_ 161:149 (1984).

35. O. S. Borgen and B. R. Kowalski, An extension of the multivariate component-resolution method to three components, _Anal.Chim.Acta_ 174:1 (1985).

36. P. J. Gemperline, Target transformation factor analysis with linear inequality constraints applied to spectroscopic-chromatographic data, _Anal.Chem_. 58:2656 (1986).

37. K. G. Jöreskog, A general approach to confirmatory maximum likelihood factor analysis, _Psychometrika_ 34:183 (1969).

38. C. Jochum, P. Jochum, and B. R. Kowalski, Error propagation and optimal performance in multicomponent analysis, _Anal.Chem_. 53:85 (1981).

39. G. E. Braun, A Monte Carlo error simulation applied to calibration-free X-ray diffraction phase analysis, _J.Appl.Cryst_. 19:217 (1986).

40. A. Lorber, Error propagation and figures of merit for quantification by solving matrix equations, _Anal.Chem_. 58:1167 (1986).

QUANTITATIVE REFERENCE INTENSITY ANALYSIS:

METHODOLOGY AND MEANS FOR VERIFICATION OF RESULTS

Briant L. Davis

Institute of Atmospheric Sciences
South Dakota School of Mines and Technology
Rapid City, South Dakota, 57701, USA

INTRODUCTION

The success of any quantitative x-ray diffraction (XRD) analytical technique depends upon good control of particle size in the sample, elimination of preferred orientation of particles, and, in the case of the reference intensity method (RIM)[1,2], the measurement of high-quality reference intensity ratios (k_i, or RIR). The method developed in our laboratory eliminates preferred orientation by use of an aerosol suspension technique[3,4] in which the sample particles are carried vertically into a glass fiber filter[5,6].

The basis of the RIM procedure for determination of weight fraction of component j in a sample of n components is Chung's self-normalizing "adiabatic" equation[1]:

$$W_j = \left| \frac{k_j}{I_j} \sum_{i=1}^{n} \frac{I_i}{k_i} \right|^{-1}$$

(1)

where I_j is the diffraction intensity for a peak from component j satisfying the constant-volume criteria (i.e., for fixed slits and "infinitely-thick" samples) and k_j is the reference intensity constant for the peak. I_i and k_i are the same quantities for all components of the sample. Thin aerosol layers collected on filter substrates require correction of the raw intensities to obtain I_i values as defined above; the procedure for this has been described previously[6].

Problems with preferred orientation and particle size effects (micro-absorption and primary extinction) can now be largely eliminated with proper care in sample preparation. The major remaining uncertainties in the RIM come from compositional variations in the components, resulting in inappropriate k_i, and difficulties with complex patterns with overlapping peaks. Under these circumstances, and while also recognizing that the RIM procedure is still in a youthful stage of development, it is important that alternate analytical methods be used frequently to verify

RIM results. The purpose of this paper is to present some procedures that may be used for such verification at the same time pointing out difficulties in using such techniques. Three such procedures will be discussed here: mass absorption analysis, polarised light microscopy (PLM), and comparison with chemical data. All x-ray work was completed using CuKα radiation of filter-loaded samples.

MASS ABSORPTION ANALYSIS

This procedure requires that an independent accurate measurement of sample mass absorption, μ_{BO}^*, be completed. Such procedures and applications have been presented elsewhere[7]. Where this is possible then a comparison of the computed mass absorption coefficient μ_{BC}^* for the sample with μ_{BO}^* may (but not always) reveal whether or not the analysis is correct. A simple example from an industrial process will be given.

Furnace firing of a mixture of carbon and amorphous silica produced SiC which was then analysed quantitatively by the RIM. The crystalline components observed were 89.3 % low cristobalite and 10.7 % SiC; mass absorption values for these components are 37.3 and 47 $cm^2 \ gm^{-1}$, respectively. The mass-weighted value for these components is 38.3, whereas the observed value from x-ray transmission measurements[7] was found to be 28.7 $cm^2 \ gm^{-1}$. This suggested that not all carbon was eliminated in the firing process. If a value of 5 $cm^2 \ gm^{-1}$ is used for carbon, the actual composition of the sample is seen to be 29 % carbon, 7.6 % SiC, and 63.4 % low cristobalite. With the proven absence of amorphous components, agreement between observed and calculated μ_B^* will generally provide verification of the XRD analysis; however, it should be pointed out that two different compounds having the same mass absorption can have any ratio whatsoever in a mixture and not affect the value of μ_{BO}^*.

POLARISED LIGHT MICROSCOPY

Davis and Walawender[8] illustrated use of this technique for analysis of coarse-grained igneous rocks, revealing very good agreement between optical, x-ray, and chemical data. For complicated mixtures, agreement will not be so good, primarily because of the following limitations in the techniques used:

(1) counting statistics (PLM)
(2) conversion of point and area data to volume/mass data (PLM)
(3) representative k_i values (RIM)
(4) peak interferences (RIM)
(5) amorphous and/or unrecognized crystalline components (RIM)

It is important to prepare the optical sample directly from the filter samples used for the x-ray analysis providing that the particle size is sufficiently large, between 5 and 10 microns diameter, to allow positive identification using polarised light properties. Where the sample load is less than about 1200 $\mu gm \ cm^{-2}$, a portion of the filter may be "cleared" in immersion oil and the particles counted directly. This permits viewing of the particles in their random orientation within the fiber matrix, a feature that is especially important when average shape factors are to be determined.

Table 1 presents an example of such an analysis where the mineralogy is rather complex. The sample is metamorphic rock from the northern Black

Table 1. Analysis of Iron-rich Greenschist,
Black Hills, South Dakota

Component-hkl	Intensity	k_i	W_i(RIM) ±error	W_i(PLM) ±error
Stilpnomelane-002	5423	3.0[a]	0.030 0.015	0.004 0.006
Muscovite-002	1959	0.57	0.062 0.011	0.016 0.006
Cummingtonite-110	539	1.31	0.007 0.011	0.003 0.008
Chlorite-002 (Fe-6)	13375	0.97	0.227 0.039	0.132 0.017
Quartz-100	18214	0.72	0.416 0.043	0.297 0.072
Goethite-110	1524	2.0	0.013 0.002	0.100 0.111
Hematite-014	2105	2.38	0.015 0.003	0.219 0.099
Limonite (Amorphous)	–	–	0.055 0.045	– –
Calcite-116	910	0.45	0.033 0.019	0.059 0.036
Siderite-014	3709	2.0	0.030 0.006	
Epidote-301, 302, 113	7459	1.1	0.113 0.017	0.163 0.049

[a] Provisional estimate

Hills, South Dakota. All phyllosilicates present are variable in composition. Other data available for this sample suggests that the carbonate
composition is also complex, with Fe, Mn, Ca, and Mg end members being
present. A computer-assisted optical measurement of shapes (equant, plate,
rod) and dimensions of 5 90-grain subsets of particles taken directly from
the filter and remounted on a flat slide was undertaken (the load was too
great to permit direct analysis on the filter). The component weight
fractions converted from volume fractions and associated standard deviation
of the subsets are labled PLM in the table. The values of μ_{BO}^*, μ_{BC}^*(RIM)
and μ_{BC}^*(PLM) mass absorption coefficients are 76.5, 70.0 and 96.0 cm^2 gm^{-1},
respectively.

Comparison of the two sets of weight fractions of Table 1 shows some
major differences. In the RIM analysis the sum of the iron oxide phases

is only 0.083 compared to 0.319 by PLM. The mass absorption data suggest that limonite or cryptocrystalline goethite was responsible for the RIM underestimation of the crystalline iron oxide content and hence the calculated sample mass absorption coefficient. For this reason the analysis shown in Table 1 includes limonite determined by mass absorption balance, where the limonite value used was 187.3 cm^2 gm^{-1}. The optical analysis more seriously overestimated the iron oxide content as shown by the large positive mass absorption error. This failing was probably due to the inability for one to observe which hematite particles (largely opaque) had a flake or plate-like shape (a common habit for hematite). This error is of the type (2) listed above. Similar type (2) error resulted from the difficulty in distinguishing between thin plates of muscovite and chlorite optically, and in determining the plate thickness. Type (3) errors and associated errors in the resulting value of μ^*_{BC} are likely to be the most serious for the RIM procedure. This exercise demonstrates how difficult it is to complete PLM analyses of highly inequant particles mounted in this manner. The mass absorption coefficients suggests that the RIM analysis is the more accurate of the two.

COMPARISON WITH CHEMICAL DATA

Table 2 presents a RIM study of a Salton Sea California drill-core sample completed independently by three analytical teams of our laboratory compared with limited chemical data obtained by atomic absorption inductively coupled plasma spectroscopy (AA/ICP). Observed and calculated mass absorption coefficients are also shown. The oxides for the RIM analyses were "reduced" from the component mineral compositions; in the case of chlorite, orthoclase, and albite, these compositions were determined from core mineral separates using electron microprobe energy dispersive analysis. Agreement is quite good, both for the oxides and for mass absorption data. Detailed component chemical data are generally not available in routine quantitative XRD work; however, use of average compositions from the published literature to obtain an estimated chemical oxide composition for the whole sample should not result in serious error except for samples containing as major components minerals with large variations in elements of strong diffracting power (e.g., Fe, Ca, K).

An alternate approach to oxide comparison is the comparison of compounds directly, using known compound compositions in an effective variance chi-square minimization (CSM) procedure[9] to evaluate their weight fractions. This method is mathematically described by

$$ \chi^2 = \sum_{i=1}^{n} \frac{[C_i - \sum_{j=1}^{p} C_{ij}W_j]^2}{\sigma_i^2 + \sum_{j=1}^{p} \sigma_{ij}^2 W_j^2} \tag{2} $$

where C_i and C_{ij} are weight fractions of n elements or oxides observed in the bulk sample and p individual mineral components, respectively, W_j are the component weight fractions to be determined, and σ_i and σ_{ij} are the errors associated with the measurements of the C_i and C_{ij}. This procedure allows iterative adjustments to the final W_i set to be computed with the final composition reflecting the influence of errors in each component evaluation. Table 3 presents the CSM matrix and results for a second sample from the Salton sea study. Errors were not available for the X-ray fluorescence method used in the element analysis (expressed as

Table 2. Analysis of Metamorphic Rock from the 5500-Foot Level, Salton Sea (Weight Percent) Continental Scientific Drilling Program

Component	Team 1	Team 2	Team 3	Mean $\pm$ SD
Albite	0.119	0.145	0.136	0.133 $\pm$ 0.013
Orthoclase	0.215	0.255	0.231	0.234 $\pm$ 0.020
Quartz	0.417	0.343	0.375	0.378 $\pm$ 0.037
Chlorite	0.229	0.232	0.241	0.234 $\pm$ 0.006
Pyrite	0.020	0.025	0.017	0.021 $\pm$ 0.004
μ_{BO}^{*} (cm^2 gm^{-1})	52.0	49.1	50.5	50.5 $\pm$ 1.45
μ_{BC}^{*} (cm^2 gm^{-1})	47.7	48.9	47.8	48.0 $\pm$ 0.67

AA/ICP and XRD "Reduced" Oxides[a]

AA/ICP		XRD Oxides			
CaO,	0.015	0.005	0.008	0.005	0.006 $\pm$ 0.002
Fe_2O_3,	0.070	0.064	0.054	0.063	0.060 $\pm$ 0.006
Al_2O_3,	0.096	0.116	0.130	0.121	0.111 $\pm$ 0.013
MgO,	0.040	0.052	0.053	0.042	0.049 $\pm$ 0.006
K_2O,	0.052	0.037	0.043	0.039	0.040 $\pm$ 0.003

[a]Electron microprobe analyses of chlorite gives 3.36 Fe mole number and Orthoclase = Or-92.8; Albite = Ab-96.1.

oxides); consequently the general trend of uncertainty of measurement with atomic number was combined with an assigned uncertainty of 15 % for Na to estimate the required variance errors. All component compositions shown in Table 3 were determined for specific mineral separates from the core by electron microprobe analysis. Three iterations of the minimization process were completed.

Table 3. CSM/RIM Analysis of Metamorphic Rock from the 7306-Foot
          Level, Salton Sea
          Continental Scientific Drilling Program

CSM Matrix (Weight Fraction)

Oxide	Core	Chlorite	Epidote	Albite	Quartz	Orthoclase
SiO_2	0.576	0.291	0.382	0.682	1.000	0.650
Al_2O_3	0.163	0.202	0.217	0.196	0	0.179
MgO	0.041	0.175	0.001	0	0	0
$FeO+Fe_2O_3$	0.073	0.199	0.158	0.001	0	0.001
CaO	0.019	0.009	0.221	0.003	0	0.0003
Na_2O	0.025	0.006	0.001	0.116	0	0.008
K_2O	0.064	0.005	0	0.002	0	0.161
H_2O	0.040	0.122	0.021	0	0	0

Modal Analyses:	Weight Fraction $\pm$ Variance Error	
Component	CSM	RIM
Quartz	0.101 ± 0.012	0.121 ± 0.029
Albite	0.169 ± 0.001	0.156 ± 0.027
Orthoclase	0.374 ± 0.002	0.421 ± 0.096
Chlorite	0.294 ± 0.001	0.256 ± 0.049
Epidote	0.081 ± 0.001	0.046 ± 0.007

The CSM/RIM comparison reveals a generally good performance by the
RIM procedure, except for epidote where the deviation from the CSM value
exceeds the sum of the error values of each method. The explanation for
this difference is probably an inaccurate epidote reference constant,
although sample correlation error (i.e., variation of epidote composition
through the sample volume) is also likely in this case. The greatest
change in component values of W_i over the three iterations was less than
3 %, indicating little influence in this case of the assigned σ_i and
σ_{ij} values.

CONCLUSIONS

Three possible procedures for verification of quantitative x-ray diffraction analysis have been discussed. Mass absorption comparison provides a simple though not infallible procedure when sample composition is simple. Polarised light methods are useful when particle size is large enough to permit identification and volume measurement by PLM, although large measurement errors can result when estimating component volumes for non-equant shapes. Perhaps the most reliable methods of verification involve chemical comparison of measured and reduced chemical data; this is particularly true when accurate individual component compositions are also known.

ACKNOWLEDGEMENTS

This research has been supported in part by the State of South Dakota and by the Department of Energy (Grant DE-FG01-85ER13407). The author is grateful to M. Spilde, K. Galbreath, B. J. Johnson, S. Simon, J. J. Papike, and C. K. Shearer of this institution and J. C. Laul of Battelle Northwest Institute for providing diffraction and chemical data on the Salton Sea core samples.

REFERENCES

1. F. H. Chung, A new x-ray diffraction method for quantitative multicomponent analysis, Adv. X-ray Anal. 17:106 (1974).
2. B. L. Davis, "Standardless" x-ray diffraction quantitative analysis, Atm. Environ. 14:217 (1980).
3. B. L. Davis, A tubular aerosol suspension chamber for the preparation of powder samples for x-ray diffraction analysis, Powd. Diff. 1:240 (1986).
4. B. L. Davis and L. R. Johnson, On the use of various filter substrates for quantitative particulate analysis by x-ray diffraction, Atm. Environ. 16:273 (1982).
5. B. L. Davis and L. R. Johnson, Sample preparation and methodology for x-ray quantitative analysis of thin aerosol layers deposited on glass fiber and membrane filters, Adv. X-ray Anal. 25:295 (1982).
6. B. L. Davis, Reference intensity quantitative analysis using thin-layer aerosol samples, Adv. X-ray Anal. 27:339 (1984).
7. B. L. Davis and L. R. Johnson, The use of mass absorption in quantitative x-ray diffraction analysis, Adv. X-ray Anal. (in press).
8. B. L. Davis and M. J. Walawender, Quantitative mineralogical analysis of granitoid rocks: A comparison of x-ray and optical techniques, Am. Mineral. 67:1135 (1982).
9. H. I. Britt and R. H. Luecke, The estimation of parameters in non-linear, implicit models, Technometrics 15:233 (1973).

CORRECTION OF COMPOSITIONAL VARIABILITY

IN THE X-RAY DIFFRACTION PHASE ANALYSIS

Agnes Griger

ALUTERV-FKI

Budapest H-1389, POB 128, Hungary

INTRODUCTION

X-ray diffraction analysis is one of the most frequent and important tools in qualifying materials, both in research laboratories and in industrial practice. In most cases materials are characterized not only by their chemical composition, but also by their crystal state and the quantity of crystalline components.

For example, in certain minerals the same chemical element may occur in different crystal structures and the structure of crystalline components strongly influences the choice of proper manufacturing process. From the point of view of alumina industry it is decisive, in which form occur the aluminium in bauxite (gibbsite, boehmite or diaspore). It is also important to know, which crystalline phases do the minor elements (Si, Fe, Ti) form, and what is the proportion of these phases.

Another example can be found for aluminium semi-products. Certain intermetallic phases formed under the effect of alloying and/or impurity elements are not desirable, whereas other phases may improve the properties of the product.

In research laboratories, quantitative phase analysis leads to the determination of physical properties of polycrystalline materials. In constructing or refining the phase diagram of binary, ternary, etc. systems, the X-ray diffraction phase analysis is the basic method.

Obviously, the accuracy requirements of phase analysis depend on the application field. Generally, in the case of industrial quality control and ensurance the semi-quantitative analysis (i.e. information concerning the main components) is satisfactory. Introducing advanced technologies, the accuracy demands of the industrial practice increase. This development challenges the research laboratories to elaborate new methods of measurements and evaluation with respect to special materials and technological processes.

The quantitative phase analysis is applied for the determination of individual crystalline constituents in multicomponent mixtures and polycrystalline materials. In mixtures, the individual constituents show a characteristic diffraction intensity pattern independently of the presence of other constituents. The intensity is proportional with the amount of the constituent. This statement of Hull[1] from 1919 was applied only in 1936 by Clark and Reynolds[2] for the purposes of quantitative phase analysis. The internal standard method, proposed by Klug and Alexander[3] in 1954 was the first well applicable method but this was very time consuming, especially for mixtures having more than two components. To overcome this difficulty was attempted in a lot of works (Bezjak[4], Black[5], Karlek and Burnett[6], Moore[7], Zoltay and Jahanbagloo[8]). Chung[9,10,11] generalized the experiences and proposed a unified method. The method can be automated knowing the Reference Intensity Ratios (introduced by Hubbard et al.[12]) of the crystalline phases. The basic method was modified for special cases (Zevin[13], Grzeta and Popović[14]) and a new calibration method was elaborated by Wölfel[15].

DISCUSSION

In the present work the credibility of results and the magnitude of errors are analyzed on the basis of the method of Chung[9,10,11]. The proportion of phase i in weight % can be obtained as

$$
f_i = \frac{I_i/k_i}{\sum \dfrac{I_j}{k_j}} , \qquad\qquad (1)
$$

where I_i is the intensity of X-ray diffracted by phase i and k_i is the Reference Intensity Ratio (I/I_c) value of phase i. If all the crystalline phases in the sample were identified, the analysis can be easily performed by using the appropriate k_i values. In order to apply this method the following conditions should be satisfied:

- the optimum particle size and the homogeneity of the sample are achieved,
- the sample is free of preferred orientation,
- the characteristic intensity values of the components are properly measured.

These conditions can be more or less satisfied by applying preparative, evaluation and correction methods.

The presumed knowledge of the k values is worth of further considerations. For constituents with simple, well-known, stable crystal structure, containing a few atoms per unit cell, the k value is available. They can be found in the catalogue of JSPDS cards[16], or they can be easily measured or they can be calculated on the basis of single crystal data (Yvon et al.[17]).

However, the structure of certain constituents are often unknown and they are not available in monophase form. Thus their k value is also unknown.

The chemical composition of the constituents may sometimes vary in a relatively wide range. As a result, the degree of crystallinity, the size of crystallites and last but not least the average atomic weight of the constituent vary. These changes may affect the actual k value.

In principle, all the difficulties can be avoided, if the k values are defined and measured for a given system, e.g. for samples originating from the same geological environment (Bayliss[18]), for samples from an equilibrium alloying ternary system (Stefániay et al.[19]), but in practice a lot of problems arise, as it is discussed below.

We shall distinguish between two cases.

(a) The variation of k as a function of the composition can be calculated or obtained by means of measuring a series of reference samples.

(b) Neither the single crystal data, nor the proper reference samples are available.

As an example for case (a) we consider goethite (α-FeOOH). It always occurs in the raw material bauxite and in the Bayer technology it is lost in the red mud. The goethite content, however, attracted the attention of the alumina industry. On one hand, goethite often contains aluminium which can be obtained by a proper technology, and, on the other hand, the particle size of goethite, influencing the speed of sedimentation, is strongly correlated with its aluminium content.

Aluminium may substitute iron in goethite lattice up to 33 mol% in solid solution form (Thiel [20]). The structure of goethite is well known, thus the effect of aluminium substitution on the k can easily be calculated. The k value decreases with increasing aluminium content, since the atomic scattering factor of iron is much higher than that of aluminium.

A series of reference samples was prepared and measured in our laboratory and calculations were also performed (Griger et al. [21]). The artificially prepared alumo-goethite samples did not entirely satisfy the criteria of ideal polycrystals, which were assumed in the calculations. Measured and calculated results somewhat differ from each other, but they equally show the above correlation between k value and aluminium content. The deviation can be explained by other changes in the physical state (decreasing degree of crystallinity) caused by aluminium substitution.

In the same series of measurements beside the k values also the dependence of the interplanar spacing d(1,1,0) on the aluminium content was determined. Afterwards in the quantitative phase analysis merely on the basis of the X-ray diffraction pattern the actual k value can be obtained from the shift of the reference line of goethite:

$$k_{(Al_x Fe_{1-x}OOH)} = k_{FeOOH} + n \ (d_{(Al_x Fe_{1-x}OOH)} - d_{FeOOH}) \ ,$$

where n is a factor known from the above sketched measurements.

The total change of the measured k values does not exceed 40 % in the full range of aluminium substitution. In usual bauxite samples of 10–12 constituents the average error of analysis is about 8 to 10 %. As the goethite content usually do not exceed 10 %, the average error does not significantly increases for the non-goethite constituents even disregarding the effect of aluminium substitution in goethite. However, as it was pointed out above, in certain cases quantitative phase analysis is made with a special accent on correctly determining the goethite content itself. For this case the correction proposed above should be applied.

The deviation of the composition of crystalline compounds from the stoichiometric composition is generally due to a non-equilibrium formation. This effect can be observed first of all for alloys. Substitution with impurity atoms or a phase formation with non-equilibrium composition occur in the stable or metastable phases formed during non-equilibrium solidification. A good example for the composition change without impurity substitution is the intermetallic compound α_H – AlFeSi formed in the ternary Al-Fe-Si system. The silicon content may vary between 7 to 9.5 % as a function of the bulk composition (Stefániay et al. [19]), whereas the iron content remains unchanged.

The actual silicon content can be determined on the basis of the changes of lattice parameters (Griger[22]). We could not measure the variation of k value as a function of the silicon content due to the lack of a proper monophase series. Calculation from the single crystal data are also impossible, as the atomic scattering factors of aluminium and silicon are close to each other. From the unchanged iron content follows, that only aluminium and silicon may substitute each other and, consequently, the molecular weight does not change drastically. Therefore the average atomic scattering factor and the k value remain also practically unchanged.

A similar situation with the lack of impurity substitution occurs in the binary system in the case of Θ-AlFe ($Al_{13}Fe_4$) equilibrium phase. The compound exists in the range of 38.6 to 41 % iron content for the equilibrium state, and the upper limit increases up to 42 % in non-equilibrium state (Griger et al.[23]). From the point of view of quantitative phase analysis the k value is interesting only at the lower and upper limits of the range. Since the molecular weight changes only slightly, one cannot think of a significant change of the k value. This was verified by the results of measurements: the k values at 25.0 %, 39.5 % and 41.1 % iron content are 0.47, 0.44 and 0.45, respectively.

However, in the ternary Al-Fe-Si system in non-equilibrium circumstances Θ-AlFe may solve silicon in a significant amount (1 to 6 %) without phase transformation (Stefániay et al.[19]). Because of reasons mentioned above, the change caused by silicon in k value cannot be estimated, since the change of the iron content in the Θ-AlFe is also insignificant.

Phase formation with impurity substitution occurs mainly for metastable compounds. These intermetallic compounds stabilize their crystal structure just by means of these impurities. For example, the compound α_c-AlFeSi often contains transition elements (Mn) in the iron positions, in the binary intermetallic phase Al_6Fe Cu and Mn may substitute Fe and in the binary intermetallic phase Al_9Fe_2 Co is the stabilizer atom. The average change of the atomic number is also small for these substitutions, thus the change of the k value is also insignificant: It was checked by means of calculations, e.g. for α_c-AlFeSi k = 2.8, while for α_c-AlMnSi k = 2.7, for Al_6Fe k = 1.25 and for Al_6Mn k = 1.22.

Now we turn to case (b), when neither credible calculations, nor adequate reference measurements can be performed. Such materials are e.g. the clay minerals (Bayliss[18]) or the intermetallic compounds containing many atoms in large cells in a ternary, quaternary, etc. system.

Let us consider again the Al-Fe-Si system. The intermetallic phases β-AlFeSi and γ-AlFeSi can be well identified on the basis of their well-known powder pattern. Nevertheless, their single crystal structures are not determined and monophase series cannot be prepared. The composition range of these phases is rather wide in the Fe-Si coordinate system (Phragmén[24], Munson[25], Stefániay et al.[19]).

In such cases the traditional method of X-ray diffraction phase analysis cannot lead to reasonable results. However, the accuracy and credibility of phase analysis can be improved by means of simultaneous application of X-ray diffraction (XRD), chemical analysis and electron probe microanalysis (EPMA). To take the change of chemical composition of components into consideration, beside of the accurate intensity measurement and the chemical analysis of the sample (Bárdossy et al.[26]), the actual composition of the phases must also be known. The results can be cross-checked by the relation (Majumdar et al.[27])

$$c_j = \sum_m a_{jm} f_m \,, \tag{2}$$

26

where the phase proportion f_m is obtained by XRD, the phase chemical composition a_{im} can be determined by EPMA and the chemical analysis yields the concentration c_j of element j.

Now, if some of the k values are not known, formula (2) gives also a basis for fitting them. Having the proper k values the phase analysis can be performed.

One has to have a series of samples with similar features and behavior, either collected from a certain geological environment or prepared in a certain alloy system (e.g. Al-Fe-Si). Denoting the samples by ℓ ($\ell=1,2,..L$) one has to minimize the expression

$$Q = \sum_{\ell=1}^{L} \left(\sum_{j=1}^{N} \left(\sum_{i=1}^{F} (a_{ji}^{\ell} \; f_i^{\ell} - c_j^{\ell}) \right) \right)^2$$

for k_i, which are related to f_i^{ℓ} by Eq. (1). Here F is the number of phases and N is the number of chemical constituents.

Of course, the known k_i values should be fixed in the minimization and Q should be minimized only for the unknown k_i values. If a "known" k_i value seems to be suspicious, we can treat it as an unknown one, etc. The analysis may include also simple assumptions concerning the dependence of k_i on the chemical composition.

This procedure is now under development in our laboratory for getting a better insight into the Al-Fe-Si system.

CONCLUSION

The composition of crystalline materials with the exception of almost ideal crystals (α-corundum, α-quartz, metals, etc.) can change in a relatively wide range and/or certain atoms can be substituted by chemically similar atoms without changing the crystal structure. In most cases these changes may be neglected from the point of view of quantitative phase analysis, but in delicate cases one has to consider them. The simultaneous application of XRD, EPMA and chemical analysis seems to be an adequate way of solving the problem.

Fortunately, the average change of atomic number is generally small and, consequently, the k values may vary only within narrow limits. In certain isomorphous substitutions, however, the difference of atomic number and the measure of substitutions are high and thus k may change up to 40-50 % which should be taken into account in applications.

In the present work the changes of atomic number and average atomic scattering factor due to the chemical composition of phase were considered as the main factors changing k values. The effects through the variation of crystalinity were neglected, but in order to diminish this limitation, k values were determined only for crystalline constituents originating from similar environments or belonging to the same alloy system. Consequently, the obtained k values are valid only for the given system.

It has to be mentioned, that the application of a complex methodology may prove to be very expensive and time consuming.

REFERENCES

1. A. W. Hull, J. Am. Chem. Soc. 41:1168 (1919).
2. G. L. Clark and D. H. Reynolds, Ind. Eng. Chem., Anal. Ed. 8:36 (1936).
3. H. P. Klug and L. E. Alexander, Quantitative Analysis of Powder Mixtures in "X-ray Diffraction Procedures", Wiley and Sons, New York (1954).
4. A. Bezjak, X-ray quantitative analysis of multiphase systems, Croatica Chem. Acta 33:197 (1961).
5. R. H. Black, Quantitative diffractometric powder analysis of multicomponent mixtures, Norelco Reporter 10:14 (1963).
6. R. F. Karlak and D. S. Burnett, Quantitative phase analysis by X-ray diffraction, Anal. Chem. 36:1741 (1966).
7. C. A. Moore, Quantitative analysis of naturally occuring multicomponent mineral systems by X-ray diffraction, Clay and Clay Min. 16:325 (1968).
8. C. Jahanbagloo and T. Zoltai, Quantitative analysis with the aid of calculated X-ray patterns, Anal. Chem. 40:1739 (1968).
9. F. H. Chung, Quantitative interpretation of X-ray diffraction patterns of mixtures I, J. Appl. Cryst. 7:519 (1974).
10. F. H. Chung, Quantitative interpretation of X-ray diffraction patterns of mixtures II, J. Appl. Cryst. 7:526 (1974).
11. F. H. Chung, Quantitative interpretation of X-ray diffraction patterns of mixtures III, J. Appl. Cryst. 8:17 (1975).
12. C. R. Hubbard, E. H. Evans, and D. K. Smith, The Reference Intensity Ratio I/I_c for computer simulated powder patterns, J. Appl. Cryst. 9:169 (1976).
13. L. S. Zevin, A method of quantitative phase analysis without standards, J. Appl. Cryst. 10:147 (1977).
14. B. Grzeta and S. Popović, Semiquantitative X-ray diffraction method for phase analysis using additions of a foreign component, J. Appl. Cryst. 18:80 (1985).
15. E. R. Wölfel, A new method for quantitative X-ray analysis of multiphase mixtures, J. Appl. Cryst. 14:291 (1981),
16. Powder Diffraction File, Inorganic Phases Alphabetical Index, International Centre for Diffraction Data, published by JCPDS (1983).
17. K. Yvon, W.Jeitschko, and E. Parthé, LAZY PULVERIX, a computer program for calculating X-ray and neutron diffraction powder patterns, J. Appl. Cryst. 10:73 (1977).
18. P. Bayliss, Quantitative analysis of sedimentary minerals by powder X-ray diffraction, Powder Diffraction 1:37 (1986).
19. V. Stefániay, A. Griger, and T. Turmezey, Intermetallic phases in the aluminium-side corner of the AlFeSi alloy system, J. Mat. Sci. 22:539 (1987).
20. R. Thiel, Zum System α-FeOOH-α-AlOOH, Z.Anorg. Allg. Chem. 326:70 (1963)
21. A. Griger, L. Bottyán, and M. Feiszl-Sajó, unpublished work (1979).
22. A. Griger, Powder diffraction data for the α_H intermetallic phases with slight variation in composition in the system Al-Fe-Si, Powder Diffraction 2:31 (1987).
23. A. Griger, V. Stefániay, and T. Turmezey, Crystallographic data and chemical compositions of aluminium-rich Al-Fe intermetallic phases, Z. Metallkunde 77:30 (1986).
24. G. Phragmén, On the phases occurring in alloys of aluminium with copper, magnesium, manganese, iron and silicon, J. Inst. Met. 77:489 (1950).
25. D. Munson, A clarification of the phases occurring in aluminium-rich aluminium-iron-silicon alloys, with particular reference to the ternary phase α-AlFeSi, J. Inst. Met. 95:217 (1967).
26. G. Bárdossy, L. Bottyán, P. Gadó, A. Griger and J. Sasvári, Automated quantitative phase analysis of bauxites, Amer. Min. 65:135 (1980).
27. A. J. Majumdar, L. S. Vallance, and C. G. G. Born, Quantitative analysis of phase composition using X-ray diffraction methods, J. Appl. Cryst. 5:343 (1972).

DILUTION AND ADDITION METHODS IN QUANTITATIVE

X-RAY DIFFRACTION PHASE ANALYSIS

S. Popović[1][2] and B. Gržeta[1]

(1) Ruder Bošković Institute,
41001 Zagreb, POB 1016
Yugoslavia

(2) Physics Department, Faculty of Science,
University of Zagreb, 41001 Zagreb, POB 162
Yugoslavia

The principles of the following doping methods in phase analysis of a
multicomponent mixture are described: (a) determination of the fraction of
a component using repeated dopings or (b) using a single doping, (c) simultaneous determination of the fractions of several components using a single
doping, (d) determination of the fraction of the dominant component, (e)
semiquantitative phase analysis using additions of a foreign component.
The methods (c), (d) and (e) have been recently postulated and developed
by the authors. The applicability of the methods is stated and the optimum
conditions to minimize systematic errors are discussed.

INTRODUCTION

In analysis of a multicomponent material by chemical and spectroscopic
methods elemental composition can be obtained, but usually great difficulties are faced in distinguishing the chemical identity of the various com-
ponents-phases present in the material and in determining the amounts of
particular phases. X-ray powder diffraction seems to be a perfect technique for the analysis of a multicomponent material. Each crystalline component of the material gives its characteristic diffraction pattern independently of the others, making it possible to identify the components of
interest. The intensities of diffraction lines of each component are proportional to its amount, except for the absorption correction, so that an
appropriate quantitative analysis can be performed.

All the methods for quantitative X-ray diffraction phase analysis developed so far are based on the principles postulated by Alexander and
Klug.[1,2] Let a system consist of several components denoted by capital
letters, and let the same notation represent their weight fractions, i.e.

$$A + B + C + \ldots + X + Y + \ldots = 1. \tag{1}$$

The integrated intensity of a selected diffraction line of a component,
say A, is related to the weight fraction of that component according to the
equation

$$I_A = K_A \frac{A}{d_A \mu} \quad , \tag{2}$$

where d_A is the density of the component A, μ is the mass absorption coefficient of the system and K_A is a factor depending on the nature of the component which is considered (in this case component A), on the chosen diffraction line and the geometry of the diffractometer.

For pure component A, since A = 1, (2) changes into

$$I_A^o = K_A \frac{1}{d_A \, \mu_A} \quad , \tag{3}$$

where μ_A is the mass absorption coefficient of the component A. From (2) and (3) it follows:

$$\frac{I_A}{I_A^o} = \frac{\mu_A}{\mu} \, A. \tag{4}$$

The absorption coefficients are not known precisely. Therefore, it is obvious from (4) that the weight fraction A cannot simply be obtained by measuring the ratio I_A/I_A^o, i.e. by a direct comparison of the diffraction pattern of the system containing component A with the pattern of the pure component A.

Analogously to (4), for another component, say B, it can be written

$$\frac{I_B}{I_B^o} = \frac{\mu_B}{\mu} \, B. \tag{5}$$

From (4) and (5) one obtains

$$\frac{A}{B} = K_{AB} \frac{I_A}{I_B} \quad , \tag{6}$$

where

$$K_{AB} = \frac{\mu_B}{\mu_A} \frac{I_B^o}{I_A^o} \quad .$$

K_{AB} is a constant for the two components which are considered (in this case A and B), for the selected diffraction lines and for a given diffractometer.

However, a direct application of (6) is not straightforward, since the absorption coefficients are not accurately known. In order to circumvent this problem semi-empirical internal-standard methods are usually used. For each component, say A, the weight fraction of which is to be determined, a calibration curve should be constructed. This curve can be constructed in such a way as to relate A/S vs. I_A/I_S, where S denotes now the internal standard. According to (6) the calibration curve is a straight line having the slope K_{AS}. The slope is obtained from the intensity measurements on a series of mixtures with known ratios A/S. To determine the fraction of the component A in the system a known fraction of the internal standard S is added to the system, the intensity ratio I_A/I_S is measured, and A is found from the previously constructed calibration curve. More details on the internal standard methods, as applied in special cases, can be found in the text-book of Klug and Alexander.[2]

In the method developed by Chung,[3,4] no calibration curve is needed, as the absorption coefficients are flushed out of the intensity-weight fraction equation. The method is based on a previous knowledge, or measurement, of relative (reference) intensities of (the strongest) diffraction lines for each pair of components that are present in the system, or rather for each component and a reference material (corundum, α-Al_2O_3). All information related to the quantitative composition of the system can be decoded directly from its X-ray diffraction pattern.

In the doping methods, described in this paper, the intensity-weight fraction equations are also free of the absorption coefficients. The doping method was introduced by Copeland and Bragg,[5] simplified by Bezjak and Jelenić,[6] and generalized by the present authors.[7,8] The generalized method involves the addition, to the original system, of known amounts of the components, the weight fractions of which are to be determined. The corresponding equations, deduced with no approximations, relate the weight fraction of the component sought to the intensities diffracted by that component and by any non-added component (already present in the system), before and after doping. This method can also be applied to systems containing unidentified components, in simultaneous analysis only for those of interest, as well as for amorphous content determination.[7,8] Two variations of this original method have also been developed.[8,9]

(a) THE WEIGHT FRACTION OF A COMPONENT OBTAINED USING REPEATED DOPINGS[5]

Copeland and Bragg[5] suggested that the system can be doped with known amounts of the component, say A, the weight fraction of which is to be determined. The system already contains an amount of another component, say B, which is considered as a reference component. A' g of the component sought per 1 g of original system is added. The weight fraction of this component is now $(A + A')/(1 + A')$, and the weight fraction of the reference component is $B/(1 + A')$. In this case (6) can be written as

$$\left(\frac{I_A}{I_B}\right)' = \frac{1}{K_{AB}} \left(\frac{A + A'}{B}\right) = (\text{constant})(A + A') , \tag{7}$$

the prime indicating the intensities measured after doping. A plot of (I_A/I_B), as a function of A' is a straight line and the quantity sought A, is the absolute value of the A'-axis intercept. From various combinations of diffractions lines of the components A and B, several estimates of A can be obtained, thus increasing the precision of the measurement.

(b) THE WEIGHT FRACTION OF A COMPONENT OBTAINED USING A SINGLE DOPING[6]

In order to avoid several consecutive dopings Bezjak & Jelenić[6] have shown that the weight fraction of any component, say A, can be determined from a single doping and from only two diffraction patterns. Namely, from (6) and (7) it follows:

$$\left(\frac{I_A}{I_B}\right)' = \frac{A + A'}{A} \left(\frac{I_A}{I_B}\right) . \tag{8}$$

On the right side of (8) are the intensities of the original sample, and on its left side the intensities of the sample doped with a known amount, A', of the component sought. B is any other component already present in the system. Equation (8) is a straight line, and its slope, $(A + A')/A$, determined from experimental data for all possible intensity ratios, yields the initial weight fraction of the component sought, A.

(c) SIMULTANEOUS DETERMINATION OF WEIGHT FRACTIONS OF SEVERAL COMPONENTS USING A SINGLE DOPING[7]

It has been demonstrated[7] that the weight fractions of several components can be <u>simultaneously</u> determined from only two diffraction patterns: the pattern of <u>the original</u> sample and the pattern of the sample doped with known amounts of all components sought. The two patterns are to be taken under the same experimental conditions. In principle, the weight fractions of all components in the system can be determined, if known amounts of all components except one are added. The equations shown below are obtained using the relation that the sum of all weight fractions after doping equals 1, instead of 1 + A', as above. The method is described for a binary system, for a ternary system with one component or two components added, and for a general multicomponent case.

Binary System

Original sample [equations (1) and (6)]:

$$A + B = 1 \tag{9}$$

$$\frac{A}{B} = K_{AB} \frac{I_A}{I_B} . \tag{6}$$

One wants to determine the weight fraction of one component, say B. A known weight fraction of this component, B_a, is added to the original sample. Then A is used as the reference component.

Doped sample:

$$A_d + B_d + B_a = 1 , \tag{10}$$

$$\frac{A_d}{B_d + B_a} = K_{AB} \frac{I_A^d}{I_B^{d+a}} . \tag{11}$$

A_d and $B_d + B_a$ are the weight fractions of the components in the doped sample and I_A^d and I_B^{d+a} the corresponding intensities. The following is also valid:

$$\frac{A_d}{B_d} = \frac{A}{B} . \tag{12}$$

There are five equations (9,6,10-12) for five unknowns, A, B, A_d, B_d and K_{AB}. After elimination it follows that the weight fractions of the components are

$$B = \frac{B_a R_{AB}}{P(1 - R_{AB})} , \qquad A = 1 - B , \tag{13}$$

where

$$R_{AB} = \frac{I_A^d}{I_B^{d+a}} \frac{I_B}{I_A} , \qquad P = 1 - B_a . \qquad (14)$$

P is a parameter, which equals 1 - the weight fraction, in the doped sample, of the added amount, or the weight fraction of the original composition in the doped sample.

Ternary System

(i)   The weight fraction of one component, say B, is to be determined.

Original sample: In addition to (6) one has

$$A + B + C = 1 , \qquad (15)$$

$$\frac{C}{B} = K_{CB} \frac{I_C}{I_B} . \qquad (16)$$

A known weight fraction, B_a, of the component sought is added to the original sample. Then, either A or C may be used as the reference component.

Doped sample: Besides (11) and (12) the following is valid:

$$A_d + B_d + B_a + C_d = 1 , \qquad (17)$$

$$\frac{C_d}{B_d + B_a} = K_{CB} \frac{I_C^d}{I_B^{d+a}} , \qquad (18)$$

$$\frac{C_d}{B_d} = \frac{C}{B} . \qquad (19)$$

It can be shown that, out of eight equations (15, 6, 16, 17, 11, 18, 12, 19) there are only seven independent equations for eight unknowns, A, B, C, A_d, B_d, C_d, K_{AB} and K_{CB}. This means that only the weight fraction of the added component, B, can be determined. If B is related to one of the two other components, either A or C, as the reference component, it follows that there are five equations for five unknowns. The problem can be, therefore, treated in terms of a binary system, regardless of the third component (or of all other components in a multicomponent system).

If B is related to A, one obtains that B is given by (13), with R_{AB} defined by (14).

If B is related to C it follows:

$$B = \frac{B_a R_{CB}}{P(1 - R_{CB})} ,$$

where

$$R_{CB} = \frac{I_C^d}{I_B^{d+a}} \frac{I_B}{I_C} .$$

In both cases the parameter $P = 1 - B_a$. It is obvious that $R_{AB} = R_{CB}$.

(ii) The weight fractions of all three components are to be determined. In this case known amounts of two components, say B and C, are added to the original sample. Then A is used as the reference component.

Original sample: In addition to (15) and (6) one has

$$\frac{A}{C} = K_{AC} \frac{I_A}{I_C} . \tag{20}$$

Doped sample (the amounts B_a and C_a are added to the original sample): Besides (11) and (12) the following can be written:

$$A_d + B_d + B_a + C_d + C_a = 1 , \tag{21}$$

$$\frac{A_d}{C_d + C_a} = K_{AC} \frac{I_A^d}{I_C^{d+a}} , \tag{22}$$

$$\frac{A_d}{C_d} = \frac{A}{C} . \tag{23}$$

There are eight independent equations (15, 6, 20, 21, 11, 22, 12, 23) for eight unknowns, A, B, C, A_d, B_d, C_d, K_{AB} and K_{AC}. It can be shown that in this case the weight fractions of the components are given as follows:

$$B = \frac{B_a R_{AB}}{P(1 - R_{AB})} , \qquad C = \frac{C_a R_{AC}}{P(1 - R_{AC})} , \qquad A = 1 - (B + C) .$$

The parameter P equals, as previously, the weight fraction of the original composition in the doped sample,

$$P = 1 - (B_a + C_a) ,$$

while R_{AB} and R_{AC} contain intensities of the added components (B,C) and the intensities of the non-added, reference component (A), before and after doping,

$$R_{AB} = \frac{I_A^d}{I_B^{d+a}} \frac{I_B}{I_A} , \qquad R_{AC} = \frac{I_A^d}{I_C^{d+a}} \frac{I_C}{I_A} .$$

Multicomponent System

The essential points in the generalized doping method can be summarized as follows. Let the system consist of N components. Its diffraction pattern is taken and a partial or complete identification of the components is performed. The prominent non-overlapping (in principle the strongest) diffraction lines of particular components are chosen and their net integrated intensities are measured. Then the system is doped by exactly known amounts of M components, the weight fractions of which are to be determined (M = 1,2,3 ..., N-1). The intensities of the chosen diffraction lines are measured again. The weight fraction of any component added can be determined from the intensities measured before and after doping. In principle, when N - 1 components are added, the weight fractions of all N components can be found. The weight fraction of any component, say X, in the original sample, is given by the equation

$$X = \frac{X_a}{P} \frac{R_{YX}}{1 - R_{YX}} \, . \tag{24}$$

Here X_a is the weight fraction, in the doped sample, of the added amount of component X, and Y is any non-added (reference) component. P equals the weight fraction of original composition in the doped sample (or 1 – the total weight fraction, in the doped sample, of all the added amounts). R_{YX} is expressed through the intensities of the components X and Y before (I_X, I_Y) and after doping (I_X^{d+a}, I_Y^d),

$$R_{YX} = \frac{I_Y^d}{I_X^{d+a}} \frac{I_X}{I_Y} \, . \tag{25}$$

One can utilize several pairs of diffraction lines of the components X and Y in order to increase the precision of the measurements. Namely, from (24) and (25) it follows:

$$\frac{I_X^{d+a}}{I_Y^d} = K \frac{I_X}{I_Y} \, .$$

A plot of I_X^{d+a}/I_Y^d as a function of I_X/I_Y is a straight line with the slope

$$K = \frac{XP + X_a}{XP} \, ,$$

from which it follows that the weight fraction of the component sought is given by the formula

$$X = \frac{X_a}{P} \frac{1}{K - 1} \, . \tag{26}$$

If a complete identification of all components is not made or is not possible, then the total weight fraction of the unidentified components follows from the determined fractions of the identified components. Analogously, if the system contains amorphous components, their total weight fraction can be found if the fractions of all crystalline components are measured.

Examples of application of this method are shown in Figure 1 and Table 1.

(d) DETERMINATION OF THE WEIGHT FRACTION OF THE DOMINANT COMPONENT[8]

A specific case of the doping method has been elaborated,[8] in which the weight fraction of a (dominant) crystalline component in the multi-component system can be determined from the measurement of diffraction line intensities of that component only.

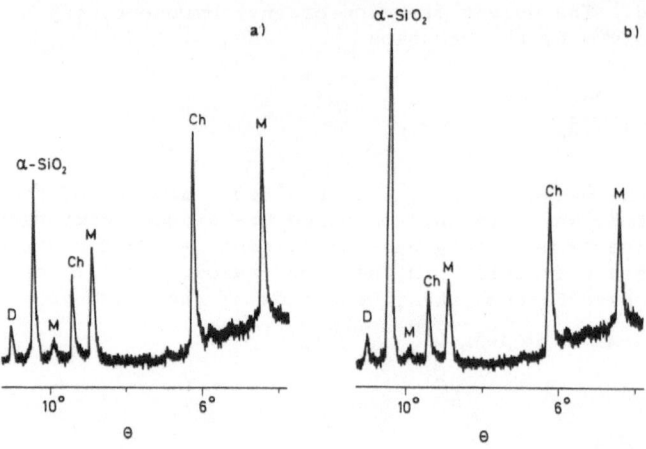

Fig. 1. Determination of the weight fraction of quartz, α-SiO_2, in industrial dust; method ($\underline{c}$): a) a part of diffraction pattern of the
original sample; b) a part of diffraction pattern of the same
sample doped with the known amount of quartz, X_a = 0.286. The
found weight fraction of quartz in the original sample: X = 0.200.
Ch-chlorites, D-dolomite, M-micas. Radiation CuKα.

Table 1. Determination of Weight Fractions in a Ternary System
by Application of the Method ($\underline{c}$)

Component	A	B	C
Compound	α-Al_2O_3	ZnO	α-SiO_2
Measured diffraction lines	102 113 204	100 102 103	100 101 112
Original sample (known)	$A = \frac{2}{4}$	$B = \frac{1}{4}$	$C = \frac{1}{4}$
Doped sample (known)	$A_d = \frac{2}{8}$	$B_d = \frac{1}{8}$ $B_a = \frac{2}{8}$	$C_d = \frac{1}{8}$ $C_a = \frac{2}{8}$
Found weight fraction	$A = 1-(B+C)$ $A = 0.487$	$B = 0.255$	$C = 0.258$

Let a system contain a component, say X, which is dominant, and let
the diffraction lines of the other components be weak. In such a case the
doping method ($\underline{c}$) is not appropriate to determine the weight fraction of X,

as the diffraction lines of any other component used as the reference component would be even weaker after doping. The problem can be solved by doping the sample by a known amount of a crystalline compound already contained in the system, or by one not contained in the system; in both cases the compound added will be used as the reference component in the doping method applied to the doped sample.

On the other hand, the following method can be used. The diffraction pattern of the investigated sample is taken (first pattern) and the net integrated intensity of a prominent non-overlapping (in principle the strongest) diffraction line of the component X is measured (I_X). Then the sample is doped by a known weight fraction, X_a, of the component X. The diffraction pattern of the doped sample is taken (second pattern), and the intensity of the chosen diffraction line of the component X is measured again (I_X^{d+a}). Finally, the diffraction pattern of the pure component X is taken (third pattern) and the intensity of the same diffraction line is measured once more (I_X^o). It is important that all three diffraction patterns are taken under the same experimental conditions. The weight fraction of the component X in the original sample is given by the equation

$$X = \frac{X_a}{P} \frac{R'}{R''} , \tag{27}$$

where

$$R' = 1 - \frac{I_X^{d+a}}{I_X^o} , \qquad R'' = \frac{I_X^{d+a}}{I_X} - 1 , \tag{28}$$

and P is the weight fraction of the original composition in the doped sample,

$$P = A_d + B_d + \ldots + X_d + Y_d + \ldots = 1 - X_a .$$

Several diffraction lines of the component X may be used in the analysis in order to increase the accuracy.

DISCUSSION ON THE METHODS (c) AND (d)[7,8]

In Eqs (24), (26) and (27) there are two factors. The first factor, X_a/P, depends on weighing. Assuming that X_a and P can be weighed with similar precision, it follows that the error in X is minimum for $X_a = P = 0.5$.

The second factor in (24), (26) and (27) contains the intensities, experimental values of which depend on the counting statistics and also on weighing. The counting statistical error is smaller for stronger diffraction lines. In the doping method (c), for a given error in R_{YX}, the error in $R_{YX}/(1 - R_{YX})$, and therefore the error in X, mathematically decreases, as R_{YX} decreases, i.e. as I_Y^d decreases (25). This means that X_a should increase. However, if X_a increases then P tends to zero, together with R_{YX}, and this may result in a large error in X. Moreover, if I_Y^d is small, the accuracy of the measurement of it may drastically decrease. Therefore, X_a must be chosen in such a way as to keep the accuracy of the measurement of I_Y^d similar to those of the other intensities. Bearing in mind the above statement on the influence of the weighing error, the value of $X_a \simeq 0.5$ appears to be most convenient. On the other hand, the counting statistical error can be minimized by averaging (a least-squares procedure) among several pairs of diffraction lines in calculation of the slope, K, of the straight line through the measured points (26).

In the method (d) [(27) and (28)], for a given error in I_X^{d+a} the error in R'/R", and therefore the error in X, mathematically decreases as I_X^{d+a} increases, i.e. as X increases. However, then both R' and P tend to zero and X may be determined very inaccurately. Therefore, it is again advisable to choose X_a close to 0.5 in order to achieve a better accuracy of X. This method is not applicable in cases where X is close to 1, as the differences between the measured intensities in (28) are small, R' and R" are both close to zero, and the accuracy of X, derived from (27), may be low.

General precautions which can increase the accuracy are as follows:
(i) Extensive grinding and mixing of the samples before and after doping are necessary to ensure sample homogenity and optimum particle size.
(ii) The preferred orientation of the crystallites has to be avoided. Its presence can be noticed: in the doping method (c) by combining different diffractions lines (from different sets of crystal planes) of the component X with the same diffraction line of the reference component Y and then repeating the procedure for different diffraction lines of Y; in the method (d) by observing different diffraction lines of the component X. If the differences among the obtained values for the weight fraction of X are out of the limits of the estimated error, with a tendency to make groups in a systematic way, depending on the Miller indices of the diffraction lines, preferred orientation is present.
(iii) The pure component added should have a similar level of crystal perfection as the same component present in the original sample.

(e) SEMIQUANTITATIVE PHASE ANALYSIS BASED ON ADDITIONS OF A FOREIGN SUBSTANCE[9]

This is a variation of the doping method, which involves the addition to the investigated system of known large (about 80-95 %) amounts of a crystalline substance X, originally not contained in the system.[9] One determines the added fraction of X, for which the intensity of a prominent non-overlapping (in principle the strongest) diffraction line of a particular component decreases to a given small detectable value. This enables the determination (estimation) of the weight fraction of that component in the original system. The method is very simple and yields the weight fractions of the major components in the system with a typical error of a few percent. The weight fractions of the minor components cannot be determined with fair accuracy in this way. The method is appropriate in cases where only a small amount of the investigated system is available.

Let a multicomponent system consist of several crystalline components:

$$A + B + \ldots + J + \ldots + N = 1 . \tag{29}$$

The diffraction pattern of such a system is a superposition of diffraction patterns of all particular components. If a known amount of the substance X, originally not present in the system, is added to the system, (29) takes the form

$$(A_d + B_d + \ldots + J_d + \ldots + N_d) + X = 1 , \tag{30}$$

where

$$A_d = A(1 - X), \quad B_d = B(1 - X), \quad \ldots, \quad J_d = J(1 - X), \quad \ldots, \quad N_d = N(1 - X).$$

The addition of substance X to the system influences the diffraction pattern: the diffraction lines of substance X appear, while the intensities of diffraction lines of the original components are weakened. Let a small but detectable weight fraction of a component, say J, Y_J, be attained by addition of the fraction X_J of substance X. Then (30) takes the form

$$(A_{d'} + B_{d'} + \ldots + Y_J + \ldots + N_{d'}) + X_J = 1 \, ,$$

where

$$Y_J = J(1 - X_J) \, , \tag{31}$$

$$A_{d'} = A(1 - X_J), \; B_{d'} = B(1 - X_J), \; \ldots, \; N_{d'} = N(1 - X_J).$$

Thus, the weight fraction of the component J in the original system follows from (31):

$$J = Y_J/(1 - X_J). \tag{32}$$

The value Y_J and the analogous values for other components (whose weight fractions are to be determined), Y_A, Y_B, $\ldots$, Y_N, are given by the corresponding detectable intensities. If the doping substance X is added to the original system in sufficiently great amounts, say X = 0.85, 0.90, 0.95, the mass absorption coefficient of the doped system is close to the one of the substance X, μ_X. Then it can be written, with a good approximation, that

$$Y_J = \frac{\mu_X}{\mu_J} \frac{I_{iJ}^{d'}}{I_{iJ}^{o}} \, , \tag{33}$$

where μ_J is the mass absorption coefficient of the component J, $I_{iJ}^{d'}$ is the intensity of the selected diffraction line i of the component J originated from the fraction Y_J in the system doped by X_J, and I_{iJ}^{o} is the intensity of the same line of pure component J. The value Y_J from (33) is used in (32) to determine the weight fraction of the component J in the original composition.

If the pure component J is not available, the intensity of its diffraction line i, I_{iJ}^{o}, in (33) is not known. Instead, one can make use of the reference intensity, I_J/I_{cor}, and the relative line intensities of the component J listed in the JCPDS Powder Diffraction File.[10] I_J and I_{cor} are the intensities of the strongest diffraction lines of the component J and corundum (α -Al_2O_3), respectively, for the binary mixture of the component J and corundum at the one-to-one weight ratio. The reference intensity values are given for a number of substances in the JCPDS Powder Diffraction File.

An example of the application of the method (e) is shown in Fig. 2. More details on systematic errors and optimum conditions in application of this method are given in the original paper.[9]

GENERAL RECOMMENDATIONS

It is supposed that a modern counter diffractometer with a monochromator is used for intensity measurements. The essential points in reducing intensity fluctuations and thus improving the counting statistics are:[2] (i) small grain (crystallite) sizes (with the order of magnitude of 10 μm; in this case the error due to the primary extinction is also small); (ii) rotation of the specimen in its own plane; (iii) relatively large divergence of the primary beam (irradiation of a larger volume of the specimen).

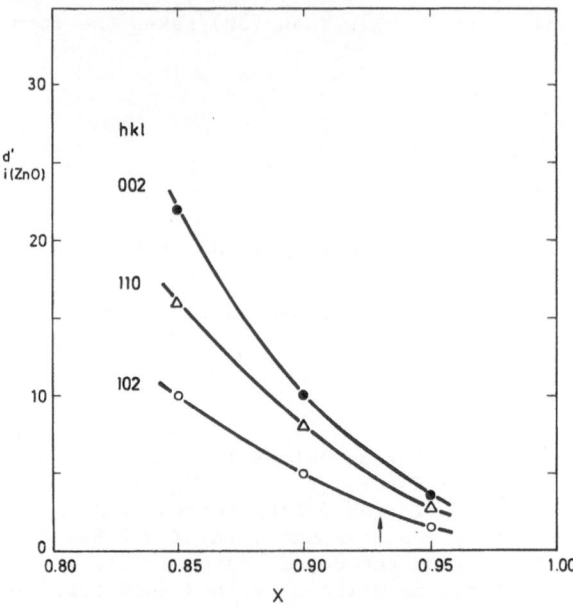

Fig. 2. Intensities of selected diffraction lines of ZnO in the ternary system α-SiO$_2$ + α-Al$_2$O$_3$ + ZnO as a function of the added weight fraction of the foreign substance X = KCl; method (e). The arrow shows the weight fraction of X, X_J, and the corresponding intensity values used in calculation of the weight fraction of ZnO (33, 32).

Quantitative analysis should be based on integrated rather than peak diffraction line intensities. The integrated intensity may be measured by recording the total counts obtained while the counter scans across the diffraction line position, followed by the background subtraction. The background is determined from measurements of counts on both sides of the given diffraction line. The measurements of the intensities have to be performed several times, with the specimen newly mounted for each measurement. The peak intensities are unreliable because: (i) the separation of the spectral doublet component $K_{\alpha_1\alpha_2}$; (ii) the height of the diffraction line decreases and its width increases as the crystallite size decreases below approximately 0.1 μm (1000 Å).

The measured intensities are also affected by the microabsorption (or particle-absorption). The change in the intensity ratios of diffraction lines of two phases in the system can be minimized by keeping grain sizes small. The microabsorption can be neglected in cases where the absorption coefficients of the phases in question do not differ greatly. Also, a difference in absorption coefficients is tolerable if the effect of microabsorption may be held constant.

The greatest source of error in quantitative analysis is the preferred orientation of the grains. Grains having characteristic, plate-like or needle-like, shapes tend to assume a preferred mode of orientation when mounted for the diffractometric analysis. Any degree of the preferred orientation results in deviations of diffraction line intensities from the true values, corresponding to the random orientation. All precautions should be undertaken to avoid the preferred orientation. If the degree of the preferred orientation is not great, rather accurate value for the weight fraction of a particular component may be obtained by averaging the data, which follow from the intensity values affected by the preferred orientation in different ways.

REFERENCES

1. L. Alexander and H. P. Klug, Basic aspects of X-ray absorption in quantitative diffraction analysis of powder mixtures, <u>Anal</u>. <u>Chem</u>. 20:886 (1948).
2. H. P. Klug and L. E. Alexander, "X-ray Diffraction Procedures",John Wiley & Sons, New York (1974).
3. F. H. Chung, Quantitative interpretation of X-ray diffraction patterns of mixtures. I., II., <u>J</u>. <u>Appl</u>. <u>Cryst</u>. 7:519, 526 (1974).
4. F. H. Chung, Quantitative interpretation of X-ray diffraction patterns of mixtures. III., <u>J</u>. <u>Appl</u>. <u>Cryst</u>. 8:17 (1975).
5. L. E. Copeland and R. H. Bragg, Quantitative X-ray diffraction analysis, <u>Anal</u>. <u>Chem</u>. 30:196 (1958).
6. A. Bezjak and I. Jelenić, The application of the doping method in quantitative X-ray diffraction analysis, <u>Croat</u>. <u>Chem</u>. <u>Acta</u> 43;193 (1971).
7. S. Popović and B. Gržeta-Plenković, The doping method in quantitative X-ray diffraction phase analysis, <u>J</u>. <u>Appl</u>. <u>Cryst</u>. 12:205 (1979).
8. S. Popović, B. Gržeta-Plenković, and T. Balić-Žunić, The doping method in quantitative X-ray diffraction phase analysis. Addendum, <u>J</u>. <u>Appl</u>. <u>Cryst</u>. 16:505 (1983).
9. B. Gržeta and S. Popović, Semiquantitative X-ray diffraction method for phase analysis using additions of a foreign component, <u>J</u>. <u>Appl</u>. <u>Cryst</u>. 18:80 (1985).
10. Joint Committee on Powder Diffraction Standards, International Centre for Diffraction Data, "Powder Diffraction File", Swarthmore, Pa., USA.

SHAPE MEMORY EFFECT ANALYSIS BY X-RAY DIFFRACTION

Zbigniew Bojarski and Henryk Morawiec

Institute of the Physics and Chemistry of Metals
Silesian University
40-007 Katowice, Bankowa 12, Poland

INTRODUCTION

Shape memory effects are closely related to the reversible martensitic transformation. Hence signal importance is attached to the study of the influence of this transformation and also martensite structure and its defects on shape memory, particularly from the aspect of alloys properties for potential applications. The X-ray method offers the possibility of direct study of the reverse martensitic transformation. The NiTi and copper-based alloys have already found a valuable practical application.

For these alloys X-ray methods do not have as broad an application as electron diffraction for structural studies of martensite and the martensitic transformation. This is due to the broadening and overlapping of peaks caused by high density of stacking faults in the martensite of copper-based alloys. However in certain cases the X-ray method is very useful. Examples are given here of the application of low and high temperature diffraction analysis for martensitic and premartensitic transformation studies, phase identification, structure analysis, Patterson analysis and X-ray topography applied for investigation of shape memory alloys. The great advantage of X-ray analysis is that it permits a quantitative relation between the phase transformation and shape recovery to be determined.

X-RAY STUDIES OF COPPER-ZINC-ALUMINIUM ALLOYS

Martensite in Cu-Zn-Al alloys is obtained by quenching from the high temperature β-phase; thus shape memory effects are only exhibited by alloys with a composition for which the high temperature β-phase exists. This phase has a body centered cubic lattice which undergoes an ordering process before the martensitic transformation occurs. Therefore the martensite inherits the long range order existing in the β_1-phase. This order may be of B2 or DO_3 type. The martensite structure most often observed in these alloys has a monoclinic structure with 9 or 18 layers of close packed planes in one period.

It is well known that the structure and properties of the Cu-Zn-Al shape memory alloys and their stability are particularly sensitive to heat treatment conditions[1-3]. X-ray studies undertaken to show the significant role played by composition and conditions of quenching as well as low temperature annealing on the process of martensite transformation and the one way shape memory effect associated with it are described.

Table 1. Composition and Transformation Temperatures of the Studied Alloys

Alloy notation	Cu % at.	Zn % at.	Al % at.	$\frac{e}{a}$	M_f $^\circ C$	M_s $^\circ C$	A_s $^\circ C$	A_f $^\circ C$
1	69.5	12.6	17.9	1.48	10	45	35	70
2	71.2	11.9	16.9	1.46	90	120	100	140

Studies were conducted on two alloys for which chemical composition and characteristic temperatures of transformation are given in Table 1. Phase changes in the quenched samples were examined by the X-ray method using a high temperature DRON-1 camera.

The β_1 martensite obtained as a result of quenching is in a metastable state and during heating undergoes reverse transformation to the ordered β_1 parent phase which subsequently suffers eutectoid decomposition $\beta_1 \rightarrow \alpha + \gamma_2$. For alloy no. 1 this sequence of transformations is shown in Fig. 1, for a sample quenched from 700 $^\circ C$ in ice-water. It is clear that during the $\beta_1' \leftrightarrows \beta_1$ reverse martensite transformation the temperature of heating of the parent phase may not exceed the temperature of eutectoid decomposition. For a sample quenched from a high temperature the temperature of reverse martensite transformation is markedly raised but during the second transformation cycle this temperature is lowered. Variations of martensite peak intensity for the first two cycles are shown on Fig. 2.

Further transformation cycles cause only a slight lowering of transformation temperature. This effect may be explained by the high concentration of vacancies frozen in during quenching from a high temperature, which are annealed out during the first reverse transformation cycle.

In certain alloys, with increase in number of thermocycles an increase in content of parent β_1-phase can be observed after the transformation shown in Fig. 3. This is presumably the result of increase in density of dislocations generated during the reversible martensitic transformation. These dislocations impede the β_1-parent phase to martensite transformation.

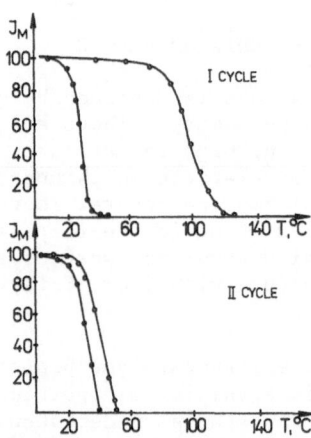

Fig. 1. The course of phase transformation in alloy no. 1 during heating of the martensite

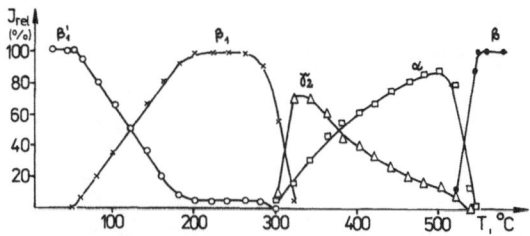

Fig. 2. The first two martensitic transformation cycles in alloy
no. 1 after quenching

The generation of dislocations in the parent during the reversible
martensitic transformation was reported by Mai et al.[4]. Their investigations were carried out on single crystals using X-ray topography.

The increase in quantity of untransformed parent phase with increasing
number of thermocycles is the reason for simultaneous decrease in degree
of shape recovery, although as can be seen from Fig. 3, this process becomes stabilized after the first few cycles.

In alloys whose composition deviates from the ideal electron ratio,
such as alloy no. 2 in Table 1, quenching in ice-water causes such a substantial rise in temperature of reverse transformation that it overlaps
with the temperature of eutectoid decomposition, as shown in Fig. 4. In
this Figure it may also be seen that only a small quantity of the martensite undergoes earlier transformation to the parent phase.

As a result of the superposition of eutectoid decomposition on the
reverse martensite transformation, during heating of deformed martensitic
samples the shape memory effect does not occur. The elimination of this
undesirable effect was found to be possible only when quenching with isothermal holding above the M_s temperature. In Fig. 5 the heating and cooling curve for alloy no. 2 after quenching with isothermal holding at
T = 200 °C is shown. A very marked lowering of temperature of reverse
transformation relative to samples quenched in ice-water may be seen.
Only the samples quenched with isothermal holding exhibit full shape recovery, while samples quenched in ice-water exhibit only 50 % shape recovery at the temperature of occurrence of the β_1 phase.

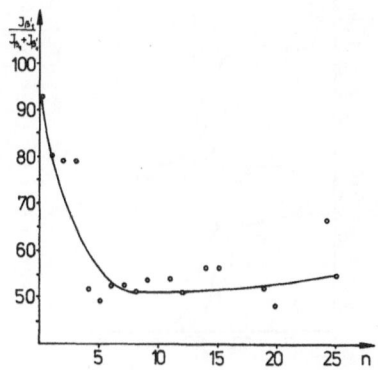

Fig. 3. Decrease in martensite content with increasing number
of martensitic transformation cycles

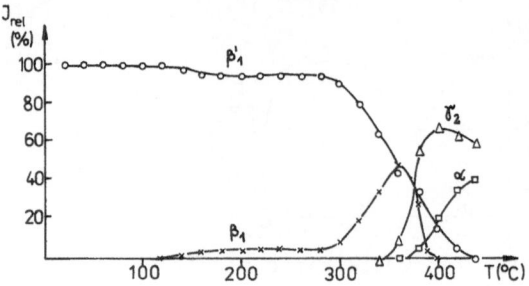

Fig. 4. The course of phase transformation in alloy no. 2 during heating of the martensite

Comparing the differences in the structure of alloy no. 2 depending on quenching conditions it may be concluded that after step-quenching with isothermal holding, the martensite exhibits greater skewing (increase of the inclination angle) of the lattice than for samples quenched in ice-water.

A further remarkable rise in the reverse martensitic transformation temperature is obtained by additional low temperature annealing of the martensite quenched in ice-water. Analysis of diffraction patterns taken from such specimens shows changes in the positions of certain diffraction peaks. Fig. 6 shows the difference between two interplanar distances determined from two peaks indexed as 040 and 320, as a function of annealing time at 200 °C. The differences between d_{040}, d_{320} decrease with increasing temperature of annealing. Interpretation of the lattice parameters showed that this low temperature annealing causes orthorhombic distortion in the monoclinic lattice of the martensite. The final effect is change of the monoclinic lattice to orthorhombic, when the 040 and 320 peaks coincide.

Delaey et al.[5] interpret this orthorhombic distortion as a result of the distortion of the regular hexagonal configuration of atoms on the basal plane of the martensite, which depends on the type of ordering of the β-phase. The highest distortion of the hexagon occurs in martensite formed from the B2 order β-phase. More regular configuration is obtained for

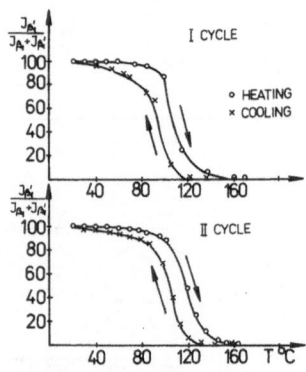

Fig. 5. The first two martensite transformation cycles in alloy no. 2 after step-quenching

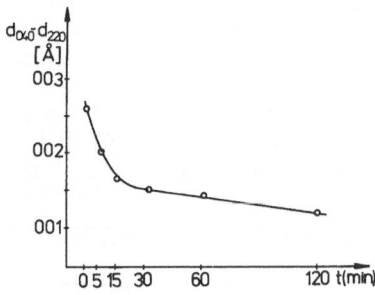

Fig. 6. Difference between interplanar distances as a function of time annealing at 200 $^{\circ}$C

martensite formed from the DO_3 type ordered β-phase, which causes the coincidence of the diffraction peaks, as mentioned above.

Thus X-ray diffraction has shown that atomic rearrangements occurring in the martensite result in modification of the lattice parameters of the martensite.

In order to explain the change in the structure of the martensite caused by ageing, Hashiguchi et al.[6] applied the Patterson diagrams to present the distribution of atoms in the lattice planes parallel to the martensite basal plane (001). X-ray measurements were made with a four circle diffractometer. Results showed that the ageing of martensite causes decrease in long range order which makes it difficult to transform disordered martensite to parent phase and thus shifts the transformation towards higher temperature.

X-RAY ANALYSIS OF NICKEL-TITANIUM ALLOYS

In these alloys the X-ray method is used principally for phase identification and transformation studies. Deviation from the equiatomic composition of the NiTi alloy is associated with the presence of Ni_3Ti or $NiTi_2$ phase. The peaks for the latter phase coincide with the martensitic structure, therefore the presence of $NiTi_2$ can only be identified in the B2 parent phase, which makes it necessary to use the high temperature attachment for the diffractometer. Because the martensitic transformation temperature is very sensitive to chemical composition and varies from the temperature of liquid nitrogen to +150 $^{\circ}$C, this equipment needs to be used for low as well as for high temperature X-ray diffractometer studies.

The high temperature parent phase in the equiatomic NiTi alloys has an ordered BCC structure (B2-type). The crystal structure of the martensite has been the subject of controversy for some time but today it is agreed that this is monoclinic distortion of the B19 structure.

The NiTi alloys exhibit premartensitic transformation with R-phase formation[7]. The R-phase transition is associated with broadening and splitting of the 110 X-ray peak on the diffraction pattern. From study of this splitting the structure of the R-phase has been determined as a rhombohedral distortion of the B2-parent phase. An example of martensitic transformation during cooling with overlapping of the R-phase transition is shown on Fig. 7 where the sequence of transformations can be seen on diffraction patterns found at various temperatures.

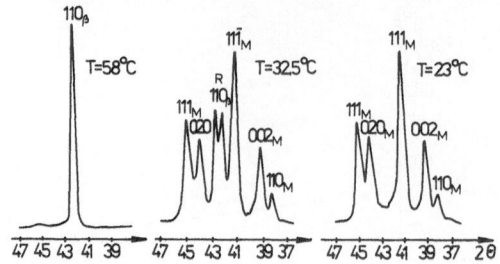

Fig. 7. Diffraction patterns illustrating the sequence of martensitic transformation in NiTi

From these studies, the sequence of transformations can be written:

B2 → B19 + B2 → B19 + R → B19

This kind of transition is appropriate for an alloy with equiatomic composition. For alloys with a higher nickel content the sequence of transformations is as follows:

B2 → R → R + B19 → B19

The R-transition is reversible with very small temperature hysteresis and is associated with shape recovery.

The experience gained at our Institute in the application of the X-ray method for analysis of shape memory alloys clearly indicates the value of this method for solving many problems in this field.

REFERENCES

1. A. Abu Arab and M. Ahlers, J. Phys. 43:C-4-709 (1982).
2. G. Scarsbrook, J. M. Cook, and W. M. Stobbs, Met. Trans. 15A:1977 (1984).
3. J. V. Humbeeck, J. Jansen, M. Wamba Nogle, and L. Delaey, Scr. Met. 18:893 (1984).
4. C. Mai, G. Guenin, M. Morin, F. Livet, and P. F. Gobin, Mat. Sc. Eng. 45:217 (1980).
5. L. Delaey, T. Suzuki, and J. Humbeeck, Scr. Met. 18:899 (1984).
6. Y. Hashigushi, H. Higuchi, I. Matsui, ICOMAT-86, Nara, Japan.
7. V. N. Hacin, Ju. I. Paskal, and V. E. Giunter, Fiz. Met. Metallov. 46:511 (1978).

FORMATION OF A SATELLITE PHASE

IN THE NEUTRON IRRADIATED REACTOR STEEL

F. Haramia

Welding Research Institute

Februárového víťazstva 71, 832 59 Bratislava, Czechoslovakia

F. Hanic

Institute of Inorganic Chemistry
Centre for Chemical Research, Slovak Academy of Sciences
Dúbravská cesta, 842 36 Bratislava, Czechoslovakia

A hexagonal satellite phase could be identified, in addition to the body-centered-cubic (bcc) iron, by X-ray powder diffraction method (CoK$_\alpha$ radiation) in a reactor steel after neutron irradiation and tensile testing. The hexagonal satellite phase is metastable, its diffraction pattern disappears from the X-ray record at room temperature after 1.5 - 2 months of relaxation. The structure of the satellite phase is closely related to the structure of the hexagonal high pressure polymorph ε-Fe. The presence of the satellite phase enhances the toughness of irradiated steel during tensile tests at room temperature.

INTRODUCTION

Iron is one of the most common and widespread metal with broad-based application. Actually, its physical properties, anticorrosive stability, chemical reactivity, formation of ferro-alloys, phase transitions, etc. are intensively investigated by physicists, chemists, metallurgists, crystallographers and others. In the forefront of interest in the physical properties of iron and its alloys at high pressure and temperatures are also engaged geophysicists, since iron is the main constituent of the Earth interior.

The present contribution provides information concerning a satellite phase observed in a pressure reactor steel after irradiation and tensile testing, when some mechanical and physical properties of the irradiated reactor steel* were investigated.

----

*Irradiation was performed in the reactor establishment of the Nuclear Research Center at Řež near Prague.

Table 1.   Chemical Composition of the Reactor Steel and its Weld Metal

Parent metal

C: 0.20,   Mn: 0.93,   Si: 0.27,   P: 0.031,   S: 0.021,   Cr: 0.17,

Ti: 0.05,   Al: 0.05,   Co: 0.01,   Mo: 0.07,   W: 0.01,   Sn: 0.10.

V: 0.002,   Mg: 0.0001.

Weld metal

C: ~0.17,   Mn: ~1.72,   Si: ~0.25,   P: ~0.022,   S: ~0.011, Cr: ~0.17

Ti: ~0.025,   Al: ~0.025.

EXPERIMENTAL

Material Characteristics

The experiments were performed on a reactor steel ČSN 13030 modified by 0.05 wt. % Al and Ti. The quantitative chemical analysis of the steel and the weld metal is given in Table 1.

Tensile Tests

Prior to irradiation and testing, prismatic steel plates were machined from welded rolled steel plates of 220 mm thickness. These plates were cut parallel to the weld joint in equal 5 mm distances. From each of these plates, several small prisms were cut out which were[1,2] then subjected to a specific heat treatment according to Table 2, in air and vacuum, at temperatures 200 ^{o}C (8 h), 500 ^{o}C (5 h) and 800 ^{o}C (2 h), some remained as welded and rolled. From these prisms, cylindrical tensile specimens were machined (Fig. 1) having 1.5 mm in diameter and 7.5 mm in length, with a definite orientation of the cylindrical axis to the rolling direction and surface. Half of the specimens were irradiated by a neutron flux $3.5.10^{22}$ n/ m^2 at actual temperatures under 100 ^{o}C (the total energy of the neutron flux was approximately 1 MeV), the rest of the specimens remained unirradiated. Irradiated and unirradiated specimens were subjected to tensile tests

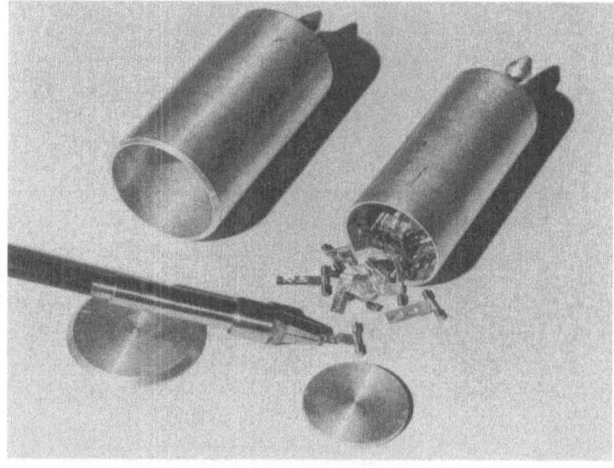

Fig. 1.   Shape of the specimens and their manipulation.

Table 2. Heat Treatment and Mechanical Properties of Investigated Reactor
          Steel

Heat treatment (before irradiation)	Yield strength (%)	Ultimate strength (%)	Elongation (%)
200 °C (8 h)			
air furnace	63.7	9.3	-12.9
vacuum chamber	65.9	2.9	-50.0
500 °C (5 h)			
air furnace	59.4	8.8	-13.5
vacuum chamber	74.6	6.8	-42.9
800 °C (2 h)			
air furnace	40.0	8.6	-10.6
vacuum chamber	44.8	3.7	-60.0
No heat treatment	64.7	6.2	-29.5

to compare their mechanical properties. The testing was performed at room
temperature in a prealigned Chevenard testing machine. The percentual averages of changes in the yield strength, ultimate strength and elongation
values of the parent metal and heat affected zone are summarized in Table 2.
In these averages, only those values were inclued, where the diffraction
pattern of the specimen was a pure α-Fe phase.

X-Ray Studies

     X-ray diffraction patterns of specimens were taken in a 90 mm photo-
graphic Debye-Scherrer powder camera using CoK_α radiation and a rotation
as well as a stationary technique. The film was centered around 90 °Θ.
Samples investigated by Debye-Scherrer method were the cylindrical tensile
specimens. When the diffraction pattern of a specimen contained after irradiation and tensile test satellite reflections, in addition to the X-ray
pattern of α-Fe, the X-ray examination was also repeated after few weeks
and months.

     The diffraction angles Θ of reflections were evaluated according to
relation:

$$\Theta = [90 - 45\, L_{hkl}/(\pi\, r_F)]^\circ,$$

where L_{hkl} is the distance between the pair of corresponding hkl lines centered round Θ = 90° and r_F is an effective radius of the camera. The lattice parameter of the cubic α-form, a_C, was refined by Nelson-Riley extrapolation method using least squares to assess linear dependence Θ vs.
0.5 $(cos^2\Theta/sin\Theta + cos^2\Theta/\Theta)$. Four or five reflections (110, 200, 211, 220,
310) were used for refinement of each parameter a_C (Fig. 2). Since the a_C
values of different specimens did not differ significantly, the weighted
average value was calculated from the large base of data, $(a_C)_{av}$ being then
2.866(2) Å. This value is in a good agreement with the parameter a_C =
(2.86638±0.00019) Å found by Donohue[3]. Therefore, the reflections of α-Fe
could be used as an internal standard for specifying and refinement of the
crystal data on the satellite phase.

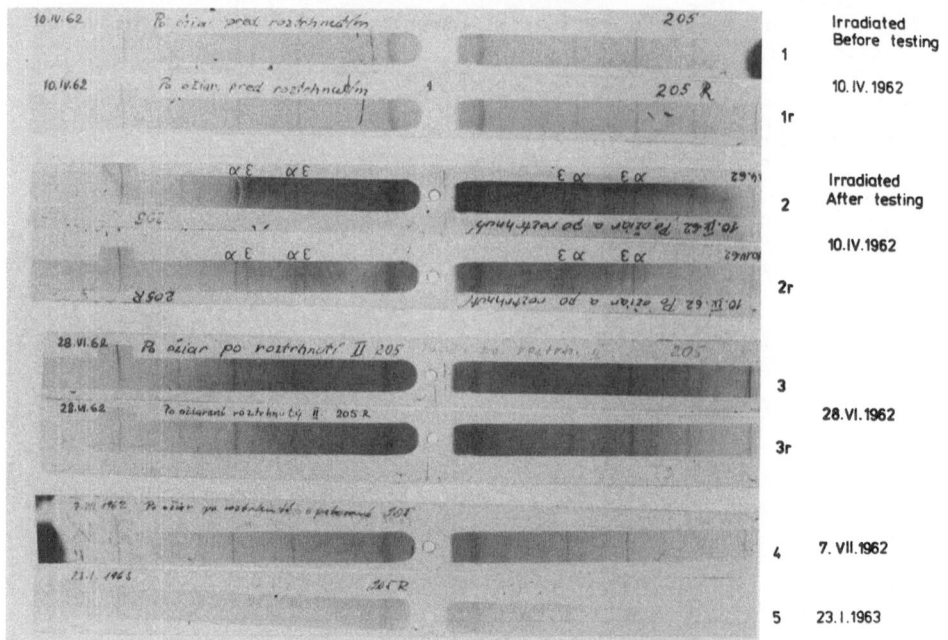

Fig. 2.   Debye-Scherrer X-ray photographs (exposure 5 min at room temperature) of irradiated specimens: 1. before tensile testing - stationary sample, 1r - rotating sample; 2. after tensile testing, appearance of satellite reflections - stationary sample, 2r - rotating
sample; 3. after 11 weeks of relaxation, disappearance of satellite
reflections - stationary sample, 3r - rotating sample; 4. after
12.5 weeks of relaxation; 5. after 9.5 months of relaxation.

The satellite lines appeared in the X-ray patterns of a part of specimens (in 50 from the total of 860 specimens), but exclusively in the samples
which were subjected to a heat treatment, irradiation and tensile tests.
Three satellite reflections appeared in the X-ray patterns in the intervals
(0.995 - 1.001) Å, (1.133 - 1.140) Å and (1.995 - 2.000) Å (Fig. 2), with
the weighted averages of d_{hkl} values: 0.9995 (15 Å, 1.137(1) Å and 2.000(2)
Å. The following information could be derived from the inspection of X-ray
patterns: (a) All satellite reflections occurred at higher diffraction angles
than those of the α-form, i.e. they define a unit cell which is smaller than
that of α-Fe; (b) The angular differences between the reflection positions
of the satellite phase and the α-form did not increase regularly from the
low to the high diffraction angles Θ. The differences $\Delta\Theta$ between the diffraction angles Θ_{110}, Θ_{211} and Θ_{220} the α-form and the corresponding satellite reflections are approximately 0.6°, 1.5° and 1.0. This requires for
the satellite phase a lower symmetry than a cubic one. Since 200 reflection
of α-Fe has no observed satellite diffraction and the intensity changes of
satellite reflections show no proportionality in comparison with the intensities of α-Fe, a tetragonal desymmetrization of the cubic 123 symmetry (α-Fe)
seems to be improbable. However, the weighted d_{hkl} values of satellite reflections are compatible with a hexagonal (hcp type) lattice having parameters a_H = 2.519(2) Å, c_H = 3.997(1) Å, V_H = 21.96 Å^3; Z = 2. Table 3 summarizes the calculated and observed crystal data on the α-form and the
satellite phase. It is interesting to note that no possible reflection of
the hexagonal phase can occur near at satellite angular position of reflection 200 of the α-Fe form.

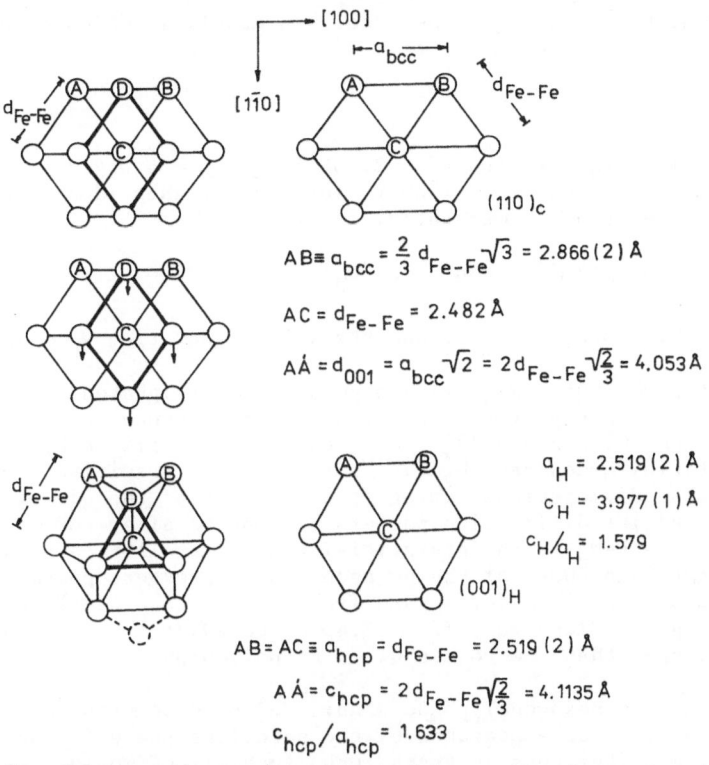

$$AB \equiv a_{bcc} = \frac{2}{3} d_{Fe-Fe} \sqrt{3} = 2.866(2) \text{ Å}$$

$$AC = d_{Fe-Fe} = 2.482 \text{ Å}$$

$$A\acute{A} = d_{001} = a_{bcc} \sqrt{2} = 2 d_{Fe-Fe} \sqrt{\tfrac{2}{3}} = 4.053 \text{ Å}$$

$$a_H = 2.519(2) \text{ Å}$$
$$c_H = 3.977(1) \text{ Å}$$
$$c_H/a_H = 1.579$$

$$AB = AC \equiv a_{hcp} = d_{Fe-Fe} = 2.519(2) \text{ Å}$$

$$A\acute{A} = c_{hcp} = 2 d_{Fe-Fe} \sqrt{\tfrac{2}{3}} = 4.1135 \text{ Å}$$

$$c_{hcp}/a_{hcp} = 1.633$$

Fig. 3.   Scheme of transformation of bcc – hcp iron structure.

RESULTS AND DISCUSSION

The volume of the hexagonal satellite phase V_H = 21.96 Å^3 differs, at room temperature, significantly from the volume V_C = 23.54 Å^3 of the cubic (bcc) α-Fe phase. The structure of the hexagonal satellite phase can be related to the structure of the high pressure hexagonal hcp polymorph (ϵ-Fe). The transition pressure $\alpha \rightarrow \epsilon$ Fe at room temperature is 11 GPa according to Akimoto[4] or 13 GPa according to Mao et al.[5]. The position of the triple point α-γ-ϵ Fe[4] is defined by p–T parameters 8.3 GPa and 440 °C[4]. However, both Akimoto[4] and Mao et al.[5] confirmed a broad p-T coexistence field for α and ϵ polymorphs and a rather sluggish transition mechanism. This is probably a well- grounded reason, why the hexagonal form of ϵ-Fe can exist at room temperature and normal pressure (10^5 Pa) as a metastable polymorph.

According to the rigid-sphere model[5], the hcp arrangement can be derived from the bcc structure by a shortening of the interatomic distances along the [001] direction of the bcc packing and by gliding of the (110) atomic planes along the [1$\bar{1}$0] or [$\bar{1}$10] directions (Fig. 3).

The nearest-neigbor interatomic distance between the iron atoms, d(Fe-Fe), in the bcc structure along the body diagonal is

$$d(Fe-Fe) = a_C \sqrt{3}/2 = 2.482 \text{ Å},$$

where a_C = 2.866 Å. This distance increases in the metastable hcp arrangement at room temperature to a value 2.519 Å, i.e. (Fig. 3)

$$d(Fe-Fe)_H = 1.015 \, d(Fe-Fe)_C = a_H.$$

The interlayer distances d_{002} in the ideal hcp model, in which $d(Fe-Fe)_C$ distances would be preserved equal to

$$a_C/\sqrt{2} = c_H/2 = 2.027 \text{ Å}.$$

The unit cell volume V_H based on $a_H = 2.482$ Å and $c_H = 4.054$ Å is 21.624 Å^3, i.e. V_H is reduced in comparison with V_C for 8.14 %, when ideal bcc $\to$ hcp transformation takes place. Generally,

$$(V_H)_{hcp} = 3\sqrt{3}\, a_C^3/4\sqrt{2} \quad \text{and} \quad (c_H/a_H)_{hcp} = 1.633,$$

while c_H/a_H in this work is 1.587, and the volume reduction $V_C \to V_H$ is 6.7 %.

At present time, two hexagonal hcp phases are designated as ε forms: one is the high pressure polymorph of iron, the second one is the iron nitride compound FeN_x with x = 0.125 – 0.5, which is stabilized by the presence of interstitial H-centres $(N_2^0)^6$. This phase can precipitate as a metastable compound on dislocations during strain ageing of the nitrogen containing steel, and its diffraction pattern disappears after some time of relaxation. Since FeN_x is an interstitial phase, its lattice parameters are always larger than those of the hexagonal high pressure polymorph. They vary from 2.693 Å to 2.762 Å for a_H and from 4.370 Å to 4.414 Å for c_H. The corresponding V_H values vary from 27.45 Å^3 to 29.16 Å^3, i.e. they are significantly larger than the volume of α Fe polymorph.

Formation of the hexagonal phase stabilized by H-centres could be a plausible explanation for appearance of the satellite phase in the irradiated reactor steel after tensile tests, when such effect would take place at very low concentrations of N in TiN_x, since then is the influence of interstitial H-centres on the lattice parameters of ε-phase negligible. Disappearance of the satellite phase could be then explained by reversible reaction $(N_2^0)_i = N_2(g) + \delta e^-$. Whether the strain-stress conditions during the tensile testing or the presence of H-centres (N_2^0); in the irradiated samples or the both effects contribute to formation of the hexagonal satellite phase, is not definitely confirmed. In any case, the hexagonal satellite phase is close to the pure ε-phase as it follows from comparison of its lattice parameters with those of ideal hcp iron form or with parameters obtained by extrapolation from the high pressure data[8] to normal pressure (10^5 Pa) and room temperature: $a_H = 2.523$ Å, $c_H = 4.044$ Å, $V_H = 22.29$ Å^3 (Table 3).

Mao et al.[5] applied an exponential form of the Murnaghan equation

$$V_p^{hcp} = V_o^{hcp} \left(1 + \frac{p}{32.5}\right)^{-0.196} \tag{1}$$

to find the hypothetical molar volume V^{hcp} of ε-Fe at zero pressure; p is the pressure in GPa and V_p^{hcp} is the molar volume of ε-Fe at pressure p. The theoretical value is $(6.72 \pm 0.06)\,10^{-6}\,m^3$/mole. The experimental value found in this work is $V_H = 6.61 \cdot 10^{-6}\,m^3$/mole.

The hypothetical pressure p at normal temperature, which would allow a volume reduction $V_H/V_{bcc} = 6.61/7.09 = 0.932$, can be calculated according to relation[5]:

$$V_H/V_{bcc} = 0.932 = (1 + p/27.5)^{-0.169} \tag{2}$$

The value of p is 14 GPa at room temperature, which approximately equals to the transition pressure $\alpha \to \varepsilon$ Fe found experimentally[4,5,8].

Table 3. Crystallographic Data on the bcc (α) and hcp (ε) Polymorphs of Iron

α-Fe: a_C (298 K, 10^5 Pa) = 2.866(2) Å;

V_C (298 K, 10^5 Pa) = 23.54 Å^3 = 7.09 . 10^{-6}m^{3}mol^{-1};

V_C (1183 K, 10^5 Pa) = 24.50 Å^3 = 7.38 . 10^{-6}m^{3}mol^{-1};
($\alpha \rightarrow \gamma$ transition)

V_C (713 K, 8.3 GPa) = 23.01 Å^3 = 6.93 . 10^{-6}m^{3}mol^{-1};
(triple point)

ε-Fe: a_H (298 K, 10^5 Pa) = 2.519(2) Å; this work;

c_H (298 K, 10^5 Pa) = 3.997(1) Å; this work;

c_H/a_H = 1.587;

V_H (298 K, 10^5 Pa) = 21.96 Å^3 = 6.61 . 10^{-6}m^{3}mol^{-1}; this work;
(metastable phase)

V_H (713 K, 8.3 GPa) = 21.02 Å^3 = 6.33 . 10^{-6}m^{3}mol^{-1}; this work;
(triple point)

a_H = 2.482 Å, c_H = 4.054 Å; c_H/a_H = 1.633;

V_H = 21.62 Å = 6.51 . 10^{-6}m^{3}mol^{-1}, rigid sphere model;

a_H (298 K, 10^5 Pa) = 2.523 Å, Mao et al.[5];

c_H (298 K, 10^5 Pa) = 4.044 Å, c_H/a_H = 1.604, Mao et al.[5];
(extrapolation from high pressure data)

V_H (298 K, 10^5 Pa) = 22.29 Å^3 = 6.71 . 10^{-6}m^{3}mol^{-1}; Mao et al.[5];

V_H (298 K, 13 GPa) = 20.87 Å^3 = 6.28 . 10^{-6}m^{3}mol^{-1}; Mao et al.[5];

γ-Fe: a_γ = (1183 K, 10^5 Pa) = 3.6467 Å, Donohue[3];

V_γ = (1183 K, 10^5 Pa) = 48.495 Å^3 = 7.30 . 10^{-6}m^{3}mol^{-1}; Donohue[3];
($\alpha \rightarrow \gamma$ transition)

V_γ = (762 K, 8.3 GPa) = 44.91 Å^3 = 6.76 . 10^{-6}m^{3}mol^{-1}; Donohue[3];
(triple point)

V_C-V_H (298 K, 10^5 Pa) = 0.48 . 10^{-6}m^{3}mol^{-1} (-6.8 %), this work;
(metastable transition)

V_C-V_H = 0.58 . 10^{-6}m^{3}mol^{-1} (-8.2 %), rigid sphere model;

V_C-V_H (713 K, 8.3 GPa) = 0.60 . 10^{-6}m^{3}mol^{-1} (-8.7 %), this work;
(triple point)

V_C-V_γ (713 K, 8.3 GPa) = 0.17 . 10^{-6}m^{3}mol^{-1} (-2.5 %), this work;
(triple point)

$V_\gamma-V_H$ (713 K, 8.3 GPa) = 0.43 . 10^{-6}m^{3}mol^{-1} (-6.4 %), this work;
(triple point)

$V_\alpha-V_\gamma$ (1183 K, 10^5 Pa) = 0.08 . 10^{-6}m^{3}mol^{-1} (-1.1 %), Donohue[3].

ε-Fe: $(d_{002})_{obs}$ = 2.000(2) Å, $(d_{002})_{cal}$ = 1.999 Å;

$(d_{103})_{obs}$ = 1.137(1) Å, $(d_{103})_{cal}$ = 1.137 Å; this work;

$(d_{004})_{obs}$ = 0.9995(15) Å, $(d_{004})_{cal}$ = 0.9993 Å.

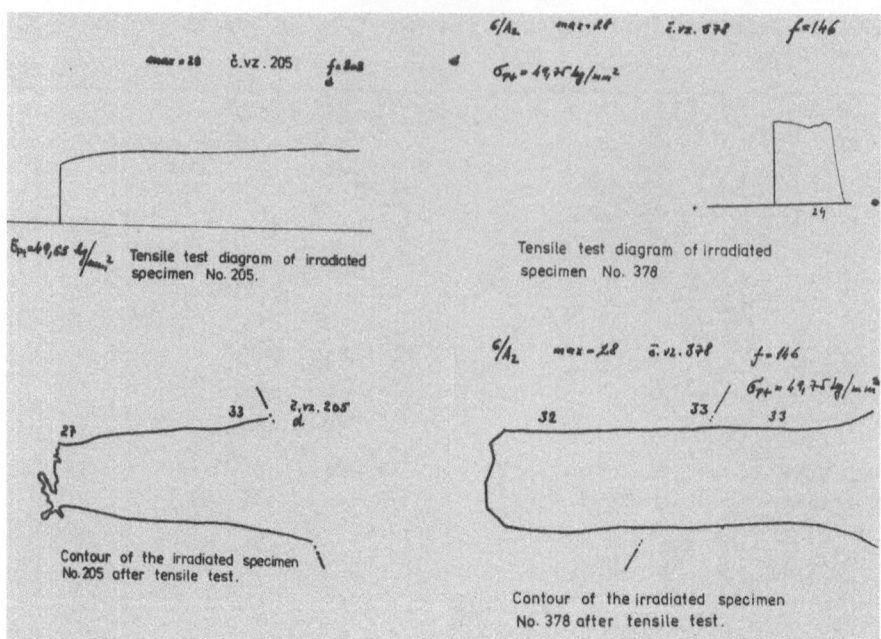

Fig. 4. Tensile diagram and contour of irradiated (a) tough, (b) brittle specimens.

Using Eqs (1) and (2), the actual volume of α, γ and ε polymorphs at 440 °C and 8.3 GPa (the triple point parameters) can be evaluated, i.e. $6.93 \cdot 10^{-6} m^3/mole$, $6.76 \cdot 10^{-6} m^3/mole$ and $6.33 \cdot 10^{-6} m^3/mole$ for the $\alpha \rightarrow \varepsilon$, $\varepsilon \rightarrow \gamma$ and $\gamma \rightarrow \alpha$ transformations, i.e. $-0.60 \cdot 10^{-6} m^3/mole$, $0.43 \cdot 10^{-6} ^3/mole$ and $0.17 \cdot 10^{-6} m^3/mole$.

The hexagonal satellite phase is unstable at normal pressure and temperature. Its diffraction lines disappear from the diffraction record of the specimen at room temperature after 1.5 – 2 months of relaxation (Fig. 2).

The specimens containing the hexagonal satellite phase showed significantly improved toughness and strength parameters (Fig. 4). This agrees with the theory of the transformation toughening of material containing dispersed metastable polymorph$_9$particles in the stable parent polymorph having different molar volumes. The marked plastic properties of the specimens containing hexagonal satellite phase, so much required in reactor steel after irradiation, brought us to the conclusion that the presence of the dispersed hexagonal satellite phase enhances the toughness of the steel and can be used in development of a steel relatively insensitive to neutron exposure.

REFERENCES

1. F. Haramia, The effect of neutron current on mechanical properties of electroslag weld joints (in Slovak), Zváranie 21:328 (1972).
2. F. Haramia, The effect of neutron radiation on some brittle fracture characteristics of the steel 15CH2 MFA used for pressure vessel fabrication (in Slovak), Zváranie 23:360 (1974).

3. J. Donohue, The Structures of the Elements, John Wiley & Sons, New York (1974).
4. S. Akimoto, High-pressure research in mineral physics, in: "Geophysics Monograph 39", M. H. Manghnani & Y. Syono, Ed., American Geophysical Union (1987), pp. 149-154.
5. H. K. Mao, W. A. Bassett, and T. Takahashi, Effect of pressure on crystal structure and lattice parameters of iron up to 300 kbar, J.Appl.Phys. 38:272 (1967).
6. J. D. H. Donnay and H. M. Ondik, Crystal Data. Determinative Tables, Volume II, Inorganic compounds, U.S. Dept of Commerce NBS and JCPDS (1973).
7. I. Hrivňák, The Theory of Mild and Micro-Alloys Steels' Weldability, ALFA, Bratislava (1969).
8. R. L. Clendenen and H. G. Drickamer, The effect of pressure on the volume and lattice parameters of Ruthenium and Iron, J.Phys.Chem.Sol. 25:865 (1964).
9. F. Kroupa, Transformation toughening of ceramics (1987), to be published.

X-RAY PHASE ANALYSIS OF Cu AND Pb ANODIC OXIDATION PRODUCTS

E. Łagiewka and A. Budniok

Institute of Physics and Chemistry of Metals
Silesian University, Bankowa 12
Katowice, Poland

INTRODUCTION

Studies on the process of anode oxidation for production of copper and lead oxides have been reported in a number of papers[1,2,3,4]. A major drawback of these methods is a lack of possibility of direct monitoring of the changes in the phase composition during the process of formation of the oxide layers. When removed from the electrolyte and subjected to a treatment such as rinsing and drying, the layer of oxidation products may also undergo structural changes. This situation prompted a decision to study the phase composition of the oxidation products of copper and lead directly during the anode electrocrystallization process. For this purpose a TURM-62 X-ray diffractometer was used with a special attachment as described in ref. 5. The use of this attachment makes it possible to obtain X-ray diffraction pattern during electrocrystallization of the oxide layer.

MATERIALS AND EXPERIMENTAL DETAILS

Anode oxidation of copper and lead was performed in a 2M NaOH solution in galvanostatic conditions at a temperature of 293 K. The electrolytic solution of volume 200 cm^3 was placed in a small vessel which forms a part of the diffractometer attachment[5]. Copper electrooxidation was carried out with a current density of 0.13 to 0.67 A/dm^2, while for lead electrooxidation a current density of 1.8 to 4A/dm^2 was used. The polycrystalline copper and lead sheets of purity 99.9 % from which anodes of diameter 50 mm were cut were used for these tests.

The copper anodes were mechanically polished first with abrasive paper and diamond paste and next electrolytically polished in a mixture of ethyl alcohol and phosphoric acid in a proportion 1 : 20, with current density D = 2.5 A/dm^2. After electrtolytic polishing the anodes were washed in distilled water and dried with filter paper. Immediately prior to the oxidation process the anodes were pickled in a 1 : 3 mixture of nitric acid and water. For the lead anodes the treatment was comprised of mechanical polishing with abrasive paper and diamond paste plus rinsing in distilled water and drying with filter paper.

After this preliminary treatment the anodes were fixed in the attachment holder and the X-ray diffractometer goniometer system was centred according to its instructions[6]. The accuracy of centering was checked by the diffrac-

Table 1. Angular Oscillation Ranges of Counter when Recording the Diffraction Reflexes of the Products of Copper and Lead Oxidation

Phase	Cu_2O	$Cu(OH)_2$	PbO yellow	PbO red	$\alpha-PbO_2$
$d_{hkl}[\text{Å}]$	2.465	2.622	3.058	3.110	3.818
Oscillation range 2Θ [degrees]	35-38	33-35	27-30	27-30	22-25

tion reflections of the substrate, i.e. copper or lead. No oxides or other phases were detected on the anodes prepared in this way. When the anode was centered as a flat specimen in the X-ray diffractometer goniometer system, it was connected to the electrical circuit of the galvanostat, the electrolyte was poured into the vessel and the current density was set at the desired value. During the whole experiment the anode rotated at a speed of 20 rpm.

A system with a TURM-62 diffractometer produced in the German Democratic Republic with the radiation from a copper-anode tube was used for measuring the X-ray diffraction reflections. The reflections were scanned in a continuous system with oscillation of the counter in ranges of angles shown in Table 1.

Diffraction reflexes from phases formed during the oxidation process of copper and lead occur close to each other in a small range of angles (Table 1). Hence it is feasible to record simultaneously at least two strongest reflections of different phases. When using a counter rate of 2°/min it was possible to record the diffraction reflexes on an average every 2 - 3 minutes. This recording rate was particularly useful in the first minutes of electrooxidation to catch the beginning of the formation of phases. The intensity of the diffraction reflex estimated as an area under the curve of intensity distribution was taken as a measure of quantity of the phases formed. The curve of the intensity distribution was plotted using the recorder tape at a tape feed rate of 36 mm/min.

RESULTS OF TEST AND DISCUSSION

Anode Oxidation of Copper

It was ascertained that during anode oxidation of copper, the phases of type Cu_2O, $Cu(OH)_2$ are formed (Figs 1, 2). Copper oxide is the principal phase of the product of electrooxidation process of copper. It is formed as the first product of electrooxidation. The beginning of its formation is associated with the density of current. For various current densities the first formation of this oxide was observed at more or less the same time (Fig. 1). If the intensity of diffraction reflexes is taken as a measure of a quantity of the phase formed, then in the current density interval $D_A = 0.13 - 0.26$ A/dm^2 there is a possibility to obtgain it in the greatest quantities, at the highest rate of formation.

A distinct drop in increment of intensity of Cu_2O reflexes per unit time was observed for the current density above $D_A = 0.33$ A/dm^2. This means, under these conditions and under the assumption that no texture occurs and

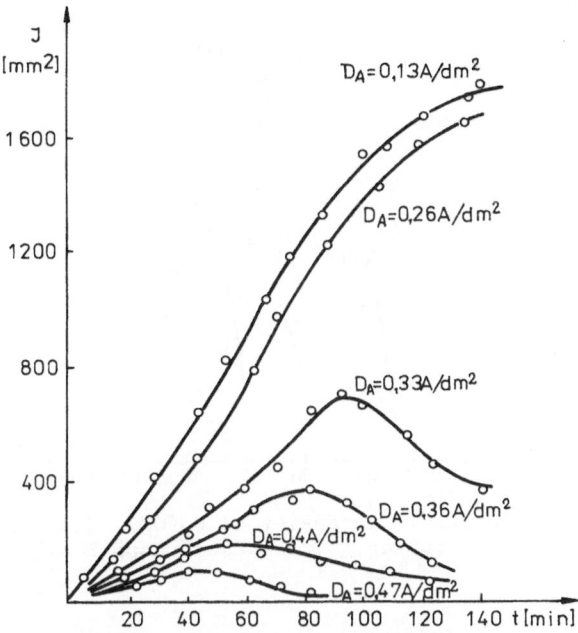

Fig. 1. Variations in intensity of Cu_2O diffraction reflexes during galvanostatic copper oxidation.

the crystallites are of approximately the same size, that the rate of Cu_2O formation may be less than the rate at lower current densities. Analysis of the full diffraction patterns for the products of copper oxidation showed that there is no change in the ratio of intensities of particular reflexes and also that their half-widths change very little. Hence it may be concluded that the Cu_2O oxide formed under the given conditions does not exhibit texturing or significant differences in size reduction of crystallites.

It was found that the intensity of copper oxide diffraction reflexes obtained above current density $D_A = 0.33$ A/dm^2 grows less after a definite time of electrooxidation (Fig. 1). The higher the current density, the shorter this time. The diffraction reflexes of copper hydroxide appeared together with the reduction in intensity of copper oxide reflexes (Fig. 2). This fact indicates that copper hydroxide is formed as a product of the electrooxidation of copper oxide.

Copper hydroxide is not a stable phase. After a certain time of duration of the electrooxidation process its diffraction reflexes also vanish and the higher the current density, the more rapidly they vanish (Fig. 2). Disappearance of the copper hydroxide reflexes did not cause the appearance of another phase. An increase in the X-ray background and a black sediment at the bottom of the vassel was observed only. It may be presumed that this is amorphous copper oxide as the last product of copper electrooxidation[1,2]. However, due to the amorphous nature of the product being formed (lack of X-ray diffraction reflexes) it was not possible to identify this phase by diffraction analysis.

These studies indicated that in the initial stage of the anode copper oxidation process there is a tendency towards the formation of copper oxide. In the range of low current densities, the tendency to form Cu_2O is found through a fairly long time of duration of the oxidation process. After the formation of Cu_2O in the initial stage, the increase in current density favours the formation of hydroxide $Cu(OH)_2$. With longer duration of the

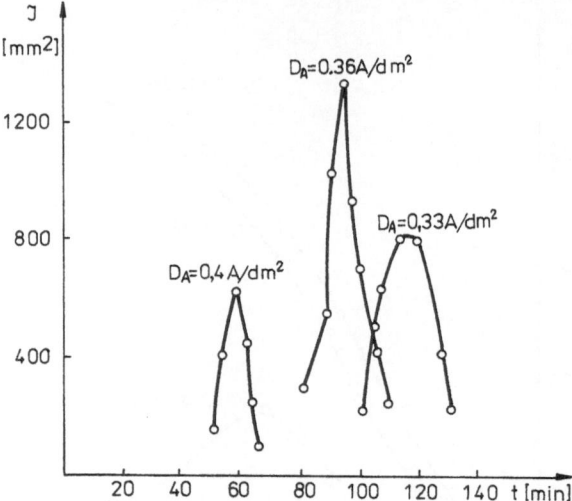

Fig. 2.   Variations in intensity of Cu(OH)$_2$ diffraction
reflexes during galvanostatic copper oxidation.

electrooxidation process, a disintegration of the hydroxide takes place,
probably to CuO, which is amorphous and only weakly adheres to the substrate.   In the most general form, the process of anode copper oxidation
may be written as:

$$Cu \rightarrow Cu_2O \rightarrow Cu(OH)_2 \rightarrow CuO \text{ (amorphous)}$$

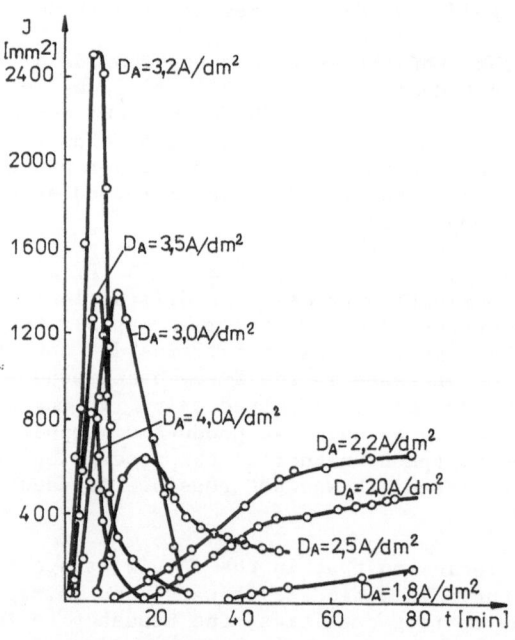

Fig. 3.   Variations in intensity of PbO$_{rhomb}$ diffraction
reflexes during galvanostatic lead oxidation.

Anode Oxidation of Lead

Our investigations showed that the process of electrooxidation of lead leads to the formation of the following phases: PbO yellow variety, PbO red variety and α-PbO$_2$. Lead oxide yellow variety crystallizes in a rhombohedral system, while lead oxide red variety crystallizes in a tetragonal system[7]. In further considerations, to stress the structural differences of these oxides they are designated as PbO$_{rhomb}$ for lead oxide yellow variety and PbO$_{tetr}$ for lead oxide red variety.

In these investigations it was found that irrespective of current density, PbO$_{rhomb}$ appears always as the first product of electrooxidation of lead. The beginning of its formation (the time to the appearance of the diffraction reflex) depends on oxidation current density, i.e. the higher the current density, the shorter is this time. Beginning from the current density D_A = 2.5 A/dm^2, a marked rise in a rate of electrooxidation was observed, evidenced by the increase in intensity of reflexes with time (Fig. 3).

The intensity of PbO$_{rhomb}$ reflexes obtained above the current density D_A = 2.5 A/dm^2 increases in the initial period, then reaches a maximum and then decreases. Simultaneously the PbO$_{tetr}$ diffraction reflexes begin to appear (Fig. 4). Intensities of diffraction reflexes from lead oxide type PbO$_{tetr}$ increase with increasing time of electrooxidation. The higher the current density, the more rapid is this growth. Lead oxide type PbO$_{tetr}$ is also formed below the current density D_A = 2.2 A/dm^2 (Fig. 4), but its rate of formation is substantially less than at higher values of current density and also the PbO$_{rhomb}$ reflexes do not attain a maximum but a stationary value only (Fig. 3). The appearance of PbO$_{tetr}$ diffraction reflexes simultaneously with the reduction in intensity of diffraction reflections from PbO$_{rhomb}$, the formation of PbO$_{rhomb}$ as the first product of electrooxidation and also the occurrence of lead in both oxides at the same degree of oxidation offers the evidence that oxide of type PbO$_{tetr}$ is not formed due to the electrooxidation process but due to the phase transformation from PbO$_{rhomb}$.

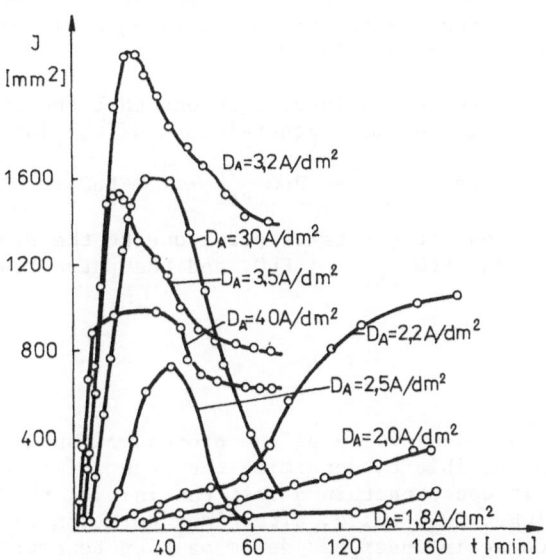

Fig. 4. Variations in intensity of PbO$_{tetr}$ diffraction reflexes during galvanostatic lead oxidation.

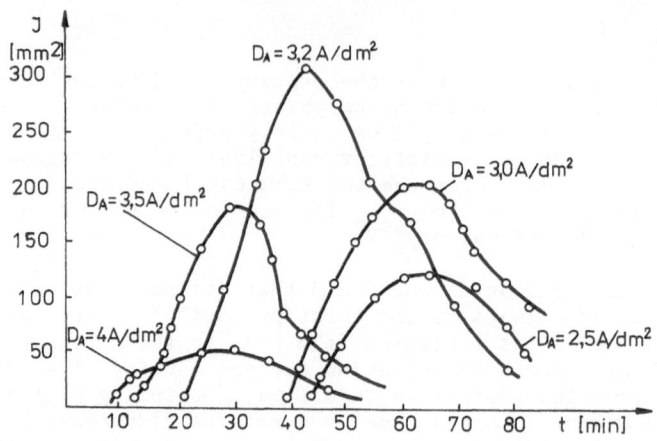

Fig. 5. Variations in intensity of α-PbO$_2$ diffraction
reflexes during galvanostatic lead oxidation.

Hence, as well as the electrochemical reaction leading to the formation of
PbO$_{rhomb}$, there also occurs a parallel phase transformation in the solid
state:

$$PbO_{rhomb} \rightarrow PbO_{tetr}.$$

An increase of lead electrooxidation in time, particularly above current
density D_A = 2.5 A/dm^2, causes a drop in intensity of PbO$_{tetr}$ diffraction
reflexes (Fig. 4). Simultaneously, the appearance of α-PbO$_2$ reflexes was
ascertained (Fig. 5). It was observed that the higher the current density,
the more rapid is the formation of α-PbO$_2$.

The reduction of PbO$_{tetr}$ diffraction reflexes in intensity, the appearance of α-PbO$_2$ reflexes, the occurrence of lead in the α-PbO$_2$ form (i.e. at
a higher degree of oxidation than in PbO$_{tetr}$) and also the fact that α-PbO$_2$
occurs after a long time duration of the electrochemical process, all considered together, suggest that α-PbO$_2$ is formed as a result of the electrochemical reaction of PbO$_{tetr}$ oxidation.

It may be concluded from these investigations that the process of lead
electrooxidation consists in the most general form of the following stages:

$$Pb \rightarrow PbO_{rhomb} \rightarrow PbO_{tetr} \rightarrow \alpha\text{-PbO}_2.$$

The transition between these stages takes place due to the electrochemical
reaction Pb $\rightarrow$ PbO$_{rhomb}$ and PbO$_{tetr}$ $\rightarrow$ α-PbO$_2$ and the phase transition
PbO$_{rhomb}$ $\rightarrow$ PbO$_{tetr}$.

CONCLUSIONS

From "in situ" X-ray examination of the process of anode oxidation of
copper and lead it was possible to determine the order of formation of the
individual products. It was ascertained that the initial product of copper
electrooxidation is always Cu$_2$O which, after long duration of the process,
passes to Cu(OH)$_2$ which again undergoes decomposition to amorphous CuO.
The rate of formation of the individual phases and the rate of their dis-
appearances increase with increasing density of current.

Investigation of the lead electrooxidation process indicated that it is associated with electrochemical reaction and with a phase transformation. The first product in the anode process of lead oxidation is always lead oxide PbO_{rhomb}. During the course of further electrooxidation this oxide undergoes a phase transformation to PbO_{tetr}. From the oxide PbO_{tetr} formed in this way, $\alpha\text{-}PbO_2$ is formed during further electrooxidation, due to the electrochemical reaction. The rate of formation of various phases is associated with the current density.

The tests conducted on the anode process of copper and lead oxidation gave an opportunity for verifying the usefulness of the attachment designed for direct determination of a phase composition of the surface layers during the time of their deposition on the electrode. Irrespective of this application, the attachment is also suitable for tracing the polymorphic transformations taking place during the deposition of layers on the electrode surface.

REFERENCES

1. A. M. Shams El Din and Em. Abb El Wahab, Electrochim. Acta 9:113 (1964).
2. D. W. Shoesmith, T. E. Rummery, D. Owen, and W. Lee, J. Electrochem. Soc. 123, 6:790 (1976).
3. Ju. D. Dunajev, Nieraztworimyje anody na osnowie swinca, Nauka, Alma-Ata (1978).
4. S. S. Popowa, A. W. Fortunatov, Elektrokhimiya 2, 4:446 (1966).
5. E. Łagiewka, A. Budniok, Surface and Coatings Technology 27:57 (1986).
6. Universal-Röntgen-Diffraktometer HZG 4/C, VEB Freiberger Präzisionsmechanik, Freiberg (1983).
7. JCPDS Powder Diffraction File, International Centre for Diffraction Data, Swarthmore, Pennsylvania (1977).

PHASE ANALYSIS OF RUSTS BY X-RAY DIFFRACTION,

IR-SPECTROSCOPY AND THERMAL ANALYSIS

J. Had, J. Balcar, and H. Pracharová

Research Institute for the Protection of Materials

Praha 9 - Běchovice, Czechoslovakia

Phase composition of corrosion products is a matter of importance in corrosion research and in protection of materials. The most important corrosion products of iron are given in Table 1.

In our institute, phase identification is provided by X-ray powder diffraction. Identification of γ-Fe_2O_3 with the help of X-ray diffraction was formerly debatable because of low intensity of lines, different from Fe_3O_4, introduced in card 4-0755 of JCPDS[1]. Later data of card 15-615 make it possible to distinguish the two phases, as proved - among others - by work of X-ray diffraction laboratory in our institute[2]. Some authors provide identification of γ-Fe_2O_3 with the help of Mössbauer spectroscopy. The others find it difficult to distinguish γ-Fe_2O_3 from Fe_3O_4 in rusts[3] even by this method. As far as X-ray diffraction is concerned, γ-Fe_2O_3 was not found in rusts. Also $Fe(OH)_2$, being an unstable intermediate product, is not usually found in rusts[2]. Green rusts I, II and δ-FeOOH are "roentgeno-amorphous", because they form too small coherent areas.

When rusts are formed under normal conditions in the atmosphere, water or earth, the most usual corrosion products identified by X-ray diffraction are α-FeOOH, β-FeOOH, γ-FeOOH and Fe_3O_4, exceptionally α-Fe_2O_3. When remnants of scales are present, FeO can be found in rusts. Significant values of inter planar spacings are given in Table 2.

In the rusts originating in water greater amounts of $CaCO_3$-calcite, in rusts from atmosphere α-SiO_2 (from dust), and in the inner layers, α-Fe from substratum can be found. When corrosion proceeds in the presence of sea water, NaCl often occurs. In other cases chlorides and sulphates which play important role in corrosion mechanism, are not usually present in such amounts to enable X-ray identification.

In our institute X-ray phase analysis is performed by diffractometer, Bragg-Brentano focusing system, CoK_{α} radiation. When only a small amount of the rust is available, Debye-Scherrer method with CrK_{α} radiation is used. Relative occurrences of the phases, which were identified in 276 samples of the rusts from carbon steels and low alloy steels are presented in Table 3.

Interpretation of phase analysis results is considerably complicated, as can be seen from works of different authors (for example refs 3,4,5,6,7,

Table 1. The Most Important Corrosion Products of Iron

Phase	Mineralogical Name	Crystal System	Interplanar Spacings		
α-Fe$_2$O$_3$	hematite	hexagonal	a= 5.04		c=13.77
γ-Fe$_2$O$_3$	maghemite	cubic	a= 8.32		
Fe$_3$O$_4$	magnetite	cubic	a= 8.396		
FeO	wustite	cubic	a= 4.29		
α-FeOOH	goethite	orthorhombic	a= 4.64	b=10.0	c= 3.03
β-FeOOH	akagaenite	tetragonal	a=10.48		c= 3.023
γ-FeOOH	lepidocrocite	orthorhombic	a= 3.88	b=12.54	c= 3.07
δ-FeOOH		hexagonal	a= 2.941		c= 4.49
Fe(OH)$_2$		hexagonal	a= 3.258		c= 4.605
Green Rust I		hexagonal	a= 3.18-3.23		c=22.5-24.2
Green Rust II		hexagonal	a= 3.17		c=10.9

8,9) and is out of range of this paper. However, one important fact is evident: presence of β-FeOOH is conditional on the chloride corrosion mechanism. From the presence of β-FeOOH in rusts conclusions may be arrived at as about higher concentration of chlorides in corrosion environment. In our institute β-FeOOH is commonly identified in rusts originated in salt spray cabinet. When corrosion products from cooling circuits of power stations (concentration of chlorides was from 20 to 50 mg Cl$^-$ (liter)) were studied, no correlation between occurrence of β-FeOOH in rusts and concentration of chlorides in cooling water was found. When influence of environment on corrosion of agricultural machines was investigated, β-FeOOH was found where chloride concentration in the rusts was higher than 0.30 % Cl$^-$.

Infra-red spectroscopy is carried out on spectrometer Shimadzu IR-435 by KBr tablet method, wave number ranging from 400 to 4000 cm^{-1}. Absorption peaks found in spectra of rusts are presented in Table 4.

Like in the paper of Misawa et al.[4], absorption peaks 880 cm^{-1} for determination of α-FeOOH and 1018 cm^{-1} for γ-FeOOH have been selected. Ratio of absorbances A_{880}/A_{1018} was calculated. The ratio increases with increasing concentration of SO$_2$ in atmosphere. Similar results were obtained in Sweden by Singh et al.[3] with the help of X-ray diffraction and Mössbauer spectroscopy. Under conditions in our country - because of high concentration of SO$_2$ in atmosphere - this dependence is distorted by wide absorption peaks of sulphates with maximum by 1115-1125 cm^{-1}.

Characterization of rusts is supplemented by thermal analysis of the corrosion products. For this purpose we use Q-derivatograph, MOM, Hungary. Parameters set on the instrument: rate of heating-up 10 $^\circ$C/min, sensitivity DTG 1/5, maximum temperature 1000 $^\circ$C, Pt covered crucible, charge 100-200 mg.

Table 2. Interplanar Spacings d Å of Corrosion Products with Relative Intensity ≥ 10

α-FeOOH 17-536	β-FeOOH 8-93	γ-FeOOH 8-98	Fe₃O₄ 11-614	α-Fe₂O₃ 13-564	FeO 6-615
	7.6^{o}*	6.26*			
	5.3^{o}		4.85*		
4.18*				3.68^{o}*	
	3.33	3.29^{o}			
			2.97^{o}		
2.69	2.62*			2.69	
	2.55^{o}*		2.53	2.51	
2.490		2.47^{o}			2.486^{o}
2.452		2.36			
	2.28				
2.192				2.20^{o}	2.153
	2.09	2.09	2.096^{o}		
	2.05				
	1.94	1.937^{o}			
	1.84	1.848		1.837^{o}	
1.721	1.74	1.732	1.712	1.691^{o}	
	1.63		1.614	1.596	
1.564		1.566			
		1.535			
	1.51	1.524	1.483^{o}	1.486	1.523^{o}*
	1.434	1.433		1.451^{o}	
	1.373	1.367	1.327	1.348	
				1.309	1.299^{o}*
		1.279		1.255	1.243

Phase and number of JCPDS card

_____ relative intensity = 100

o relative intensity ≥ 50

* values characteristic for identification

Record of the derivatograph consists of four curves (all being functions of time): temperature T, weight TG, weight deerivation DTG and curve of differential thermal analysis DTA. For interpretation of the records it is necessary to consider all curves simultaneously. DTG curve is more illustrative for determination of mass decreases than TG curve, which enables to associate DTA peak with either mass change or phase change.

For thermal analysis of corrosions products the substantial area is that up to 400 °C, in which rusts release H_2O. At first stage humidity decreases and physically bonded water is released. This stage is finished

Table 3. Occurrence and Relative Occurrence of Phases
in 276 Samples of Rust

Phase	Occurrence	Relative occurrence (%)
α-FeOOH	174	63.0
β-FeOOH	53	19.2
γ-FeOOH	96	34.8
α-Fe_2O_3	22	8.0
Fe_3O_4	180	65.2
FeO	31	11.2
α-Fe	22	8.0
$CaCO_3$	52	18.9
NaCl	3	1.1
unidentified	12	4.3
amorphous sample	3	1.1

Table 4. IR Absorption Peaks of Rusts

Wave number $[cm^{-1}]$	Phase
470	predominantly amorphous phases, partially also γ-FeOOH
585-595	Fe_3O_4, amorphous phases, β-FeOOH
740	γ-FeOOH
790	α-FeOOH
880	α-FeOOH
1018	γ-FeOOH
1115-1125	sulphates
1460	probably organic impurity
1630-1640	water absorbed on tablet
2840+2910	ν_{C-H}
3130	ν_{O-H}
3410	ν_{OH}

between 180-210 °C under given experimental conditions, especially speed of heating-up. Decomposition of oxyhydroxides and releasing of H_2O with corresponding endothermic effect follow. By measuring this peak on TG curve it is possible to evaluate the sum of all forms of FeOOH in sample. This stage is finished at temperature range 305-330 °C. When amorphous oxyhydroxides in rusts are present, distinguished exothermal peaks on DTA curve in temperature range 300-450 °C are found. High content of amorphous oxyhydroxides used to be typical for initial stage of corrosion. Crystallinity increases and DTA curve becomes flatter.

Corrosion products from water surroundings contain often carbonates, which can be easily determined by derivatography. Carbonates are characterized by exothermic peak on DTA curve and loss of weight on TG curve at temperature about 800 °C.

REFERENCES

1. Powder Diffraction File of the JCPDS - International Center for Diffraction Data, Swathmore, Pennsylvania.
2. K. Volenik, M. Seberini, J. Neid, A. Mössbauer and X-ray diffraction study of nonstoichiometry in magnetite, Czech. Phys. B25:1063 (1975).
3. A. K. Singh, T. Ericsson, L. Häggström, Mössbauer and X-ray diffraction phase analysis of rusts from atmospheric test sites with different environments in Sweden, Corrosion Science 25:931 (1985).
4. T. Misawa, T. Kyuno, W. Suetaka, S. Shimodaira, The mechanism of atmospheric rusting and the effect of Cu and P on the rust formation of low alloy steels, Corrosion Science 11:35 (1971).
5. S. Yamaguchi, H. Sawamura, Sulfid als Korrosionsprodukt an Stählen, Werkstoffe und Korrosion 17:654 (1966).
6. Z. Takehara, A. Saito, S. Yoshizawa, Corrosion of carbon steel through a thin film of solution, Corrosion Science 16:91 (1976).
7. T. Misawa, K. Asami, K. Hasimoto, S. Shimodaira, The mechanism of atmospheric rusting and protective amorphous rust on low alloy steel, Corrosion Science 14:279 (1974).
8. J. E. Hiller, Phasenumwandlungen im Rost, Werkstoffe und Korrosion 17:943 (1966).
9. T. Misawa, K. Hashimoto, S. Shimodaira, The mechanism of formation of iron oxides and oxyhydroxides in aqueous solutions at room temperature, Corrosion Science 14:131 (1974).

DYNAMICAL X-RAY DIFFRACTION STUDY

ON THE PHASE TRANSFORMATION IN RAPIDLY QUENCHED $Au_{71}Sn_{29}$ ALLOY

Norbert Mattern

Academy of Sciences of the GDR, Central Institute of Solid
State Physics and Materials Research
8027 Dresden, Helmholtzstr. 20, GDR

INTRODUCTION

Materials can be obtained in a non-equilibrium state by means of rapid
quenching as e.g. amorphous alloys, metastable phases, supersaturated solutions. These alloys undergo phase transformation into the stable state by
heat treatment. In the case of rapidly quenched $Au_{71}Sn_{29}$ drastic changes of
the mechanical properties already at room temperature are observed. The
alloy shows elastic behaviour after its formation, but it becomes brittle
after about 10 minutes. The aim of this work was to investigate the structural development of rapidly quenched $Au_{71}Sn_{29}$ as a function of time.

EXPERIMENTAL

Ribbons of the $Au_{71}Sn_{29}$ alloy with 30 μm in thickness and 15 mm in width
were prepared by means of the melt-spinning technique cooling down to -190 °C
X-ray measurements were performed by a HZG-4A goniometer of VEB Carl Zeiss
Jena and a position-sensitive detector system at room temperature[1]. The
radiation used was CoK_{α}.

In the stable state the alloy $Au_{71}Sn_{29}$ consists of the tetragonal phase
AuSn (NiAs-type) and the hexagonal close-packed ζ_2-phase (Mg-type) with
11-16 at.% Sn corresponding to the phase diagram. This phase composition
is observed in the X-ray diffraction pattern of the alloy 2 days after its
formation as shown in Fig. 1. Some reflections of a SnO-surface film are
also visible in the diagram. The lattice parameters were determined to a
(0.432 ± 0.001)nm, c=(0.552 ± 0.001)nm for the AuSn phase and a=(0.29 ± 0.001)nm,
c=(0.478 ± 0.001)nm for the ζ-phase, respectively. The X-ray diffraction diagram immediately after the rapid quenching process shown in Fig. 2a indicates
no reflection of the ζ-phase. Besides the peaks of the AuSn phase we observe maxima which can be indexed by a cubic face-centered lattice with
a=(0.339 ± 0.001)nm. The γ-brass phase which was found in splat cooled Au-Sn
alloys with 28-31 at.% Sn by Giessen[3] was not detected.

To investigate the phase transformation the change of the (220)-reflec-
tion of the f.c.c. phase ($2\Theta = 78,5°$) to the (110)-reflection of the h.c.p.
phase ($2\Theta = 76.2°$) was recorded by means of dynamic X-ray diffraction measurements using a position-sensitive detector. The time of measuring was
120 s for each diagram. The obtained diffraction patterns are shown in
Fig. 2. Up to 8 minutes we only observe in the measured region of 2Θ the

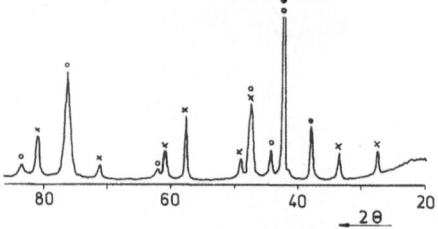

Fig. 1. X-ray diffraction diagram of rapidly quenched $Au_{71}Sn_{29}$ after 48 hours at room temperature (x:AuSn, o: ζ-phase, •: SnO_2).

(220)-reflection of the f.c.c. phase (Fig. 2a-c). After 12 minutes three overlapping new reflections arise at lower values of 2Θ (Fig. 2d). The behaviour of the integral intensity of the (220)-reflection and the sum of the three overlapping peaks in dependence on time is given in Fig. 3. The intensity of the (220)-reflex increases at first. This could be caused by a transformation of an amorphous phase to the f.c.c. lattice. The decrease of the (220)-intensity and the increase of the sum intensity of the three new reflections after 12 minutes indicate the phase transformation from the f.c.c. phase to a more complicated phase X of unknown structure. In the case of a simple phase transformation from the metastable f.c.c. lattice to the stable h.c.p. ζ-phase we would expect only one peak at $2\Theta = 76.2°$. After about 25 minutes the integral sum intensity of the three reflections remains constant, but as it can be seen from Figure 2e-k, the intensities of the components are changing as a function of time. The integral intensities of the single components were estimated by fitting the diffraction intensities by three pseudo-Voigt functions[4]. The obtained results are given in Fig. 4. The second component at $2\Theta = 76.2°$ which corresponds to the (110)-reflection of the h.c.p. ζ-phase increases in dependence on time while the other components decrease. The growth of the (110)-reflection of the ζ-phase lasts up to some days.

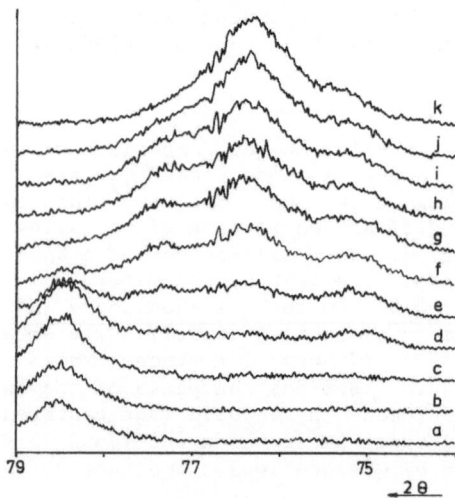

Fig. 2. X-ray diffraction patterns of rapidly quenched $Au_{71}Sn_{29}$ at room temperature in dependence on time t after ribbon formation:
a:t = 2 min., b:t =   5 min., c:t =   8 min., d:t = 12 min.,
e:t = 17 min., f:t = 25 min., g:t = 30 min., h:t = 40 min.,
i:t = 90 min., j:t = 150 min., k:t = 240 min.

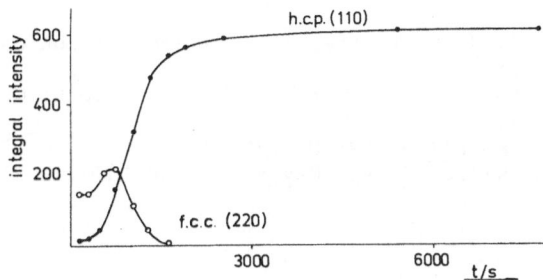

Fig. 3. Integral intensity of the (220)-reflection of the f.c.c. phase and sum intensity of the triple reflections of the h.c.p. phase as a function of time.

CONCLUSION

From the dynamic X-ray diffraction investigations we draw the following conclusions. As-quenched ribbons of the $Au_{71}Sn_{29}$ alloy consist of a metastable f.c.c. phase and the stable phase AuSn. The metastable phase transforms at room temperature to a second metastable phase X of unknown structure. The phase transformation starts after 10 minutes and is finished after about 25 minutes. The metastable phase X then transforms to the stable h.c.p. ζ-phase. This transition starts after about 15 minutes and is finished after about two days. The embrittlement of the as-quenched $Au_{71}Sn_{29}$ alloy is probably connected with the phase transformation of the f.c.c. to the X-phase. The lattice parameter of the metastable f.c.c. phase of the as-quenched state is smaller than that of pure Au or a solid solution of Sn in Au lattice. The transformation from the f.c.c. to the X-phase is probably connected with a volume extension, which causes internal stresses. The mean atomic volume increases from 0.0159 nm^3 for the f.c.c. phase to 0.0175 nm^3 for the h.c.p. ζ-phase.

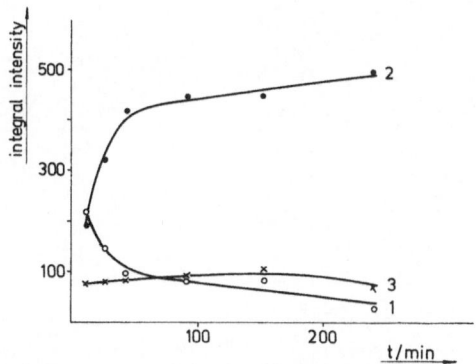

Fig. 4. Time dependence of the integral intensity of the components at $2\Theta = 77.5$ (1)°, $2\Theta = 76.2$ (2)°, $2\Theta = 75.2$ (3)° of the triple reflection.

REFERENCES

1. K. H. Kleinstück, K. Richter, Preprint 05-11-84, Technical University Dresden, GDR (1984).
2. M. Hansen, "Constitutions of Binary Alloys", New York: McGraw Hill Book Co., N.Y. (1958).
3. B. C. Giessen, Z.f.Metallkde 59:805 (1968).
4. R. A. Young, D. B. Wiles, J.Appl.Cryst. 15:430 (1982).

PROCESS KINETICS STUDIED BY X-RAY DIFFRACTION

Tomáš Havlík and Milan Škrobian

Department of Non-Ferrous Metallurgy
Metallurgical Faculty, Technical University
Švermova 9/A, Košice, Czechoslovakia

INTRODUCTION

An analysis can be regarded as a process of changing the degree of uncertainty with respect to the unknown values. The goal of an analysis is
then to gain a maximum decrease of uncertainty. This implies that the decrease of uncertainty also describes the inadequacy of the analysis and
may be considered as a measure for the amount of information obtained[1,2,3].
X-ray diffraction is a very powerful analytical technique since it enables
to obtain a lot of information on the structure of matter: its phase composition, molecular structure, and real structure. The information content
of X-ray diffraction analytical techniques can further be extended when they
are applied to the examination of processes.

As a process represents a dynamic evolution of the state of matter, it
can be described e.g. in terms of the change of the concentrations of the
relevant phases in course of time, i.e. $dc/d\tau$. This can be determined by
the quantitative X-ray diffraction analysis of the phase composition of the
system under study executed in accurate time intervals under controlled conditions, which enables in its turn to calculate the kinetic characteristics
of the process (e.g., process temperature dependence, activation energy,
frequency factor, order of reaction, diffusion characteristics, the influence of structure upon a given process, kinetic equation, etc.). Furthermore, the use of X-ray diffraction phase analysis offers a number of other
advantages, e.g. it immediately discovers the presence of intermediate
phases, which are "parasitic" in given process. That is impossible by means
of other methods. Intermediate phases may be the source of great difficulties for the mentioned study.

Amont a great number of papers concerning X-ray diffraction phase analysis only a few studies deal with process kinetics[4,5,6]. As a consequence
of development of new equipments, ancillaries and precision methods one can
assume that such papers will appear more and more frequently to get new results in development of new technologies and materials. As a practical
example of application of the above mentioned consideration the X-ray study
of thermal decomposition of calcite has been chosen.

Thermal decomposition of calcium carbonate

$$CaCO_3 = CaO + CO_2 \qquad (1)$$

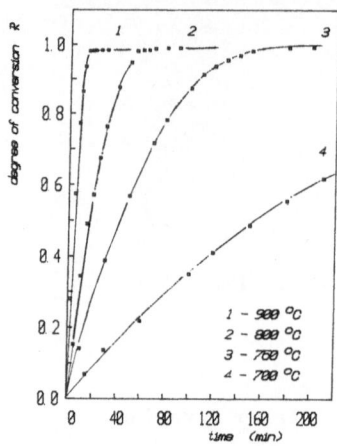

Fig. 1. Conversion versus time for thermal calcite decomposition.

has been a subject of interest of many research workers because of its great
practical utilization.

Even earlier works showed[7] that in case of decomposition of various
types and shapes of calcium carbonate the thickness of calcined layer raises
linearly with time, and so decomposition rate is checked by chemical reaction
at the phase interface of $CaCO_3$ - CaO. Kinetics of the process depends mainly on the area of interface, on difference of partial pressures of CO_2 between its equilibrium and actual value, and on heat conductivity[9].

Many authors (e.g. ref.[8,9]) have studied thermal decomposition of calcium carbonate from many aspects in order to acquire knowledge of influence
of such parameters as the shape of particles, temperature, partial pressures,
etc. Various methods have been used, e.g. temperature measuring inside limestone and on its surface during the course of calcination, partial pressures
measuring, calcination rate measuring by means of gravimetry, study of porosity, study of calcination in fluidized bed, microscopic studies, etc.[10].

Beside determination of influence of technological parameters also the
laws of kinetics and process mechanisms have been studied. The values of
activation energies and the range of temperatures within the course of calcination of calcium carbonate obtained by various authors are summarized in
Table 1.

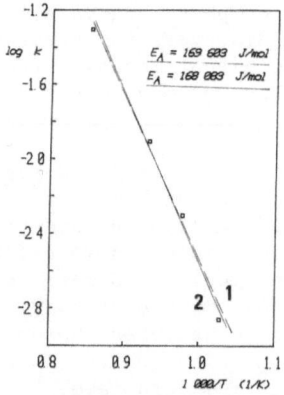

Fig. 2. Arrhenius plot for calcite thermal decomposition. 1 - calculated
from model kinetic equation, 2 - calculated at R = 0.5.

Table 1. Values of the Published Activation Energies

$E_A/kJ\ mol^{-1}$	Range of Temperature	Literature
169.84	750 - 900	8
167.45	950	11
150.7 - 163.3	775 - 850	11
205	661 - 740	12

EXPERIMENTAL

Calcium carbonate $CaCO_3$ chemically and mineralogically pure representing mineral calcite was used in this work in a form of very fine powder. Isothermal decomposition was carried out in electric chamber resistance furnace. Heating and regulation were controlled by thyristor and Pt - PtRh 10 thermocouple.

Specimens of weight of 5 g were placed in equal corundum crucibles, that were placed in preheated furnace. In given time intervals (from 2 to 200 minutes) the specimens were taken out one after another from the furnace and after cooling them, they underwent X-ray diffraction phase analysis in order to determine quantitative contents of calcite and lime in the specimens.

X-ray diffraction phase analysis was carried out by X-ray diffractometer DRON - 2,0 (USSR) by means of Cu K_α radiation with graphite monochromator diffracted beam. The calibrating relationship as a function of weight portion of individual components against the ratio of absolute intensities of the calcite 1 0 4 plane and lime 2 0 0 plane was used for quantitative phase analysis.

RESULTS AND DISCUSSION

The temperatures for thermal decomposition of calcite were chosen 700, 750, 800 and 900°C. Fig. 1 shows the relation between the time of calcination and the degree of conversion:

$$R = (W_o - W_t)/W_o \qquad (2)$$

where W_o is an amount of studied phase for $t = 0$ and W_t is the amount of studied phase at time t.

The Arrhenius plot gives the activation energy $E_A = 168.08\ kJ\ mol^{-1}$. This value is in considerable correspondence with the published data[8,11].

The kinetic equation describing the thermal decomposition of calcite was obtained in the form:

$$\log k = 6.294075 - 8856/T. \qquad (3)$$

These results can be approximated easily by model kinetic equation

$$1 - (1 - R)^{1/3} = k \cdot t \qquad (4)$$

describing the contracting sphere model and is again in considerable correspondence with earlier works. The value of activation energy $E_A = 169.6\ kJ\ mol^{-1}$ was obtained by means of this model.

REFERENCES

1. K. Eckschlager, I. Horsák, and Z. Kodejš: "Vyhodnocování analytických výsledků a metod", SNTL Praha (1980) (Czech text).

2. J. Šesták, "Měření termofyzikálních vlastností pevných látek", Academia, Praha (1982) (Czech version).

3. J. Fiala, Z. Melichar, Informační obsah rentgenové difrakční fázové analýzy, Chem. Listy 78:54 (1974) (Czech version).

4. C. Barriga, J. Morales, and J. L. Tirado, Changes in the kinetics of the vaterite - calcite transformation with temperature and sample crystallinity, J.Mater.Sci. 21:947 (1986).

5. H. Chen, Diffusivity measured by X-ray diffraction, J.Metals 10:36 (1986).

6. M. A. Kipnis, D. A. Agievskii, Metodika kontrolya stepeni fazovykh prevratschtsenii s pomostschyu rentgenografii, Zavodskaya laboratoriya 10:24 (1984) (Russian version).

7. E. C. Furnas, in: M. E. Brown, D. Dollimore, A. K. Galwey: "Reactions in the Solid State", Elsevier, New York (1980).

8. T. R. Ingraham, P. Marier, in: M. E. Brown, D. Dollimore, A. K. Galwey, "Reactions in the Solid State", Elsevier, New York (1980).

9. E. T. Turkdogan, R. G. Olsson, H. A. Wriedt, and L. S. Darken, Calcination of limestone, Trans.Soc.Min.Eng. AIME 254:9 (1973).

10. Z. Asaki, Y. Fukunaka, T. Nagase, and Y. Kondo, Thermal decomposition of limestone in a fluidized bed, Met.Transaction 5:381 (1974).

11. F. Habashi, "Principles of Extractive Metallurgy", vol. 1, Gordon and Breach, New York (1975).

12. D. Beruto, A. W. Searcy, in: M. E. Brown, D. Dollimore, A. K. Galwey, "Reactions in the Solid State", Elsevier, New York (1980).

X-RAY DIFFRACTION ANALYSIS OF 12 CaO . 7 Al_2O_3 POLYMORPHS

A. Derdacka-Grzymek, Z. Konik, J. Iwanciw,
M. Pyzalski, and A. Stok

Interbranch Institute of Building and Refractory Materials
Academy of Mining and Metallurgy
Cracow, Poland

INTRODUCTION

As it results from the literature data, the problem of 12 CaO . 7 Al_2O_3 polymorphism has not been fully explained until now. The authors have obtained some new informations concerning 12 CaO . 7 Al_2O_3 polymorphs and polymorphic transitions[1,2,3]. These data allow to improve the production of self-disintegrating sinter consisting of 12 CaO . 7 Al_2O_3 and calcium orthosilicate – the semi-product in J. Grzymek's alumina technology[4,5].

12 CaO . 7 Al_2O_3 SYNTHESIS IN AIR AND IN VACUUM

12 CaO . 7 Al_2O_3 was synthesized in a superkanthal furnace at 1823 K in air atmosphere from the raw mixture of CaO/Al_2O_3 weight ratio equal to 0.94. Then the samples were cooled slowly in the furnace or quenched rapidly in liquid nitrogen. The same raw mixture has been heated in the "Setaram" thermoanalyser in a vacuum of $3.10^{-2} - 2.10^{-4}$ Tr at 1823 K and subsequently cooled in a vacuum. The samples were analysed by XRD and subjected to optical microscopy and SEM observations.

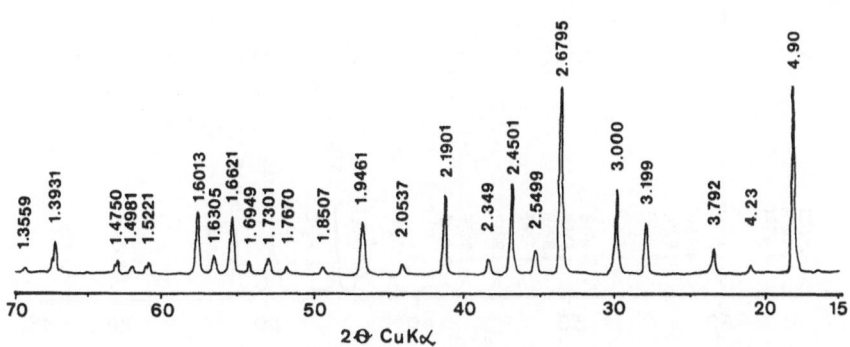

Fig. 1. X-ray diffraction pattern of α 12 CaO . 7 Al_2O_3.

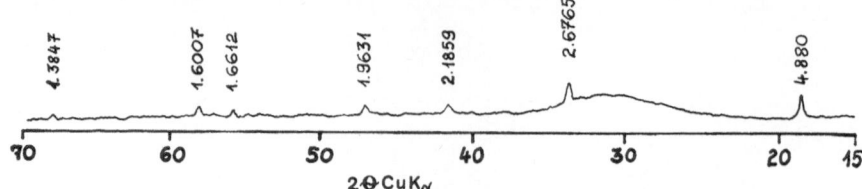

Fig. 2. X-ray diffraction pattern of aluminate glass.

XRD Studies of 12 CaO . 7 Al$_2$O$_3$ Polymorphs

The XRD studies have been made using Philips diffractometer with CuK$_\alpha$ radiation and graphite monochromator in the range of 15 – 70° 2θ with the following parameters: voltage – 50 kV, current – 25 mA, time constant – 2 sec, slits – 1°/0.2 mm, goniometer speed – 1°/min, chart speed – 2 cm/min. XRD pattern of the sample synthesized and cooled in air (Fig. 1) is typical for the isotropic, regular 12 CaO. 7 Al$_2$O$_3$ phase and the "d" values correspond well to the literature and catalogue data. This phase is further determined as α 12 CaO . 7 Al$_2$O$_3$. As it results from the XRD pattern, the sample quenched in liquid nitrogen (Fig. 2) consists of aluminate glass with regular 12 CaO . 7 Al$_2$O$_3$ phase nuclei. The peaks appearing on the pattern give some idea about the growth of regular 12 CaO . 7 Al$_2$O$_3$ crystals. 12 CaO . 7 Al$_2$O$_3$ crystallizes very quickly and therefore the obtaining of aluminate glassy phase optically homogeneous is practically impossible. The sample synthesized and cooled in vacuum, as it results from the XRD patterns (Fig. 3), is not isotropic, regular 12 CaO . 7 Al$_2$O$_3$. The set of "d" values is different from the one for the isotropic 12 CaO . 7 Al$_2$O$_3$ and does not correspond to the mixture of the other calcium aluminates.

Microscopic Studies

The samples have been additionally subjected to the observations using optical microscope "Reichert MeF - 2" and scanning electron microscope Tesla BS 300. The observations of polished sections revealed, that the α – 12 CaO . 7 Al$_2$O$_3$ sample synthesized and cooled in air consisted of optically anisotropic and isotropic phases. Isotropic phase forms colourless crystals, irregular or isometric, characteristic for the regular 12 CaO . 7 Al$_2$O$_3$ phase. Anisotropic phase forms predominantly skeleton structures consisting of irregular or prismatic crystals. The samples synthesized and cooled in vacuum, γ 12 CaO . 7 Al$_2$O$_3$, is built of coarse, plate-like or elongated crystals. SEM observations confirmed this microstructure.

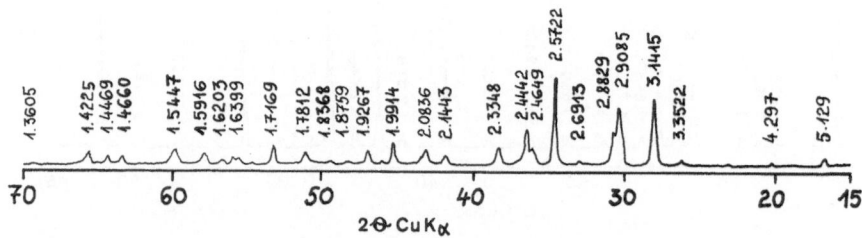

Fig. 3. X-ray diffraction pattern of γ 12 CaO . 7 Al$_2$O$_3$.

TEMPERATURE OF $\gamma \rightarrow \alpha$ 12 CaO . 7 Al_2O_3 TRANSITION DETERMINATION

As it results from the studies, the mixture of CaO/Al_2O_3 weight ratio equal to 0.94 can give two different 12 CaO . 7 Al_2O_3 phases or aluminate glass, depending on the conditions of the synthesis. γ – 12 CaO . 7 Al_2O_3 synthesized in vacuum has been heated at 1473 K in air and subsequently cooled in air. The sample thus obtained has been subjected to XRD studies and α 12 CaO . 7 Al_2O_3 has been found. One can therefore state that α and γ are the polymorphs of the compound 12 CaO . 7 Al_2O_3 and the temperature of transition $\gamma \rightarrow \alpha$ is equal to 1473 K. It should be mentioned that the γ 12 CaO . 7 Al_2O_3 sample heated at lower temperatures for example 873 K does not reveal any change (Fig. 4).

As it results from the literature data, regular 12 CaO . 7 Al_2O_3 phase is probably stabilized by OH or H_2O and therefore can be obtained in the presence of water vapour. The thermogravimetric measurements of α 12 CaO . 7 Al_2O_3 revealed the 1.5 % weight loss mainly at 1473 K. One can presume that during the thermal treatment of anisotropic γ 12 CaO . 7 Al_2O_3 at 1473 K in air, OH^- groups or H_2O particles can incorporate to the structure and the structural rearrangement thus takes place.

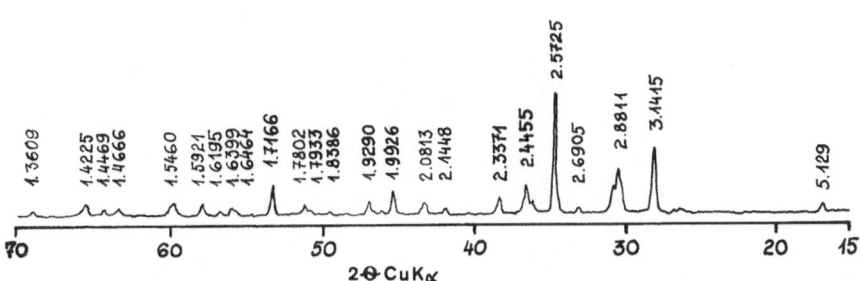

Fig. 4. X-ray diffraction pattern of γ 12 CaO . Al_2O_3 heated at 873 K.

CONCLUSION

1. Optically anisotropic γ 12 CaO . 7 Al_2O_3 phase can be produced at vacuum from the raw mixture of CaO/Al_2O_3 weight ratio equal to 0.94. This phase reveals different XRD pattern than the isotropic, regular 12 CaO . 7 Al_2O_3.

2. Anisotropic optically γ 12 CaO . 7 Al_2O_3 transforms at 1473 K in air into the isotropic, regular α 12 CaO . 7 Al_2O_3.

3. γ and α phases are the polymorphs of 12 CaO . 7 Al_2O_3. The temperature of $\gamma \rightarrow \alpha$ transition is 1473 K.

REFERENCES

1. J. Grzymek, A. Derdacka-Grzymek, Z. Konik, M. Gawlicki, and A. Stok, The complex production of aluminium oxide and iron from laterite raw materials applying the calcium aluminates polymorphism, in: "Light Metals 1985", H. O. Bohner, ed., AIME New York (1985).

2. J. Grzymek, A. Derdacka-Grzymek, Z. Konik, A. Stok, M. Pyzalski, and J. Iwanciw, Some physicochemical properties of 12 CaO . 7 Al_2O_3 phase in relation to Al_2O_3 production from self-desintegrating sinters, in: "Light Metals 1987", R. D. Zabreznik, ed., AIME, Denver (1987).

3. J. Grzymek, A. Derdacka-Grzymek, Z. Konik, M. Pyzalski, A. Stok, and J. Iwanciw, The effect of synthesis and cooling conditions on phase properties of 12 CaO . 7 Al_2O_3, in: "Proceedings of 8th International Congress on Chemistry of Cement", Rio de Janeiro, Vol. 4:401 (1986).

4. J. Grzymek, A. Derdacka--Grzymek, Z. Konik, W. Grzymek, Methods for obtaining iron, alumina, titania and binders from metallurgical slags and from red mud remaining in the Bayer method, in: "Light Metals 1982", J. E. Andersen, ed., AIME, Dallas (1982).

5. A. Derdacka-Grzymek, J. Grzymek, M. Gawlicki, Z. Konik, and A. Stok, The complex technologies to get aluminium oxide, iron, titanium oxide and cementing binders from clay materials, Mineral.Records 41:201 (1986).

X-RAY ANALYSIS OF MINERAL SUBSTANCES OF BLOOD VESSELS

Andrzej Szytula and Eugeniusz Rokita

Institute of Physics, Jagellonian University

Reymonta 4, 30-059 Cracow, Poland

INTRODUCTION

The artery wall mineralization is defined as the deposition of inorganic compounds under normal and pathologic conditions. The results of IR and X-ray diffraction measurements of the human artherosclerotic artery are reported.

MATERIALS AND METHODS

The investigations were carried out on human aorta and skeleton muscle samples. The material was excised from cadavers (adults ranging in age from 16 to 69 years). An identical procedure was used to examine all samples. After dissection, the material was freeze-dried and ground down in an agate mill. The composition of the powder was determined by the application of the proton induced X-ray emission (PIXE) method and of the infrared (IR) spectroscopy. Apparently the IR spectra of dry aorta samples contained only bands resulting from the presence of organic molecules, and this is why the organic molecules had to be removed in order to get information about an inorganic material. It was achieved by the inceneration for 8 h at 450 $^{\circ}$C. The ashed material was examined by PIXE method and IR spectroscopy.

The PIXE measurements were performed using the 2.7 MeV proton beam from the C-48 cyclotron in the Institute of Nuclear Physics in Kraków. For analysis, the previously described[1-2] set-up was used. The targets were prepared in the form of pellets and irradiated using external beam technique. Concentrations of the following elements were measured: P, S, Cl, K, Ca, Fe, Cu, Zn, Br, Sr and Pb.

The infrared spectra were obtained in the 4000 - 400 cm^{-1} region from a Specord 75 IR spectrometer (Carl Zeiss Jena) using compressed KBr pellets. Approximately 1 mg of ash was needed to obtain reliable IR spectra.

RESULTS

The elemental composition of human aorta homogenates are given in Table 1. The age of patients is also presented in the table. Since mineral deposits in a pathologically changed aorta wall contain mainly calcium salts, the correlation of Ca concentration with age of patients was found. Con-

Table 1. Concentrations of Elements [μg/g Dry Mass] in the Human Aorta [A] and Skeleton Muscle [M] Samples

Sample	P	S	Cl	K	Ca	Fe	Cu	Zn	Br	Sr	Pb	Age
A-27	9300	7800	10200	4900	2900	191	10.9	97.3	61.2	2.15	1.96	37
A-28	8700	8100	8050	4300	3400	162	15.3	90.6	79.5	2.58	1.51	34
A-30	8100	8850	12000	5840	2000	92	14.1	131	85.1	1.97	2.41	16
A-32	11300	8020	7960	4360	9400	147	19.2	184	68.6	6.69	11.20	69
A-33	9500	8200	9320	4710	3260	126	10.4	84.6	93.2	1.90	1.67	30
A-34	12500	6840	9100	4050	6300	103	17.0	107	74.2	3.81	2.85	60
A-35	10400	7410	8900	5300	9100	224	15.9	160	59.7	4.43	7.21	62
M-29	5200	6720	5180	11700	365	250	12.1	148	151	1.2	0.8	16
M-31	6130	6430	6590	12300	540	381	9.4	135	170	1.2	0.8	69

centrations of other elements determined in this study do not demonstrate correlation with age. The elemental composition of ashed human aorta samples are shown in Table 2. The quantitative results obtained by such method of material preparation can be only considered as crude estimations[1].

Fig. 1A shows IR absorption spectrum of a dry human aorta sample containing an atherosclerotic plaque. Many different bands, being in good agreement with previously reported results[3], may be easily seen in the spectrum. The bands are mainly due to such substances as lipids, lipoproteins, nucleic acids and polysaccharides. We cannot recognize any bands characteristic for an inorganic material. Therefore, the IR spectra of dry aorta samples are completely useless for the investigation of inorganic deposits within an aorta wall. To get any information about an inorganic material the organic matrix has to be removed. It may be done by an enzymatic digestion or inceneration. We used the latter procedure.

Table 2. Concentration of Elements [μg/g Ashed Mass] in the Human Aorta [A] and Skeleton Muscle [M] Samples

Sample	P 10^3	S	Cl	K	Ca 10^3	Fe	Cu	Zn	Br	Sr	Pb
A-27	82	–	–	–	27	5800	120	1310	–	27.4	70.6
A-28	89	–	–	–	33	4100	251	1080	–	36.7	45.2
A-30	74	–	–	–	17	3300	196	1420	–	20.8	84.3
A-32	105	–	–	–	96	3700	165	1970	–	7.31	197
A-33	91	–	–	–	41	2600	110	986	–	22.4	37.1
A-34	98	–	–	–	74	2300	234	1250	–	41.5	61.4
A-35	112	–	–	–	91	7000	182	2000	–	39.6	110
M-29	87		–	–	3.2	6400	134	2430	–	6.14	5.05
M-31	93	–	–	–	4.6	7800	97	1935	–	6.94	11.6

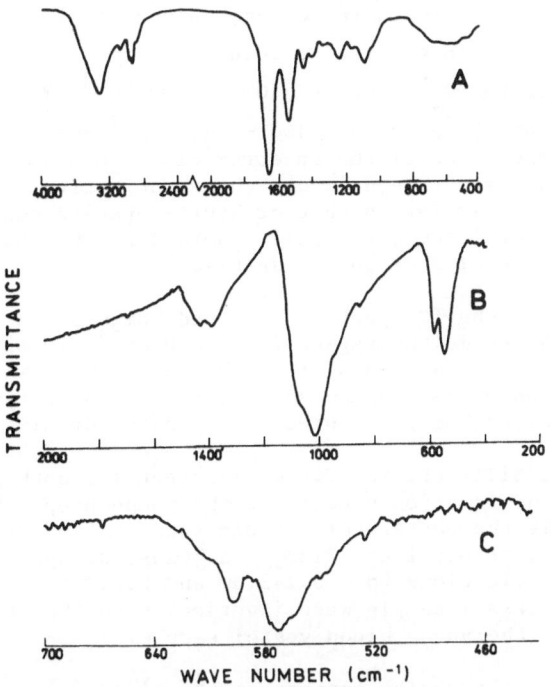

Fig. 1. IR spectra of various samples

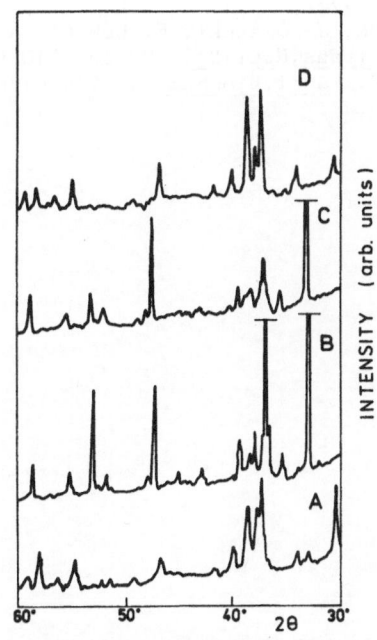

Fig. 2. X-ray diffraction patterns
A - hydroxyapatite standard, B,C - young persons,
D - old persons

The IR spectrum of incenerated human aorta samples as well as a synthetic hydroxyapatite in the region 2000 – 400 cm^{-1} are presented in Fig. 1B. The bands due to PO$_4^{3-}$ group at 1100 – 1000 cm^{-1}, 960 cm^{-1} and 600 – 550 cm^{-1} as well as CO$_3^{2-}$ group at 1460 – 1410 cm^{-1} and 875 cm^{-1} can be recognized. The comparison of the spectrum of a human aorta sample with the spectrum of a hydroxyapatite delivers the quantitative proof that the "apatic" structures similar to that of hydroxyapatite constitute the inorganic phase of mineralized aorta wall. Moreover, the bands due to CO$_3^{2-}$ ion confirm the presence of a carbonate apatite.

In the Fig. 1C the IR spectrum of ashed human aorta samples and a synthetic hydroxyapatite in the region 700 – 400 cm^{-1} are presented. The splitting of 600 – 550 cm^{-1} band from a broad singlet, which is characteristic for an amorphous calcium phosphate, to a doublet typical for the fully crystaline mineral can be considered as an indicator of crystallinity.

The obtained diffractograms were different for both groups (Fig. 2). The X-ray powder diffraction pattern of the young people blood-vessels was not indexed, while the pattern of the old ones was indexed on the basis of the strongest reflections identified as a hexagonal hydroxyapatite structure. Two weak reflections (d = 0.312 nm and 0.221 nm) in the X-ray pattern of the old blood-vessel sample were identical with the strongest reflections in the pattern of the young blood vessel sample.

It shows that the ageing process in the blood-vessels may be investigated through the structural transformations analysis.

REFERENCES

1. J. Gałuszka, L. Jarczyk, E. Rokita, A. Strzałkowski, M. Sych, Nucl.Instr. Meth. B3:141 (1984).
2. K. Durek, L. Jarczyk, J. Oszacki, E. Rokita, A. Strzałkowski, M. Sych, A. Orlsan, IEEE Trans.Nucl.Sci. NS-30:1310 (1983).
3. F. S. Parker, Ans.R., Anal.Biochem. 18:414 (1967).

OPTICAL DIFFRACTION OF COMPUTER SIMULATED PATTERNS

OF TRANSITIONS IN TWO DIMENSIONAL REGULAR LATTICE

J. Sołtys and W. Kołyniak

Institute of Physics, Jagellonian University

Reymonta 4, 30-059 Cracow, Poland

Methods of investigations of ordering or decomposition processes in crystalline structures are often based on diffraction of X-rays, electrons or thermal neutrons. Diffraction patterns are difficult to interpret and calculation of structural parameters is rather complicated. Due to the universality of wave diffraction laws the possibility of application of optical diffraction method has already been noticed by W. L. Bragg and M. Wolfke[1,2]. In the present work the applicability of optical diffraction method for verification of structure models and comparative analysis of patterns obtained by other diffractional methods was studied.

A computer simulation of order-disorder transitions in processes of ordering and decomposition of binary alloy for the two dimensional lattice was performed. The distribution of A and B type atoms was randomly generated in two dimensional regular lattice. Two processes are possible: ordering, when it is energetically preferable for the A-type atoms to have nearest neighbours of the B-type, and decomposition, when forming clusters of the atoms of the same type is more preferable. The transition's probability is given by Boltzman's formula. Let $N(dE)$ denote a number of pairs of atoms which, after being permuted, cause energy change equal to dE. After the given structure is randomly chosen, $N(dE)$ distribution is approximately Gaussian and after ordering one obtains Boltzman distribution, except for the case dE=0. A number of pairs of such a kind is considerable for concentration $\neq$ 50 %, what means that not all of the atoms may occupy energetically preferable positions because there exist the excess of atoms of the one type and lack of the second type.

In case of decomposition the situation is different. From the moment of cluster forming the reverse process is less probable. Although the atoms on the surface are in the positions energetically unpreferable, the probability of coagulation of the clusters is low. Because of that, the final energetical distribution is asymmetric, deformed towards the energetically preferable locations, in spite of the random cluster distribution in lattice.

Figure 1 shows atom distribution in chosen successive stages of the process of ordering of atoms in two dimensional structure. The distribution of A and B type atoms in the successive stages of the computer modelling were printed, and then, applying photographical techniques, masks of these structures were made. These masks were then applied in the process forming optical diffraction patterns.

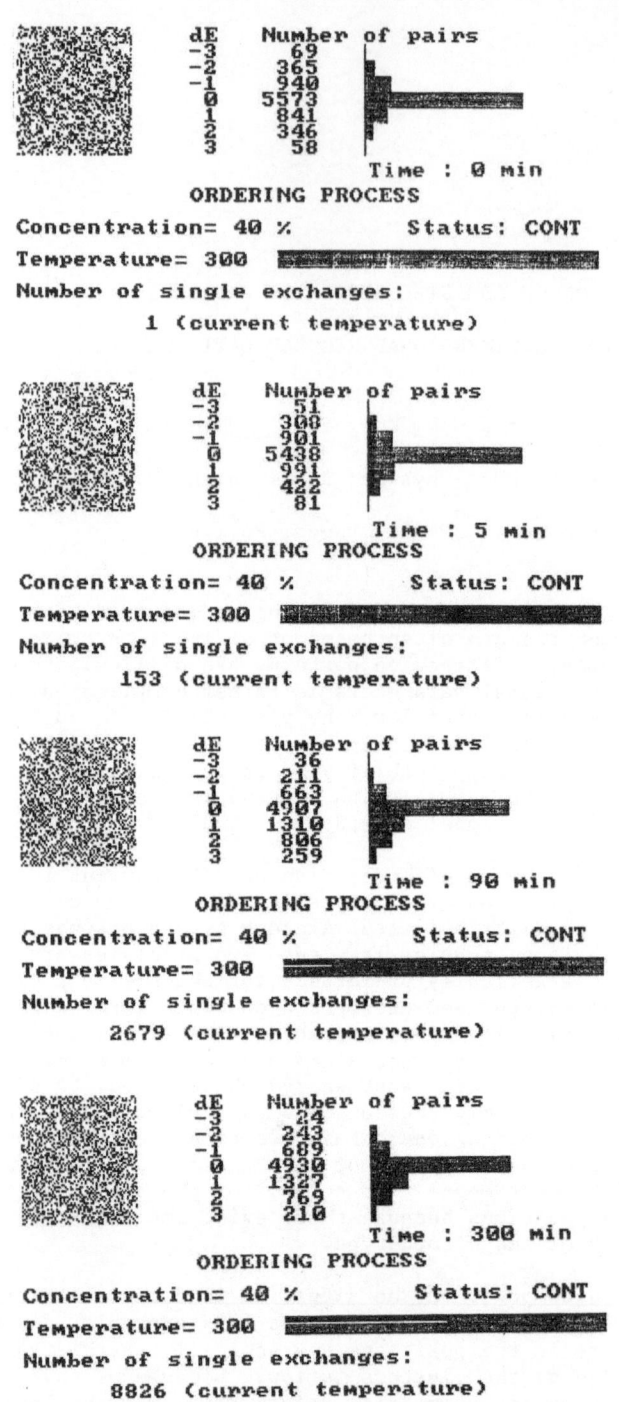

dE   Number of pairs
-3   69
-2   365
-1   940
 0   5573
 1   841
 2   346
 3   58
                    Time : 0 min
        ORDERING PROCESS
Concentration= 40 %   Status: CONT
Temperature= 300
Number of single exchanges:
        1 (current temperature)

dE   Number of pairs
-3   51
-2   300
-1   901
 0   5438
 1   991
 2   422
 3   81
                    Time : 5 min
        ORDERING PROCESS
Concentration= 40 %   Status: CONT
Temperature= 300
Number of single exchanges:
        153 (current temperature)

dE   Number of pairs
-3   36
-2   211
-1   663
 0   4907
 1   1310
 2   806
 3   259
                    Time : 90 min
        ORDERING PROCESS
Concentration= 40 %   Status: CONT
Temperature= 300
Number of single exchanges:
        2679 (current temperature)

dE   Number of pairs
-3   24
-2   243
-1   689
 0   4930
 1   1327
 2   769
 3   210
                    Time : 300 min
        ORDERING PROCESS
Concentration= 40 %   Status: CONT
Temperature= 300
Number of single exchanges:
        8826 (current temperature)

Fig. 1. The stages (a, b, c, d) of the ordering process in two dimensional
         structure (computer simulation).

For the investigation of diffractional patterns optical diffractometer
working with He-Ne laser as the source of light was constructed. Diffraction patterns were registered on a photographical plate. In case of dis-

ordered and totally ordered structures a comparison of obtained results and results of theoretical calculations of diffractional distributions was made. The diffraction patterns from the following processes were examined:

1. Ordering process in an alloy containing atoms showing great or small values of scattering factors.

2. Order and disorder in an alloy containing atoms exhibiting similar scattering factors.

3. Order and disorder in alloy containing atoms of different scattering factors, when the atoms of each kind are not identical but accidentally deformed.

The following conclusions were drawn from the obtained results:

1. Intensities of superstructural reflections gradually increase during the ordering process.

2. The shape of scattering objects does not affect the information about the state of order yielded by optical diffraction.

3. If all components of a given kind are not identical (in molecular crystals it may result from molecule deformation, e.g. caused by thermal vibrations), the optical diffraction may also differentiate between the order and disorder states; however, apart from fundamental reflections.

The results of model diffraction study may be qualitatively compared to the diffraction patterns of real crystalline structures.

REFERENCES

1. W. L. Bragg, Nature 143 (1939); 149 (1942).
2. M. Wolfke, Phys.Zeitschr. 21 (1920).

II. DIFFRACTION ANALYSIS OF AMORPHOUS MATERIALS GLASSES AND POLYMERS

INVESTIGATION OF THE MICROSCOPIC MECHANISM

OF THE HIGH STRENGTH OF AMORPHOUS ALLOYS

Liu Rang-su

Department of Physics, Hunan University

Changsha, China

INTRODUCTION

From the viewpoint of statistical thermodynamics and by means of the "model of double-layer structural units (MDLSU)" proposed by the author recently, we investigated the specific process of formation and the mechanism of structure defects of amorphous alloys formed by melt quenching, and reasonably explain the microscopic mechanism whereby amorphous alloys possess high strength.

Amorphous alloys possess various excellent physical properties, especially their striking ultra-high strength at temperatures below the glass transition temperature, T_g. In order to improve their strength and other mechanical properties, we have to understand more thoroughly the relations between the macroscopic properties and microscopic structures. In this paper, we have studied the process of formation of amorphous structures formed by melt quenching from the point of view of statistical thermodynamics, and propose a forming mechanism of structure defects (network of micro-dislocation elements). By means of the "model of double-layer structural units (MDLSU)" proposed by the author recently[1,2], we obtain the concrete distribution of structure defects. Thus, we can reasonably explain the microscopic mechanism by which amorphous alloys possess high strength; furthermore, we point out the ways to enhance the strength of amorphous alloys.

The Forming Process and Structure Defects of Amorphous Alloys

From the viewpoint of statistical thermodynamics, the system of melting metal-metalloid alloys can be considered as a gas system which consists of non-interactive and independent atoms and, by the action of a certain potential energy, the gas system would be transformed into a liquid system in which the atoms are interactive and dependent; then, by rapid quenching, the liquid system would be transformed into the amorphous solid.

When the system is in the gaseous state, it would obey Maxwell-Boltzmann distribution law whether in energy or in velocity. In general, the metal atoms have more weight than the metalloid. In the case of only two elements, there will be two different Maxwell velocity distribution curves. According to their velocity, we can divide the atoms of the system into four regions, as shown in Fig. 1.

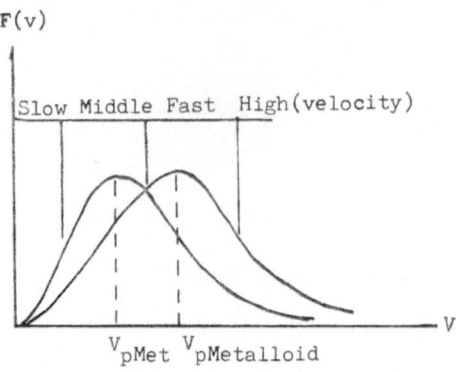

Fig. 1. Maxwell's velocity distribution curves of metal and metalloid
         atoms in melt state.

When the system is cooling, the atoms can be condensed to transform
into a liquid by the action of a potential between atoms. Consider that,
in the same potential, the atoms with the same or nearby the same velocity
would have a larger collection probability than those with more different
velocity. Since the metalloid elements have higher electronegativity than
the metal, a few of the metal atoms and a few of the metalloid atoms with
slower velocities would be gathered at first and many small aggregates would
be formed which possess chemical short range order (CSRO) with the metalloid
atoms in the center and the metal atoms in the nearest neighbourhood.

When the temperature decreases near T_g, the cooperative collective effect of all atoms occurs; many of the atoms with middle and fast velocities
would rapidly form aggregates of CSRO to fill most of the space. Then they
attach to each other, but, in the end, a very few of the atoms with high
original velocity would be ejected into the gaps left by a great number of
small aggregates. Meeting the condition of CSRO and being consistent with
topological short range order (TSRO) as much as possible, the atoms with
high original velocity would be frozen and the whole system would transform
into an amorphous structure.

It is the last few atoms that fill in the larger space and decrease
the density of amorphous alloys; these give rise to the commonly called the
"free volume". Since the atoms in the "free volume" can move more freely,
the arrangement of atoms is quite irregular and they can form the particular polyhedra corresponding to three types of the five polyhedra in amorphous
structure of the Bernal hard sphere model. We can consider the space where
the few atoms locate as the structure defect, as shown in Fig. 2.

The Structure Defects in the New Model

According to the results of a great number of experiments, the amorphous
structure is characterized by long-range randomness but short-range order
still exists, and the range of the latter is about 5 - 6 atomic distances[3].
The range of the simple tetrahedra and octahedra, however, is of only 1 - 2
atomic distances, which cannot reflect the short-range order in experiments.
From this, the author proposed a new model of amorphous structure recently[1,2].

In the new model, we supposed that the double-layer structure is
within the Bernal polyhedra (mainly within the tetrahedra and octahedra
and their combining units), i.e. they themselves are the area of short-range

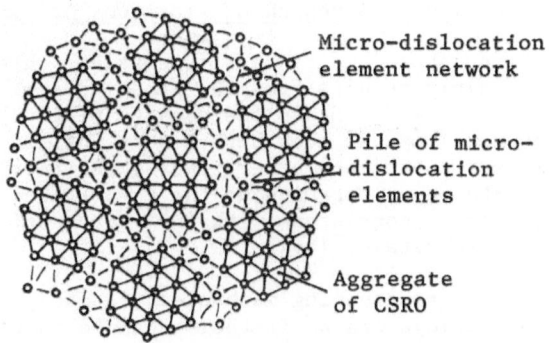

Fig. 2. Schematic diagram of the microscopic structure of amorphous metal-metalloid alloys and the "micro-dislocation element network" (in plane).

order, and the whole amorphous structure is made of these double-layer polyhedra with short-range order by connecting triangular plane with dense-packing. This is the "model of double-layer structural units (MDLSU)" of amorphous alloys formed by melt quenching, as shown in Fig. 3.

According to the MDLSU, we can consider that the double-layer combining unit in Fig. 3(c) consists of one double-layer octahedron (DLO) and two double-layer tetrahedra (DLT), and the DLO is a small aggregate of spherical shells which consist of a metalloid atom as the center, 12 metal atoms as the nearest neighbours, and 6 metalloid atoms as the next nearest neighbours. As mentioned above, both the DLT close to the DLO would be formed from freezing a few atoms with high velocity. It is obvious that the DLT has serious distortion, particularly at the tips of DLT. This is the place where the structure defects of amorphous alloys locate. In three-dimensional space, these distorted areas would construct a random network connected to each other. By the aid of the concept of "dislocation" in crystals, the distorted area of several atomic distances can be considered as the "micro-dislocation element"; therefore, the random network formed by the distorted areas can be considered as the random "micro-dislocation element network" and the severely distorted area, namely the "junction" of the network, can be considered as the "pile of micro-dislocation elements", as already shown in Fig. 2.

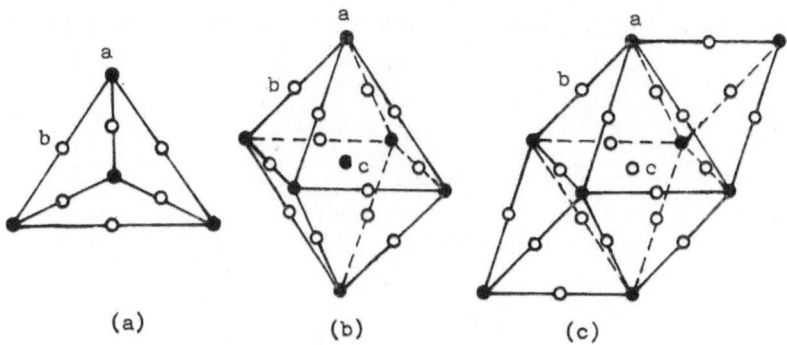

Fig. 3. The model of double-layer structural units
● metalloid atom; o metal atom;
(a) double-layer tetrahedron (DLT);
(b) double-layer octahedron (DLO);
(c) double-layer combining unit (1DLO + 2DLT).

Microscopic Mechanism of High Strength of Amorphous Alloys

From the mentioned above, we can easily consider that the high strength of amorphous alloys mainly results from the random distribution of "micro-dislocation element entworks". Since the density of the "micro-dislocation elements" in amorphous alloys is much higher than that in crystal alloys, and they interlock very tight each other, therefore, amorphous alloys possess very high strength, even approaching the theoretical strength. This strengthening mechanism is consistent with the microscopic mechanism of deep cold work hardening in crystal alloys.

According to this strengthening mechanism, we can predict that the strength of amorphous alloys can be further enhanced in two ways. First, increase as much as possible the content of the metalloid used as the nucleating center, in order to form more small aggregates of atoms, but the metalloid content can not be infinite; according to the MDLSU, its maximum is about 30.47 at %. Second, the cooling speed of rapid quenching must be increased to decrease the volume of the small aggregates of atoms. Both the ways may make the "micro-dislocation element network" have a more dense, homogeneous and random distribution so that amorphous alloys possess higher strength.

DISCUSSION

1. In fact, the first way mentioned above[4] has widely been used in various preparations of amorphous alloys and it has been shown that the strength of amorphous alloys do increase with the increase of the content of metalloid. For the second, even though it has not yet been proved experimentally, it is worthy of future study.

2. Using the above mentioned strengthening mechanism, it is possible that the strength of amorphous alloys can be calculated quantitatively. As an example, the strength of the amorphous Fe-Si-B system has been calculated and the result will be published elsewhere[5].

REFERENCES

1. Liu Rang-su, Kexue Tongbao (Science Bulletin), 31:1167 (1986).
2. Liu Rang-su, Kexue Tongbao (Science Bulletin), 32:1527 (1987).
3. Y. Waseda, H. Okazaki, and T. Masumoto, J.Mat.Sci. 12:1927 (1977).
4. T. Masumoto, Sci.Rep.RITU 26A:246 (1977).
5. Liu Rang-su, Scientia Sinica (Series A) 31 (1988), to be published.

PARTIAL STRUCTURE FACTORS FOR AMORPHOUS $Fe_{75}B_{25}$
DETERMINED BY DIFFRACTION OF POLARIZED NEUTRONS

W. Matz

Central Institute for Nuclear Research
Rossendorf, Dresden, GDR

J. Kulda and P. Mikula

Institute for Nuclear Physics
Řež, Czechoslovakia

INTRODUCTION

In the case of amorphous metals the understanding of structure is mostly limited to the short range order. The information on the real space structure gained from the diffraction experiment is further limited, because only radial distribution functions can be determined. This is a one-dimensional structural picture of the real three-dimensional atomic arrangement.

This relatively low information content can be considerably enhanced, if it is possible to obtain partial structure factors and hence partial radial distribution functions. The latter can be achieved by combining experiments with different radiations or in the case of ferromagnetic amorphous metals by the diffraction of polarized neutrons.

THEORETICAL BACKGROUND

Suppose that in a binary alloy AB only the A-atoms carry a magnetic moment. The diffraction of polarized neutrons with spin parallel and anti-parallel ($\pm$) to a saturating magnetic field $\vec{H}$ applied to the sample gives different elastic coherent cross sections $\frac{d\sigma}{d\Omega}$ if $\vec{H}$ is perpendicular to the scattering vector $\vec{Q}$ (perpendicular to the plane of incident and scattered neutrons).

$$\vec{H}_\perp \vec{Q}: \quad \frac{d\sigma^\pm}{d\Omega} = c_A^2(b_A \pm p_A)^2 S_{AA}(Q) + 2c_A c_B(b_A \pm p_A)b_B S_{AB}(Q) + c_B^2 b_B^2 S_{BB}(Q) \qquad (1)$$

The $S_{ij}(Q)$ are the partial structure factors of interest for the binary amorphous alloy. The c_i denote the concentrations and the b_i the coherent neutron scattering lengths. The magnetic scattering length $p(Q)$ is defined by the magnetic moment μ and the neutron magnetic form factor $f(Q)$.

$$p(Q) = 0.270 \; \mu \; f(Q) \times 10^{-12} \; cm \qquad (2)$$

A third equation is obtained for the pure nuclear scattering which can be measured by applying the magnetic field $\vec{H}$ parallel to $\vec{Q}$.

$$\vec{H} \parallel \vec{Q}: \quad \frac{d\sigma}{d\Omega}^{(n)} = c_A^2 b_A^2 S_{AA}(Q) + 2c_A c_B b_A b_B S_{AB}(Q) + c_B^2 b_B^2 S_{BB}(Q) \qquad (3)$$

RESULTS

Amorphous ribbons of the composition $Fe_{75}B_{25}$ were studied by diffraction of polarized and unpolarized neutrons. The sample was made from boron enriched with ^{11}B in order to reduce neutron absorption. The neutron diffraction experiments using polarized neutrons were performed with the SPN-100 spectrometer at the research reactor of the UJF Řež (mean power 6.5 MW). During the measurements a magnetic field of 0.35 T was applied to the sample. The corrected experimental results are given as the points in Fig. 1.

The partial structure factors $S_{ij}(Q)$ will be obtained by the solution of the system of linear equations (1,3). The direct estimation of the partial structure factors from the experimental data was not possible, because the elements of the inverse coefficient matrix has rather big values and magnify the experimental uncertainties considerably. Furthermore, there are two additional parameters, the magnetic moment μ and a width parameter W (ref.1) for the parametrized magnetic form factor of iron.

$$f_{Fe}(Q) = 0.385 \exp(-0.0852.W.Q^2) + 0.623 \exp(-0.0321.W.Q^2) - 0.017 \qquad (4)$$

So we adopt an iterative procedure for the determination of the partial structure factors with the following steps:

(i) Normalization of the corrected intensities for a given set of μ and W.

(ii) Calculation of $S_{FeFe}(Q)$ and $S_{FeB}(Q)$ from each 2 of the 3 experiments, neglecting $S_{BB}(Q)$.

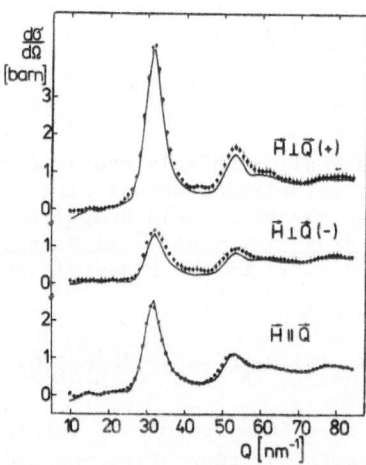

Fig. 1. Coherent elastic cross sections for amorphous $Fe_{75}B_{25}$ determined by diffraction of polarized ($\vec{H} \perp \vec{Q}$) and unpolarized neutrons ($\vec{H} \parallel \vec{Q}$). The points are the experimental data and the lines represent cross sections, recalculated from partial structure factors.

(iii) Smoothing of the partial structure factors $S_{ij}(Q)$ at a graphical display. The partial structure factors have to fulfil the sum rule and the inequalities[3]:

$$\int_0^\infty Q^2[S_{ij}(Q)-1]dQ = -2\pi^2\rho_o \tag{5}$$

$$c_i + c_i^2(S_{ii}(Q)-1) > 0 \tag{6a}$$

$$[c_i + c_i^2(S_{ii}(Q)-1)] \cdot [c_j + c_j^2(S_{jj}(Q)-1)] > (c_ic_j)^2(S_{ij}(Q)-1)^2 \tag{6b}$$

(iv) Recalculation of the elastic coherent cross sections with the smoothed partials $S_{ij}(Q)$ and computation of χ^2.

The recalculated cross sections are displayed in Fig. 1 as full lines. The choice of a optimal data set according to the χ^2-minimum was stronger affected by the pure nuclear scattering ($\vec{H} \parallel \vec{Q}$) in account of the smaller experimental errors. The smoothed partial structure factors for the atomic pairs Fe–Fe and Fe–B are shown in Fig. 2.

The partial structure $S_{FeFe}(Q)$ reproduces the main features of the total structure factors, because in all cases the Fe–Fe contribution to the cross section is the largest one. A sharp main maximum at 31.5 nm^{-1} and a second maximum with a shoulder are characteristic. In the partial $S_{FeB}(Q)$ a peak at 35 nm^{-1} and a broad second maximum centered around 65 nm^{-1} are observed. The double maximum below 25 nm^{-1} was observed in the pure nuclear scattering experiment ($\vec{H} \parallel \vec{Q}$) and appears enlarged as a consequence of the smaller prefactors of $S_{FeB}(Q)$.

The determination of the partial structure factor for the metalloid--metalloid-pair $S_{BB}(Q)$ was not possible because of the small weighting factor. On the other hand, the magnetic data of iron in amorphous $Fe_{75}B_{25}$ could be deduced from this set of experiments: W = 0.95$\pm$0.05 and the magnetic moment μ = (1.77$\pm$0.03)μ_B. The width parameter W < 0 indicates a tendency of electron localization on iron in the amorphous alloy. This result was also found for other amorphous metal-metalloid alloys[1,4].

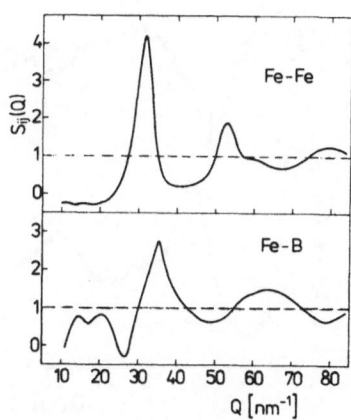

Fig. 2. Smoothed partial structure factors for the atomic pairs Fe–Fe and Fe–B determined from the polarized neutron diffraction experiments.

COMPARISON TO THE TWO-BEAM EXPERIMENT

The partial structure factors for the atomic pairs Fe-Fe and Fe-B of amorphous $Fe_{75}B_{25}$ were independently determined by combining neutron (n) and X-ray diffraction (X).[6] The measured total structure factors S(Q) are connected by the following equations with the partials $S_{ij}(Q)$:

$$S_n(Q) = 0.685 \cdot S_{FeFe}(Q) + 0.285 \cdot S_{FeB}(Q) + 0.03 \cdot S_{BB}(Q) \qquad (7a)$$

$$S_X(Q) = 0.883 \cdot S_{FeFe}(Q) + 0.113 \cdot S_{FeB}(Q) + 0.004 \cdot S_{BB}(Q) \qquad (7b)$$

Neglecting the contribution $S_{BB}(Q)$ these two experiments gave the results displayed in Fig. 3. The general form of the two sets of partials in Fig. 2 and Fig. 3 is similar but in detail there are significant differences. The main maximum of $S_{FeFe}(Q)$ in Fig. 2 is about 30 % higher compared to the curve in Fig. 3 and the minimum around Q = 40 nm^{-1} is not so deep in Fig. 2. For $S_{FeB}(Q)$ the experiment with polarized neutrons gave a main maximum which is narrower than that in Fig. 3. The data from the combination of neutron and X-ray diffraction seem more reliable, because of the smaller experimental uncertainties.

CONCLUSIONS

The determination of partial structure factors for metal-metalloid alloys by diffraction of polarized neutrons at a small reactor gives results with high uncertainties. The metalloid-metalloid contributions can not be determined.

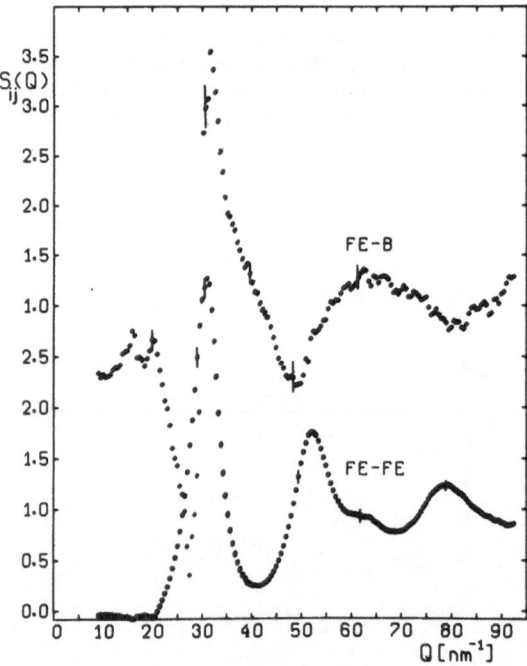

Fig. 3. Partial structure factors of the Fe-Fe- and Fe-B-pairs of amorphous $Fe_{75}B_{25}$ determined from the combination of neutron and X-ray diffraction experiments[5].

The derivation of partial structure factors is only possible, if additional information (sum rules, inequalities) is used. The recalculation of cross sections and comparison to experimental data after smoothing the partial structure factors is important.

The diffraction of polarized neutrons provides information on the magnetic characteristics of the metal atom in the amorphous alloy. Neutrons are a local probe as compared to most other magnetosensitive techniques.

REFERENCES

1. J. Bletry and J. F. Sadoc, Determination of the three partial interference functions of an amorphous cobalt-phosphorus ferromagnet by polarized-neutron scattering, J.Phys.F. 5:L110 (1975).
2. E. J. Lisher and J. B. Forsyth, Analytical approximations to form factors, Acta Cryst. A27:545 (1971).
3. Y. Waseda, The structure of liquids, amorphous solids and solid fast ion conductors, Prog.Mater.Sci. 26:1 (1981).
4. Wu Guan, N. Cowlam, H. A. Davies, R. A. Cowley, D. McK Paul, and W. G. Stirling, Structural investigation of an FeB metallic glass by polarised neutron diffraction, J.Physique 43:C7-71 (1982).
5. W. Matz, H. Hermann, and N. Mattern, On the structure of amorphous $Fe_{75}B_{25}$, J.Non-cryst.Solids, 93:217 (1987).

NEUTRON-DIFFRACTION INVESTIGATION OF THE SHORT-RANGE ATOMIC ORDER

IN TELLURITE GLASSES

S. Neov, I. Gerasimova

Institute of Nuclear Research and Nuclear Energy

Sofia 1784, Bulgaria

P. Mikula

Institute of Nuclear Physics

250 68 Řež, Czechoslovakia

The present neutron-diffraction investigation aims to establish the effect of the second oxide component on the short-range atomic order in tellurite glassy systems: TeO_2-MoO_3, TeO_2-V_2O_5, TeO_2-Al_2O_3 and TeO_2-P_2O_5. The experiments were carried out at λ = 0.1061 nm on the IRT-2000 reactor in Sofia and at λ = 0.0803 nm on the VVR-S pile in Řež. In both cases the maximum value of the scattering vector Q = $4\pi \sin\Theta/\lambda$ attained was 100 nm^{-1}. Experimental radial distribution functions (RDF), which are Fourier transforms of the neutron interference functions, were interpreted by comparison with the quasicrystalline model RDF's, calculated by using the structural diffusion model.

The binary TeO_2-MoO_3 system possesses wide concentration glass formation limits, between 12.5-58.5 mole % of MoO_3. The TM1, TM2 and TM3 compositions containing 20, 33.3 and 50 mole % MoO_3, respectively, were investigated. RDF's shown in Fig. 1, calculated without using any modifying function, allow us to trace the effect of MoO_3 on the short-range atomic order in the glasses. The gravity center of peak P1 shifts toward the shorter distances as the MoO_3 content increases: R_1=0.192, 0.190 and 0.187 nm for samples TM1, TM2 and TM3, respectively. The increment of the short-distant component in P1 is due to the contribution of the shorter Mo-O bonds in the MoO_6 octahedron, four of which are located in the 0.17-0.195 nm range. In the crystalline α-TeO_2, the basic structural units building up the spatial network are TeO_4 groups, which can be considered as trigonal bipyramids with one unoccupied equatorial position. Unlike β-TeO_2, where the binding of the coordination polyhedra is effected through a mutual edge, the TeO_4 groups in α-TeO_2 are bound vertex - to - vertex by means of an oxygen atom in an axial-equatorial position, respectively, as shown in Fig. 2. One of the axial Te-O bonds which is not equivalent in strength to the remaining ones of the TeO_4 bipyramid undergoes elongation of about 10 % in the glasses[1]. In this case, the Te coordination can be regarded as 3+1. In TM1 composition the coordination number (c.n.) of Te is 3.3 as calculated from the area under P1. Increasing the MoO_3 content stimulated the transition to a triple Te coordination: a reduction of the c.n. of Te being observed in TM2 and TM3 samples.

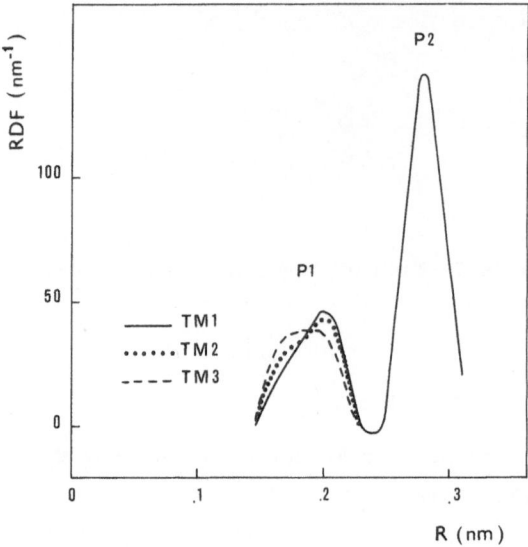

Fig. 1. RDF's for TM1, TM2 and TM3 glass samples.

Simultaneously, changes in the MoO_6 octahedron are also registered. To explain the experimentally observed area under P1 one has to assume that the coordination state of Mo atoms is also reduced, MoO_5 and MoO_4 groups being probably formed.

The $2TeO_2 \cdot V_2O_5$ glass corresponds to the $Te_2V_2O_9$ compound, in which the Te is in a triple coordination with respect to oxygen. In the RDF obtained, the partial V-O distributions are missing, since the coherent scattering amplitude of the V nucleus is too small. The coordination number value, 3.35, at $R_1 = 0.196$ nm shows that 65 % of the Te atoms in the glass are in triple coordination. The newly-formed groups, in which a Te atom is in vertex of the TeO_3 pyramid, are bound into chains characteristic of the $Te_2V_2O_9$ compounds.

In a $4TeO_2 \cdot WO_3$ glass one observes a reduction of the first coordination peak P1 formed by the Te-O and W-O distributions with respect to the model RDF. The shift of P1 towards the shorter distances: $R_1 = 0.195$ nm relative to $R_{model} = 0.2$ nm indicates that the reduction of the coordination number of Te leads to the formation of TeO_3 pyramids, in which the Te-O distances are on the average shorter than those in the TeO_4 bipyramids.

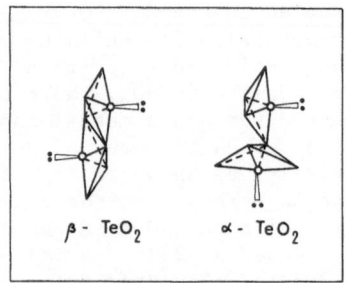

Fig. 2. Binding scheme in α-TeO_2 and β-TeO_2.

The obtained area and position of the RDF peaks show that the irregular WO_6 octahedron is retained in the glassy state.

The mutual correlation of the glass-forming TeO_2 and P_2O_5 was studied in detail earlier[2]. On passing from 8 to 26 mole % P_2O_5 a rapid reduction is observed of the contribution of O-O and Te-O distribution characteristic for α-TeO_2 in the RDF. Instead of peak P1 with R_1 = 0.195 nm, characteristic of glasses rich in TeO_2, a new maximum appears at R_1 = 0.16 nm, due to the P-O distribution in P_2O_5.

In experiments with very cold neutrons by the time of flight method, the concentration and size of the regions with density different from the average one for the samples were determined. While clusters were not observed at a low P_2O_5 content, their concentration was 6.5×10^{14} cm^{-3} at 20 mole % of P_2O_5 and their diameter was about 10 nm. In a sample with 25 mole % of P_2O_5 their concentration increased to 1.2×10^{16} cm^{-3}, while the diameter was about 5 nm. These results confirm our conclusion made earlier[2] that at P_2O_5 concentration of over 26 ± 5 mole %, there is a stable immiscibility in the TeO_2-P_2O_5 glassy system.

The effect of Al_2O_3 on the short-range order of TeO_2 is investigated on the basis of three compositions containing 7, 10 and 13 mole % of Al_2O_3. The first maximum P1 on the RDF shifts towards the shorter distances in all the three cases, while its area remains constant and equal to the area under P1 of the quasicrystalline model RDF. Therefore, in the TeO_2-Al_2O_3 glassy system the coordination number of Te is 4 at low Al_2O_3 content, i.e. the TeO_4 bipyramid is retained. The shift of P1 is explained by the contribution of the shorter (R = 0.19 nm) Al-O distances in the AlO_6 octahedron.

Fig. 3 shows a RDF obtained by means of superimposing experimental RDF's of the investigated glasses (except for that of $2TeO_2 \cdot V_2O_5$). Peak P1 is smaller in area and is shifted towards the smaller values of R_{Te-O}, a fact due to the general trend towards a transition to the 3+1 coordination of the Te atom. The lack of any marked P3 peak on the experimentally obtained RDF's within the 0.38-0.4 nm range is characteristic of all tellurite systems investigated so far and is a consequence of the deformation of the O_{ax}-Te-O_{eq} angle in glassy state.

This work was sponsored by the Science Committee of Bulgaria, Res. Contract No 7.

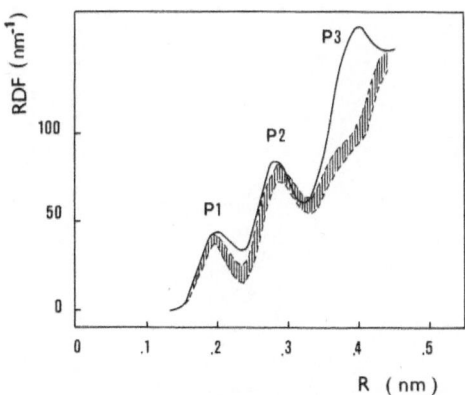

Fig. 3. Comparison between theoretical RDF for pure α-TeO_2 and RDF's for the glasses studied.

REFERENCES

1. S. Neov, I. Gerasimova, K. Krezhov, B. Sydzhimov, and V. Kozhukharov, Atomic arrangement in tellurite glasses studied by neutron diffraction, Phys.Stat.Sol.(a) 47:743 (1978).
2. S. Neov, I. Gerasimova, V. Kozhukharov, and M. Marinov, The structure of glasses in the $TeO_2-P_2O_5$ system, J.Mater.Sci. 15:1153 (1980).

SUPERMOLECULAR STRUCTURE OF THE "LADDER-TYPE"

STYRENE-MULTIMETHACRYLATE COPOLYMERS

Stanisław Rabiej and Andrzej Włochowicz

Textile Institute, Technical University of Łódź

Findera 32, Bielsko-Biała, Poland

Supermolecular structure of the "ladder-type" styrene-multimethacrylate copolymers were investigated using SAXS, WAXS and DSC methods. The structural studies were preceded by detailed measurements of the copolymers composition (IR method), the average molecular weights and the molecular weight distributions (GPC method).

CHARACTERIZATION OF MATERIAL

Styrene-multimethacrylate copolymers were synthesized by the method developed by Połowiński[1,2]. In the first stage, fractions of p-cresol-formaldehyde resin were obtained by polycondensation. The selected fractions (matrices) of known composition and molecular weight were estrified with methacryloyl chloride in the presence of triethylamine. Multimethacrylates were obtained as a product of the reaction. The copolymerization of multimethacrylate and styrene was carried out in benzene at 90 °C using AIBN as an initiator. Macromolecules of ladder-type copolymers are built of alternating multimethacrylate blocks with fixed length and polystyrene blocks with variable uncontrolled lengths. The term "ladder-type copolymers" is connected with the constitution of the multimethacrylate blocks, for which the system of bonds and atoms resembles a ladder (Fig. 1).

A series of styrene-multimethacrylate copolymers were used in our studies. The homopolymers, i. e. polystyrene and multimethacrylate homopolymer (polymerized multimethacrylate) were also investigated. Total mass-average molecular weights measured by gel permeation chromatography ranged from 2.10^4 to 3.10^4 (Table 1). The weight fractions of polystyrene determined by IR spectroscopy lay between 26 % and 69 %.

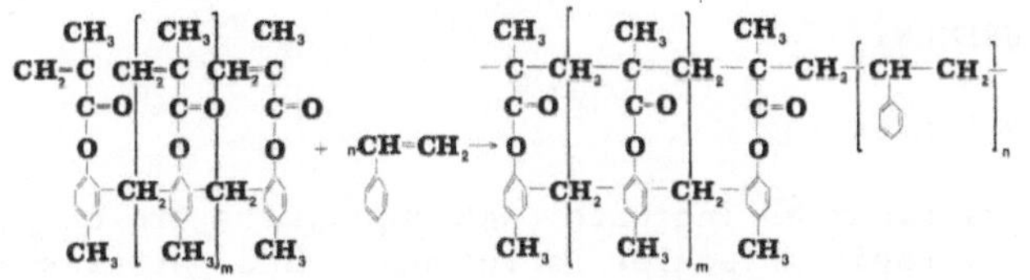

Fig. 1. Scheme of copolymerization of multimethacrylate and styrene.

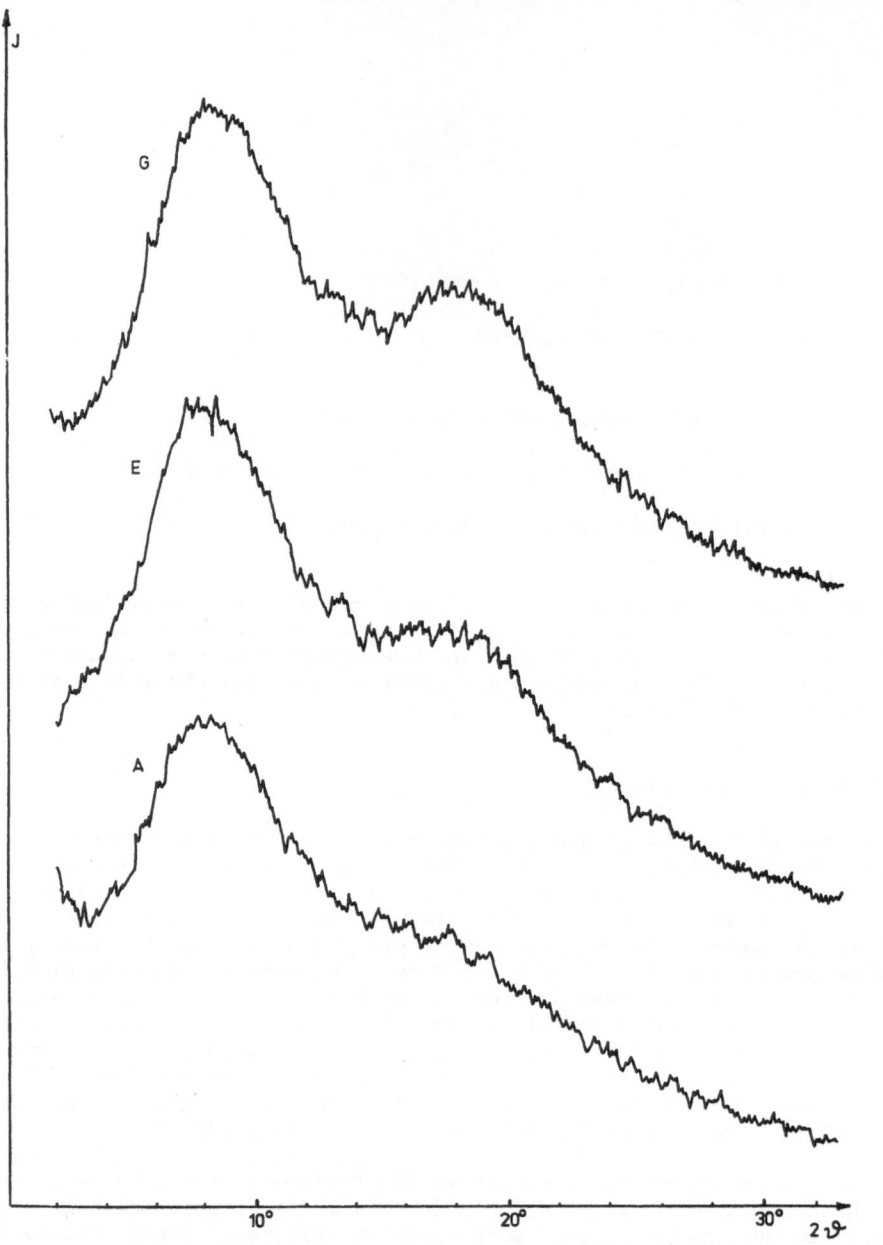

Fig. 2. WAXS intensity distributions for samples A (74 % MM), E (53 % MM) and G (31 % MM), (MM - multimethacrylate)

RESULTS OF MEASUREMENTS

Wide Angle X-Ray Scattering

WAXS scattering curves indicate that copolymers are completely amorphous independent of their molecular structure. The patterns exhibit two broad, diffuse maxima. The maxima are located at 8° and 18° (Fig. 2). These positions are close to those observed for amorphous polystyrene (10°, 19°).

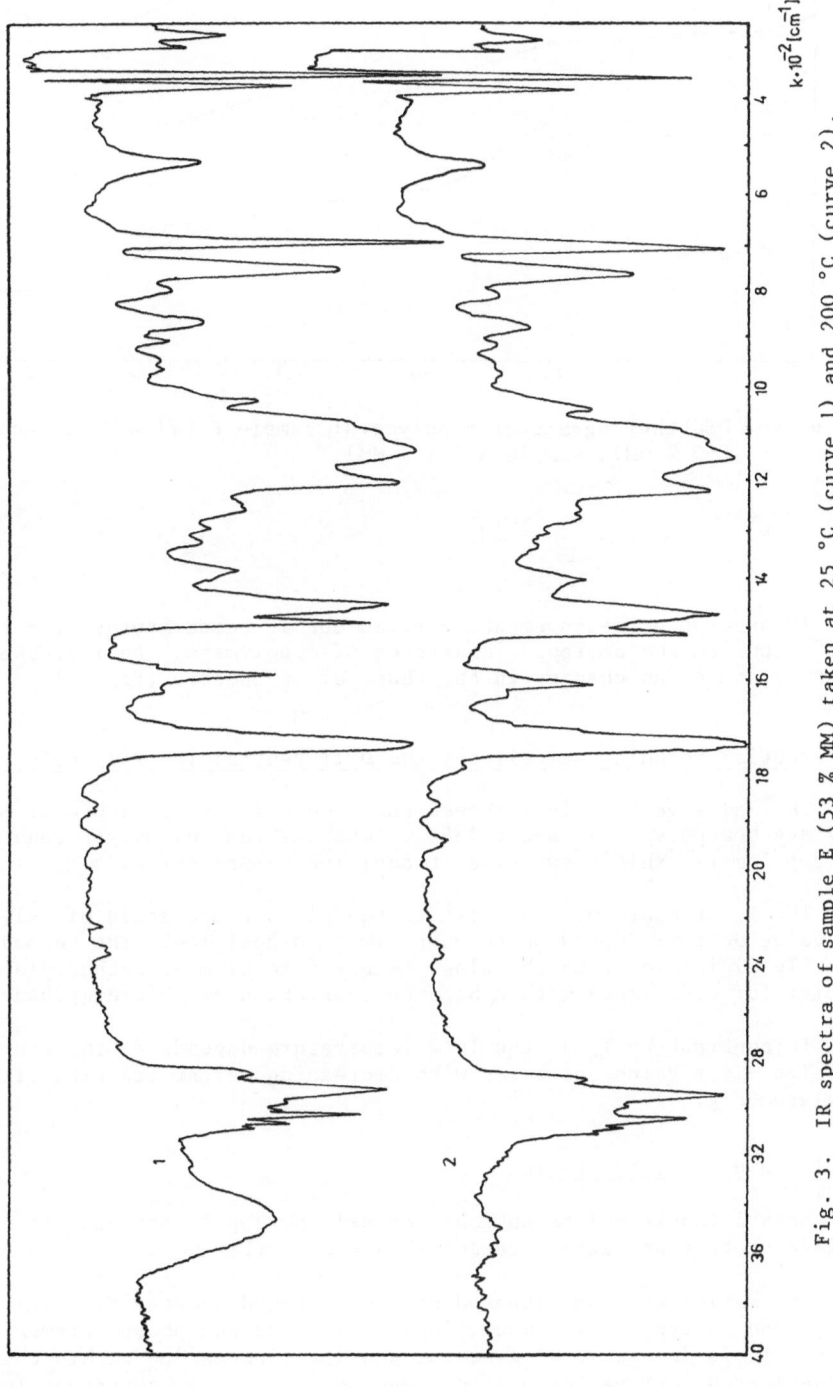

Fig. 3. IR spectra of sample E (53 % MM) taken at 25 °C (curve 1) and 200 °C (curve 2).

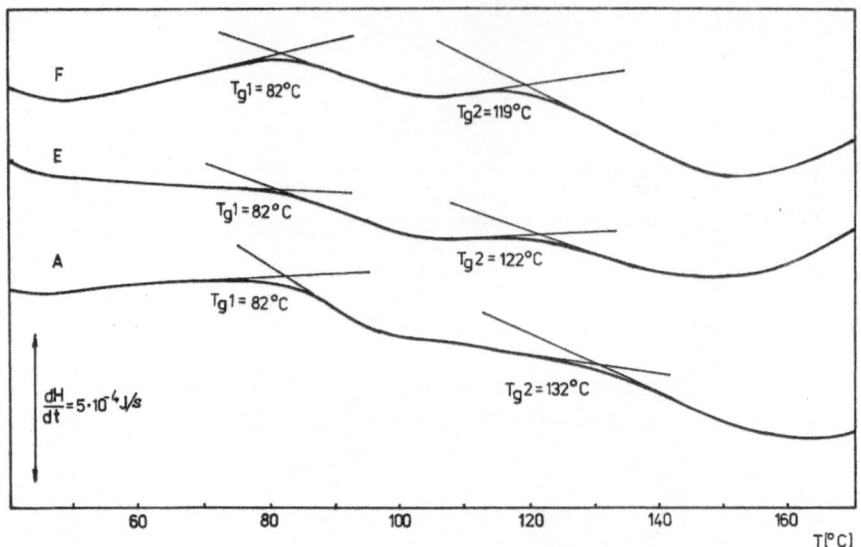

Fig. 4. DSC thermograms of copolymers: sample F (41 % MM), sample E (53 % MM), sample A (74 % MM).

IR absorption measurements carried out at temperatures from 25° to 200 °C confirm the amorphous character of copolymers. An increase in temperature causes no changes in the shape of IR spectra (Fig. 3).

Differential Scanning Calorimetry and Differential Thermal Analysis

Thermal investigations showed that the glass temperature of multimethacrylate homopolymer is about 138 °C, whereas that of polystyrene is 88 °C. The copolymers exhibit two glass transition temperatures.

The T_{g1} temperature is close to the glass temperature of polystyrene. Its value does not depend on the copolymer composition. The second temperature ($T_{g}2$) is closer to the glass temperature of multimethacrylate, in particular for copolymers with a high concentration of this component.

In contrast to $T_{g}1$, the $T_{g}2$ temperature depends on the copolymer composition - its value decreases with decreasing weight fraction of multimethacrylate (Fig. 4).

Small Angle X-Ray Scattering

SAXS intensity distributions for all the copolymers exhibit clearly visible scattering maxima between 2° and 2.5° (Fig. 5).

The maxima are superimposed on a background scattering. This kind of background is typical for many single-phase and two-phase polymeric materials. It is due to local electron density fluctuation within the phases. The desmeared SAXS patterns after subtraction of the background for two copolymers are presented in Fig. 5. (Desmearing of the SAXS intensity distributions and quantitative analysis of the SAXS data were performed with the aid of a computer program FFSAX-5 elaborated by Vonk.)

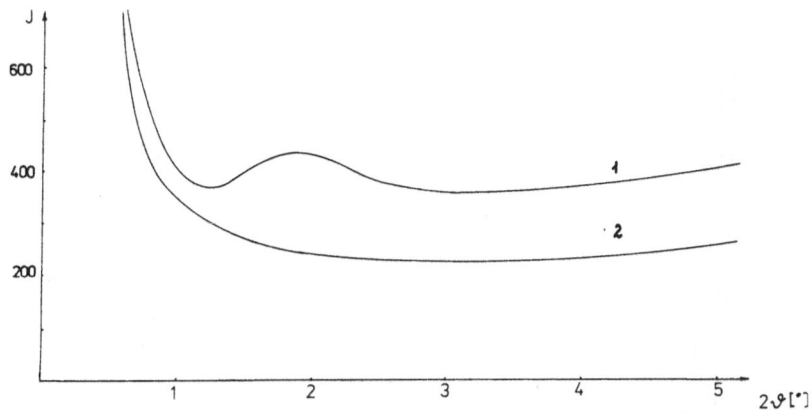

Fig. 5. Smeared SAXS intensity distributions: curve 1 - for copolymer F (41 % MM), curve 2 - for blend made of homopolymers (50 % PS and 50 % MM).

The scattering maxima indicate that the copolymers are composed of two phases different in their electron densities. However, the maxima appear at relatively large scattering angles. Generally at large scattering angles the interparticle interference can be neglected - for this reason, the maxima are most probably associated with intraparticle interference. They are due to individual particles, microdomains that exist in the volume of a sample. Assuming that microdomains are monodisperse spheres, their radii were estimated.

The scattering intensity distribution for a spherical particle of radius R is given by:

$$I(h) = (\rho_2 - \rho_1) \cdot V^2 \cdot [3(\sin hR - hR \cos hR)/(hR)^3]^2 \qquad (1)$$

$h = 4\pi \cdot \sin\vartheta/\lambda$, 2ϑ - scattering angle, V - volume of the particle, ρ_1, ρ_2 - electron densities of the particle and surrounding medium.

Table 1. Results of Measurements

Sample	% MM	N	$\overline{M}$	R [Å]	E [Å]
A	74	4	36 800	33	8
B	70	4	–	34	–
C	71	4	31 400	32	9
D1	63	4	29 100	37	9
D2	62	5	–	35	–
D3	65	6	–	35	–
D4	63	8	21 700	36	10
E	53	4	29 700	38	8
F	41	4	16 600	40	10
G	31	10	23 600	42	9
H	44	8	–	49	–

% MM - weight fraction of multimethacrylate, R - radius of the microdomain, E - thickness of the transition layer, N - number of monomers in MM blocks, $\overline{M}$ - weight-average molecular weight

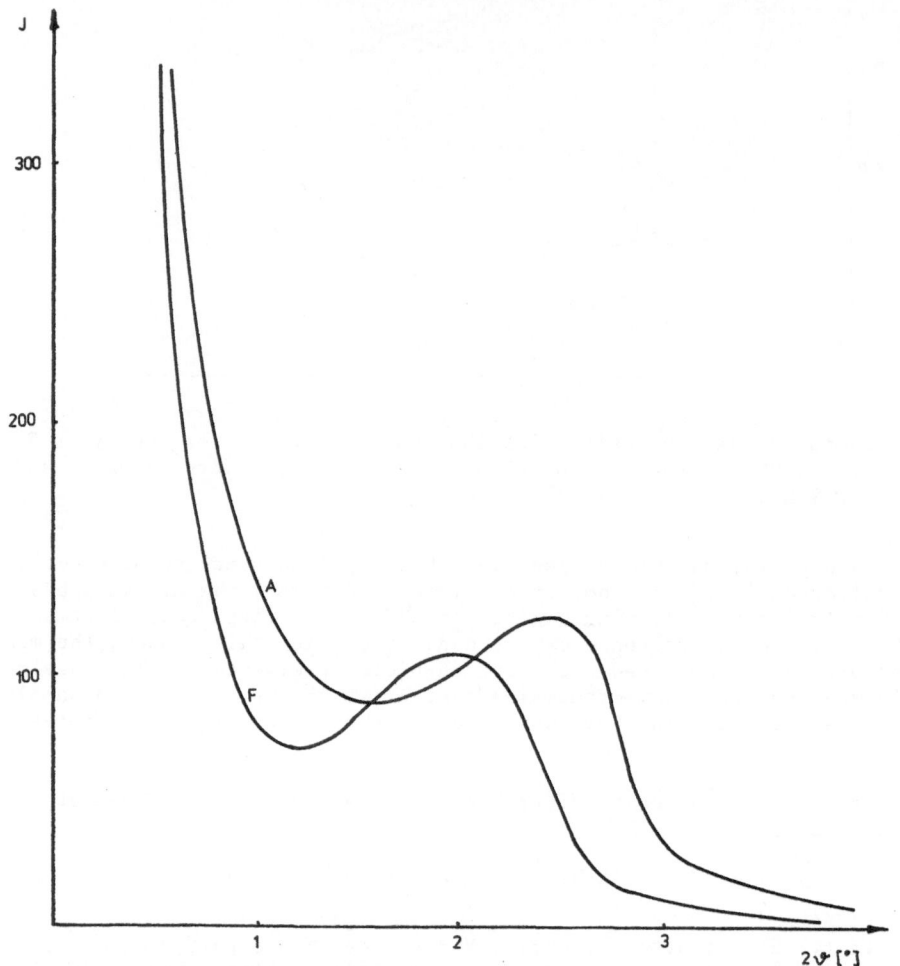

Fig. 6. Desmeared SAXS intensity distributions after subtraction of the
background for samples A (74 % MM) and F (41 % MM).

The intensity distribution exhibits scattering maxima due to intraparticle interference. The first maximum is located at an angle satisfying
the following relation

$$(4 \pi \cdot \sin \vartheta/\lambda) \cdot R = 5.765 \qquad (2)$$

Using equation (2) the radii of the microdomains were calculated (Table
1). The radii range from 33 Å to 42 Å and they increase with increasing
polystyrene weight fraction in copolymers.

The thickness of the transition layer between the microdomain and surrounding medium (an interphase) was calculated by the method of Ruland[4].
According to him, the observed intensity (after subtraction of the background)
can be expressed at large scattering angles as:

$$I(h) = (const.) \cdot h^{-4} \exp(-\sigma^2 \cdot h^2) \qquad (3)$$

where σ is related to the thickness E of the transition layer by:

$$E = 2 \sqrt{3} \cdot \sigma$$

114

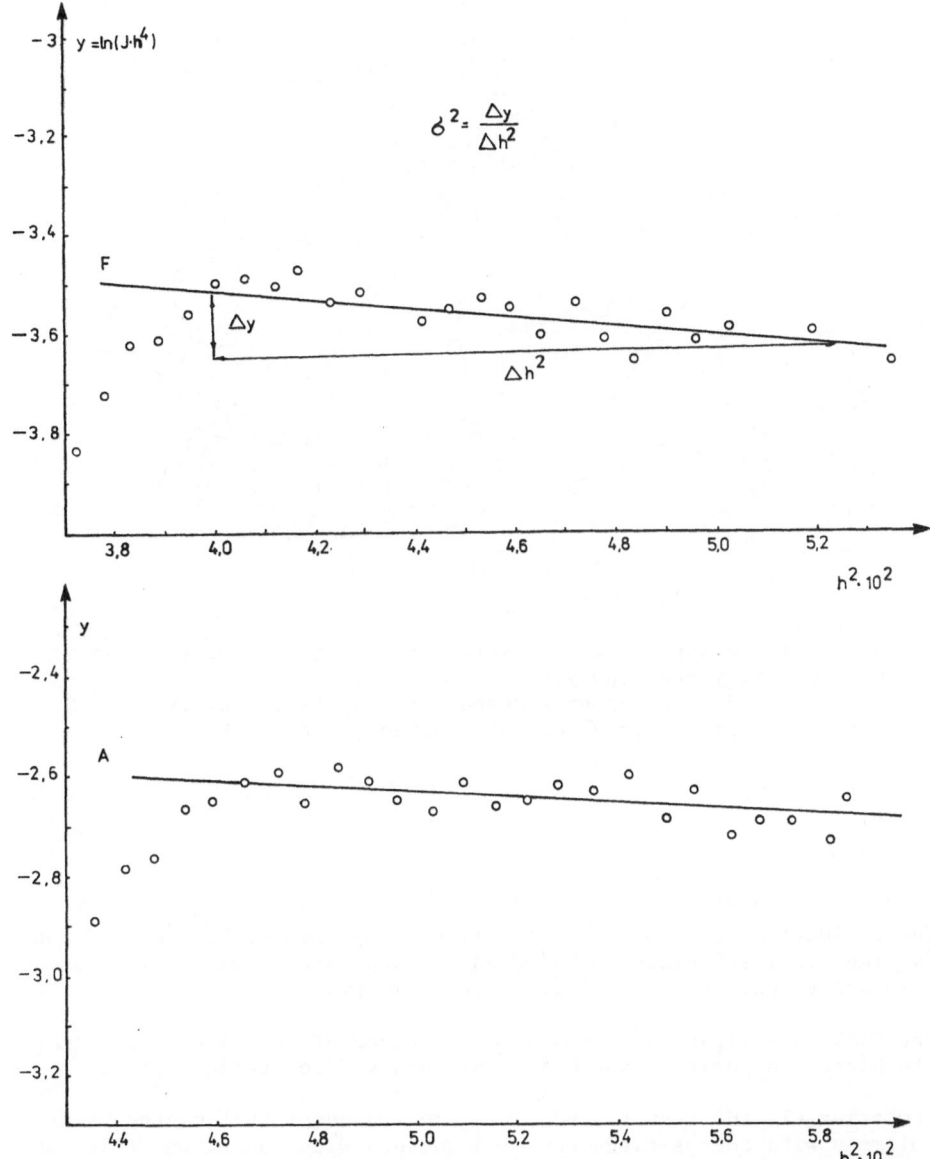

Fig. 7. The plots for evolution of the transition layer thickness E for samples A (74 % MM) and F (41 % MM).

Thus a plot of $\ln[I(h) \cdot h^4]$ vs. h^2 gives the value of σ (Fig. 7). The thickness calculated by this method is practically independent of the copolymer composition, its value is about 9 Å (Table 1).

SAXS investigations were carried out not only at room temperature but also at temperatures up to 200 °C. An increase in temperature results in progressive decrease in SAXS reflection height and slight shift towards higher angles. On the other hand, the background intensity distinctly increases with increasing temperature.

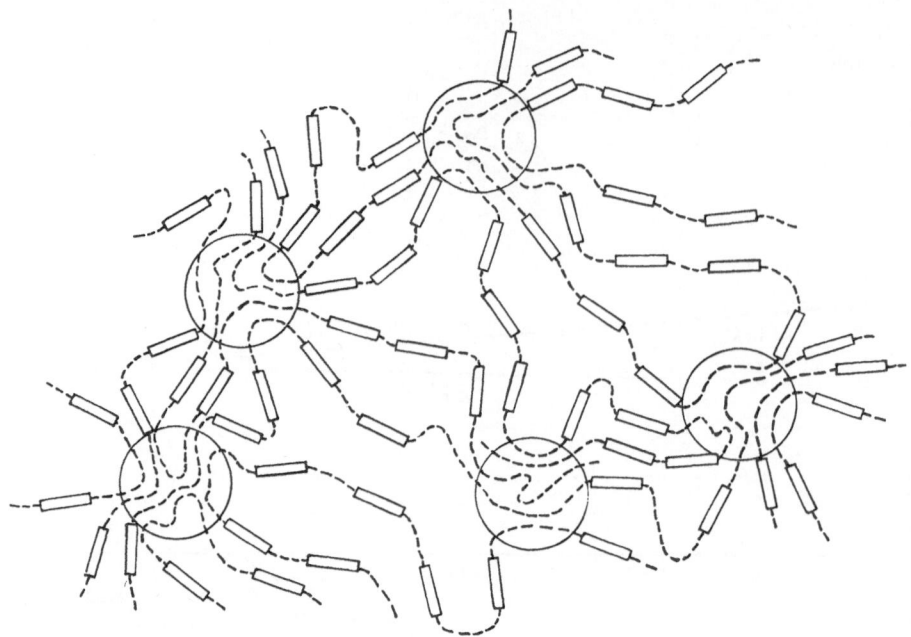

Fig. 8. Model of the supermolecular structure of the ladder-type styrene -
multimethacrylate copolymers
- the dashed line segments stand for polystyrene blocks,
- the rectangles stand for multimethacrylate blocks.

The reflections disappear completely at temperatures between 150 and
175 °C. The results indicate on a gradual disappearance of the microdomain
structure and transition to a single-phase structure.

The SAXS investigations showed no influence of the size of multimethacrylate blocks on position and height of observed scattering maxima.

Comparing all the results and taking into account that microdomains exist
in copolymers with compositions covering quite a wide range, we arrive at the
following conclusions. Ladder-type copolymers are systems in which a partial
microphase separation take place. The microdomains are formed from polystyrene blocks, while the continuous matrix contain both multimethacrylate and
polystyrene blocks. A likely model of the supermolecular structure of the
copolymers is shown in Fig. 8. The reasons for this kind of supermolecular
structure lie in the essential differences between the chemical structures
of polystyrene and multimethacrylate blocks.

REFERENCES

1. S. Połowiński, Polimery 17:409 (1972).
2. S. Połowiński, Europ.Polym.J. 14:563 (1978).
3. C. G. Vonk, J.Appl.Cryst. 8:340 (1975).
4. W. Ruland, J.Appl.Cryst. 4:70 (1971).

THE STRUCTURE EVALUATION OF POLYPROPYLENE FIBRES

BY COMPUTATIONAL RESOLUTION OF WAXS

Václav Marcian and Jozef Karniš

Research Institute for Man-Made Fibres

059 21 Svit, Czechoslovakia

INTRODUCTION

Methods of wide-angle X-ray scattering (WAXS) are advantageously used for obtaining detailed information on molecular as well as supermolecular structure of synthetic fibres. Crystallinity is one of basic characteristics of supermolecular structure of fibre-forming polymers, determining their physical and mechanical properties. Even if it is possible to use many other methods, the WAXS is frequently used for crystallinity evaluation because its results are influenced neither by the presence of additives, molecular orientation nor porosity.

In contrast to others, the WAXS analysis simultaneously gives the information on structure arrangement - crystallite size, orientation, polymorphism, unit cell dimension etc. By exact evaluation of diffraction data obtained by suitable techniques it is possible to gain some basic information on the structure of regions with the lower arrangement level - noncrystalline regions.

Classic ways of the WAXS quantitative analysis are mostly based on the graphical separation of equatorially scanned diffraction profiles into areas for crystalline peaks as well as for scattering by noncrystalline matter of the polymer. Earlier, the graphical methods used for quantitative analysis of WAXS employed the approximation of noncrystalline scattering by either melted polymer diffraction profile or by solid amorphous polymer one. Many simple graphical procedures[1,2,3] led into Hermans-Weidinger method[4]. The subjective estimation of the noncrystalline background line in the diffraction profile is the main problem of the analysis.

This problem is fundamental for such polymers as polypropylene (PP). Noncrystalline scattering occurs in the same region of angles as the diffraction by crystallites and frequently overlaps with it. The shape of the noncrystalline profile line cannot be determined objectively, because of the impossibility to prepare the perfectly noncrystalline standard sample in solid state. In addition, the phase problem solution is complicated by means of crystalline reflection widening as a result of changes in arrangement as well as in crystallite region size. For these reasons the simple graphical methods cannot be used for low-crystalline polypropylene samples with crystallinity of 20 % and below.

EXPERIMENTAL

Two methods for quantitative characterization of crystalline structure in PP fibres based on WAXS data processing are presented. The first relatively simple method, suitable for the routine crystallinity as well as crystallite size evaluation, consists of computational substracting of noncrystalline scattering from total diffraction intensity in a chosen range of diffraction angles. In this procedure the rules for determination of the noncrystalline background line as proposed by Hermans-Weidinger in their graphical method have been used. To increase the objectivity the additional criteria have been introduced, describing the noncrystalline background shape:

- shape of noncrystalline profile is approximated by scattering of melted isotactic polypropylene,

- position of the maximum of noncrystalline profile is 16 °2Θ, as found by extrapolation of temperature changes in the profile position shift of melted polymer to room temperature,

- noncrystalline profile line is identical with the diffraction profile in region of angles from 8 to 12 °2Θ ,

- noncrystalline profile does not intersect the sample diffraction one,

- noncrystalline profile is always in touch with the diffraction profile in minimum between 220 and 060 reflections as well as between 111, 041, 131 and 060 ones,

- noncrystalline profile is in touch with the diffraction profile in absolute minimum. It can be the minimum between either 110 and 040 reflections or 130 and 111, 041, 131 ones.

Crystallinity is calculated by the simple relation

$$X = A_c/(A_c + A_n),$$

where A_c and A_n are areas for crystalline as well as noncrystalline components.

The program for noncrystalline profile construction was written in BASIC and calculations of structural parameters have been performed on the IT-10 computer.

The latter method uses the improved version of previously developed computer program[5]. The procedure consists of the computational resolution of diffraction scans into eleven individual peaks represented as Gaussian and Pearson VII functions from crystalline planes as well as the noncrystalline components.

Physical Aspects of WAXS Resolution

WAXS patterns of semicrystalline isotactic polypropylene samples with the high content of α-monoclinic modification are considerably complicated. The diffraction scan is characterized by 6 strong crystalline reflections at least, indexed as 110, 040, 130, 111 041 131, 060 and 220, recorded in the regions of angles from 8 to 32°2Θ. For uniaxially oriented samples such as fibres the reflections 110, 040, 130, 060 and 220 are located on the equator, whereas the mixed one is in a common position.

Owing to the bimodal orientation of crystallites the frequent case of 110 as well as 130 reflection splitting can be observed[6]. As a rule, for oriented fibres the anisotropic component of noncrystalline scattering concentrates on the equator as a result of diffraction by extended tie chains.

The weak isotropic component can be considered as the amorphous background.

In some cases the weak crystalline peak for γ-modification can be seen between 110 and 040 reflections.

Mathematical Model

The evaluated mathematical model describes the integral intensity of the diffraction profile as the sum of the linear background, six symmetric profiles for single crystalline reflections at least and one asymmetric curve for noncrystalline scattering.

$$I = P + Q \cdot x + \sum_{i=1}^{6} f_i(x) + f_{am}(x)$$

The approximation by means of both the three-parameter Gaussian function

$$f(x) = I_o \cdot \exp(-4z^2)$$

as well as by the flexible four-parameter Pearson VII one

$$f(x) = I_o/[1 + 4z^2(2^{1/m} - 1)]^m$$

have been used for the mathematical description of single reflections, where parameter $z = (x - x_o)/H$. I_o, x_o H and m are the profile intensity, position, width and form-factor, respectively.

The Pearson VII function enables modelling of the profile shape by the form-factor change between Gaussian function ($m \rightarrow \infty$) and Cauchy one ($m = 1$).

The asymmetric profile for noncrystalline scattering has been modelled as the graphic sum of two or three symmetric profiles with different parameters. So, 24 parameters for crystalline reflections as well as 8 ones for the noncrystalline profile approximation are needed at least, in case of eight-curve diffraction profile resolution. The calculation with so large number of parameters appeared unstable and did not give the best fit. For this reason, simplifying assumptions have been done leading to a reduced number of parameters without a change of a physical sense:

- linear background parameters are determined by least-squares method

- crystalline reflections are represented as three-parameters Gaussian functions

- noncrystalline scattering is resolved into Pearson VII profiles with a different width as well as the intensity in maximum, but with the same form-factor

- constant ratio among integral intensities for crystalline reflections can be found

- program for the nonlinear regression by the least-squares method is used for the search of parameters of mathematical functions.

In order to accelerate the WAXS profile resolution process the minimal/ maximal limits of curve parameters are predetermined as well as the required calculation accuracy. The parameters approximation is included into the data input.

All data input are applied by a dialogue via terminal display. Some physically interesting quantities can be calculated:

crystallinity
crystallite size
intermolecular distances

The computer program in FORTRAN IVa performed on HP 1000 computer enables the graphical record of resolution by means of HP 7225A plotter.

DISCUSSION

The 060 reflection is very weak, so that the parameter establishment is not reliable in some cases. This is manifested mainly by the shift of the intensity maximum in the lower diffraction angles direction and by either intensity increase or unreally great widening. By this is negatively influenced the shape as well as the position of noncrystalline maximum. The resolution stability has been increased by setting the reasonable limits of parameters of 060 reflection. The stability can be increased also by description of the mixed crystalline reflection by a sum of two peaks.

However, the most important source of the resolution instability is the nonsymmetric profile approximation for noncrystalline scattering.

Owing to the scattering intensity fluctuations the 110 crystalline profile is frequently either expressed or suppressed. The calculation stability has been gained by limit narrowing for both the peak intensity and the position, without the physical sense failure.

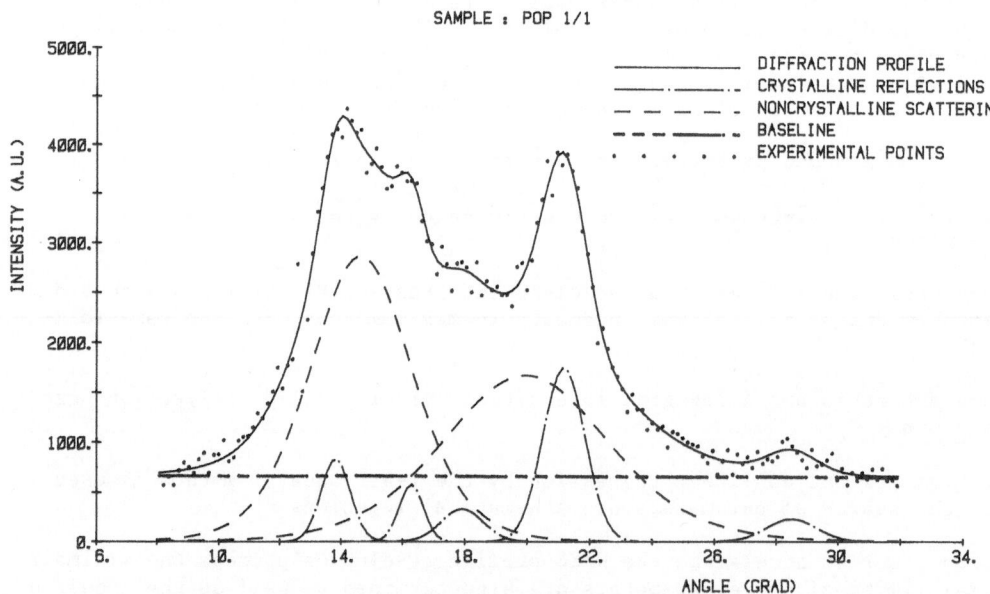

Fig. 1. WAXS profile analysis of low crystalline polypropylene sample (POP 1/1).

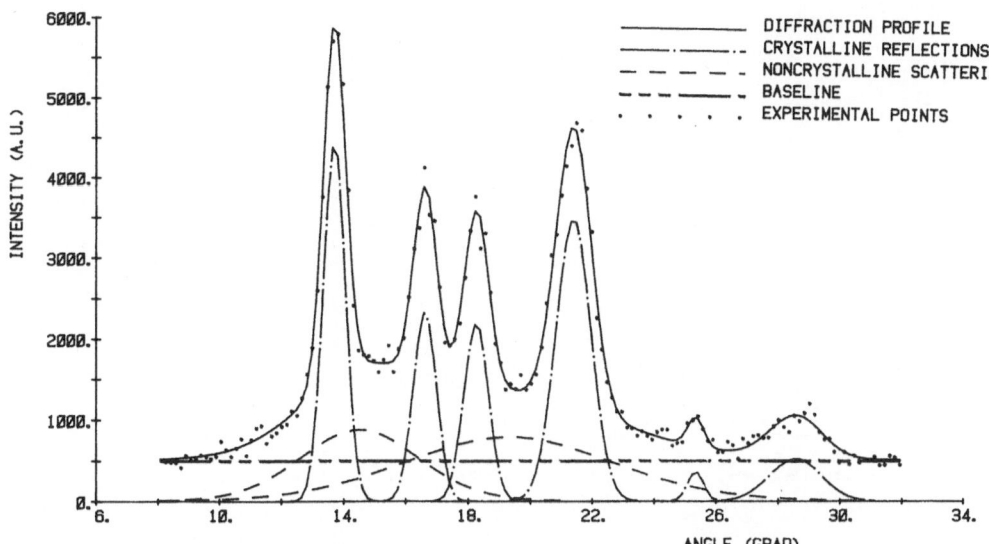

Fig. 2. WAXS profile analysis of highly crystalline polypropylene sample
(POP 11/1).

CONCLUSION

In contrast to conventional graphic methods the WAXS computational resolution one brings the advantage of the high objectivity in the noncrystalline profile estimation and by this the increased accuracy of crystalline
structural parameters calculation. It enables almost noncrystalline sample
evaluation considering the content of noncrystalline phase and the peak
widening caused by the crystallite size and quality (compare Figs 1, 2).

The basic version of the computer program in FORTRAN makes possible
very precise determination of the crystalline peak positions, which can be
used for calculation of the mesomorphic region content.

The sensitivity of the measurement itself as well as the data processing
objectivity have been found as limiting factors for attainign of the high
accuracy of structural parameters determination.

REFERENCES

1. M. A. Martynov and K. A. Vylegzhanina, "Rentgenografia polimerov",
   Chimiya, Leningrad (1972).
2. G. Farrow, The measurement of crystallinity in polypropylene fibres by
   X-ray diffraction, Polymer 2:409 (1961).
3. F. Rybnikář, Rentgenografické měřeni krystalinity lineárního polyetylenu
   a izotaktického polypropylenu, Chem.vlákna 18:162 (1968) (in Czech).
4. A. Weidinger and P. H. Hermans, On the determination of the crystalline
   fraction of isotactic polypropylene from X-ray diffraction, Makromol.
   Chem. 50:98 (1961).
5. V. Marcian, Některé možnosti využití röntgenografických metod pro hodno-
   ceni struktury nekrystalických oblastí syntetických vláken, Chem.
   vlákna 35:22 (1985) (in Czech).
6. V. Marcian, Morphology of Porous Polypropyelen Fibres, in: Proceedings
   of 17th Europhysics Conference on Macromolecular Physics, Walter de
   Gruyter, Berlin (1986).

III. REAL STRUCTURE OF CRYSTALLINE MATERIALS

PROBLEMS IN DIFFRACTION ANALYSIS

OF REAL POLYCRYSTALS

P. Klimanek

Section of Metallurgy and Materials Technology
Academy of Freiberg, POB 47
9200 Freiberg, GDR

INTRODUCTION

It is well known that X-ray or neutron diffraction phenomena of real
polycrystalline materials are often strongly influenced by the microscopical
structure (phase content, grain size and orientation, spatial arrangement
of different phases) of the scattering object volume. Methodological problems arising from this fact in the diffraction analysis of polycrystals have
been treated with regard to systematic errors of intensity measurements and,
of course, in connection with special applications of X-ray or neutron scattering for microstructure characterization. In most of the considerations
it was implicitly assumed that all crystallites of a given phase have similar
structure (i.e. perfect crystal structure and lattice disorder). Such a pre-
sumption is necessary (and possible) in investigations of constitution –
related structure characteristics (e.g. electron density, thermal lattice
vibrations, short-range order in solid solutions) which in principle require
a knowledge of the single-crystallite scattering and therefore must usually
be performed with carefully prepared (structurally homogenized), fine-grained
powder specimens. However, in the examination of process-related structure
parameters (e.g. dislocation densities due to various modes of plastic deformation, substructure characteristics associated with heat treatment, precipitation processes or phase transformations), which more and more becomes
the dominating problem of diffraction analysis in materials research, no
homogenization treatment of the crystallites structures can be carried out in
general. That means, in real polycrystalline objects the existence of crystallite fractions with significantly different lattice disorder must be expected and accepted as an inevitable and sometimes regular (but nevertheless
undesired) feature of the microscopical structure which has to be incorporated into a physically realistic interpretation of diffraction.

The consequences of a structural inhomogeneity for diffraction investigations have firstly been considered for single-phase materials with random
distribution of the differently distorted grains and could be confirmed in
experiments with both model objects (powder specimens) and real materials
(martensitic structures)[1,2,3]. More recently considerations concerning the
diffraction phenomena of polycrystals with a non-random distribution of
structurally different crystallites were performed, too[4,5,6]. The present
paper gives a brief survey of this work. Moreover, some aspects of microstructure characterization by means of X-ray profile analysis are discussed
which should be particularly of interest in studies of polycrystalline ob-

jects with substructure gradients due to plastic deformation or heat treatment and also for the interpretation of diffraction data due to thin layers and coatings.

GENERAL CONSIDERATIONS

The Structures of Real Polycrystals

The classical structure model used in the treatment of X-ray or neutron scattering by polycrystals (e.g. refs [7,8,9,10]) is a spatial arrangement of monocrystalline particles (crystallites, grains) which are assumed to be

- identical in both the perfect lattice structure and the lattice disorder,
- uniform in geometry (size and shape),
- completely randomly oriented in the space, and
- very small in comparison with the scattering object volume.

The diffraction phenomena of such an ideal (single-phase) polycrystal are representative for the structure of each of its crystallites and can easily be derived from the spatial intensity distribution of the single-crystallite scattering by means of a simple orientation averaging. A similar correlation between the crystallite structures and the diffraction effects is also obtained, if the scattering object has a multiphase grain structure which can be described as a random mixture of ideal polycrystals. A material of this kind may therefore be called an ideal phase mixture. In every case the microscopical structure of a real polycrystal is significantly different from that of the model objects just mentioned and must be characterized by means of statistics. In this connection the following types of crystallite configurations may be distinguished[4,5]:

(i) statistically (structurally) homogeneous polycrystals, in which the deviations from the ideal crystallite arrangements are small and, in particular, the grains of one phase have (with the exception of small fluctuations) similar lattice disorder and substructure, respectively;

(ii) statistically (structurally) inhomogeneous polycrystals where crystallites of different phases and/or with different lattice disorder are nearly randomly (i.e. isotropically) distributed in the space, and

(iii) so-called layer structures as a special form of structurally inhomogeneous polycrystals, in which structurally different grains are successively (i.e. anisotropically) arranged approximately parallel to a selected plane of the scattering object.

A statistically homogeneous crystallite ensemble is approximately realized in a fine-grained powder specimen formed by fillings, for instance. Single-phase polycrystals with high degree of (statistical) structural inhomogeneity are represented by mixtures of recrystallized and plastically deformed grains as occurring in hot working or heat treatment of metals and alloys, for example. Structures of the type (iii) are often caused by deformation or temperature gradients even in chemically homogeneous materials.

In the traditional diffraction analysis, which is mainly concerned with statistically homogeneous polycrystals, the correlation between the microscopical structure of an object and its scattering effects is characterized by a sample statistics being determined by the mean grain size (grain size statistics[11,12]) and, more completely, by the orientation distribution (texture[13,14]) of the crystallites. Both quantities can be enclosed into a generalized axis-distribution Ω ($\vec{b},\vec{e}$) as defined in the quantitative texture analysis[13]. The function Ω ($\vec{b}$, $\vec{e}$) describes the distribution of volume fractions, in which the direction $\vec{b}$ (e.g. the normal direction of a family of lattice planes (hkl)) of a structure-related, crystallographic axis system

$\{\vec{b}_1,\vec{b}_2,\vec{b}_3\}$ is parallel to the direction $\vec{e}$ of a macroscopically defined sample-related axis system $\{\vec{e}_1,\vec{e}_2,\vec{e}_3\}$. In investigations of X-ray or neutron diffraction peaks the quantity $\Omega(\vec{b},\ \vec{e})$ determines the reflection probability of the grains in dependence on the orientation $\vec{g}=\vec{e}$ of the diffraction vector $\vec{g} = (\vec{s}-\vec{s}_0)/\lambda$.

In order to characterize the microscopical structure of inhomogeneous crystallite ensembles the introduction of a generalized sample statistics is necessary. In single-phase materials it encloses[1,5,6]

- a lattice-disorder statistics $w(\rho)$ describing the volume fractions $w_j=W(\rho_j)$ of crystallites with significantly different content ρ of lattice defects,
- a set of reflection statistics $\Omega_j(\vec{b},\vec{e})$ associated with the volume fractions w_j, and
- a "position statistics" which describes the spatial arrangement of the structurally different grains.

In multiphase systems the distribution $w(\rho)$ and the corresponding quantities $\Omega(\vec{b},\vec{e})$ must be defined for each phase.

On practical conditions it cannot be expected that the lattice-disorder statistics is represented by a simple distribution function $w(\rho)$, but for theoretical considerations such an assumption is necessary. Basic types of the distribution $w(\rho)$ being of interest are shown in Fig. 1. However, since in practice absorption of the radiation takes place, the lattice-disorder statistics $W(\rho)$ really operative in a diffraction experiment can significantly be different from the true distribution $w(\rho)$. It depends on the position statistics of the structurally different grains, the correct characterization of which is obviously a very difficult problem. In the polycrystal structures introduced above limiting cases of the position statistics are realized.

Diffraction Peaks due to Real Polycrystals

The intensity distribution due to X-ray or neutron scattering by an ideal polycrystalline object containing N crystallites is obtained by averaging the intensity distribution $I_{sc}(\vec{g})$ of its single-crystallite scattering over all orientations of the diffraction vector $\vec{g}$ with respect to the crystal

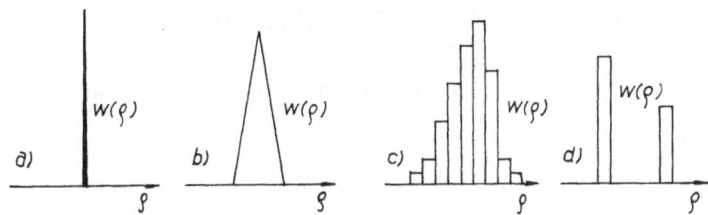

Fig. 1. Basic types of the lattice-disorder statistics in polycrystalline materials
a) ideal polycrystal, b) structurally homogeneous polycrystals, c,d) structurally inhomogeneous polycrystals

lattice. The result is a centrosymmetrical function of the variable $|\vec{g}| = 2 \sin\Theta/\lambda$ which can explicitly be calculated by integration of $I_{sc}(\vec{g})$ over the surface elements dS_g of the spheres $|\vec{g}| = $ const. of the reciprocal space[8,9]:

$$I_{pc}(g) = (N/4\pi g^2) \int I_{sc}(\vec{g})\, dS_g \tag{1}$$

Practical diffraction analysis of polycrystals is usually based on the reflection peaks associated with crystallographically equivalent nodes $|\vec{h}| = $ const. of the reciprocal lattice (and families $\{hk\ell\}$ of lattice planes, respectively). In the case of an ideal crystallite ensemble the intensity distribution of such a diffraction peak is represented by the average

$$I_{pc}(g - h) = (Np_{hk\ell}/4\pi h^2) < \int I_{sc}(\vec{g} - \vec{h})\, dS_g >|\vec{h}| = \text{const.} \tag{2}$$

where $I_{sc}(\vec{g}-\vec{h})$ denotes the intensity distribution of a single-crystal reflection and $p_{hk\ell}$ is the multiplicity of the scattering lattice planes $\{hk\ell\}$.

The parameters (i.e. integrated intensity, peak position, line width and shape) of the function $I_{pc}(g-h)$ are unambiguously related to the single-crystallite structure of the scattering object. For this reason, the reflections of ideal (and also of structurally homogeneous) polycrystals can be called diffraction singlets.

In order to obtain the intensity distributions of diffraction peaks due to real polycrystals the following modifications of the scattering law are necessary[1,4]:

(i) $I_{sc}(\vec{g})$ must be replaced by the weighted sum of the scattering functions $I_j(\vec{g})$ corresponding to the crystallite fractions N_j or w_j, respectively, of the lattice-disorder statistics.

(ii) The contributions of the surface elements dS_g to the resulting intensity distribution must be weighted by the reflection statistics $\Omega_j(\vec{b},\vec{e})$ for each crystallite fraction.

(iii) The contributions of the crystallite fractions N_j to the diffraction must be weighted with respect to the spatial distribution (position statistics) which can formally be done by introduction of direction-dependent absorption factors $A_j(\vec{g})$.

That means, the intensity distribution of X-ray or neutron scattering by a real polycrystal becomes anisotropic in the space and has to be defined as a function of both the magnitude $g = 2 \sin\Theta/\lambda$ and the direction $\vec{e}$ of the diffraction vector $\vec{g}$ within the sample-related axis system $\{\vec{e}_1, \vec{e}_2, \vec{e}_3\}$. Formally, it can be expressed by the formula

$$I_{pc}(g,\vec{e}) = \sum_j (N_j A_j(\vec{g})/4\pi g^2) \int I_j(\vec{g})\, \Omega_j(\vec{b},\vec{e})\, dS_g \tag{3}$$

The intensity distribution of a diffraction peak is given then by

$$I_{pc}(g-h,\vec{e}) = \sum_j (N_j A_j(\vec{g}) p_{hk\ell}/4\pi h^2) < \int I_j(\vec{g}-\vec{h}_j)\, \Omega_j(\vec{b},\vec{e})\, dS_g >|\vec{h}| = \text{const.} \tag{4}$$

which represents a weighted sum of partial reflections and may be therefore called a diffraction multiplet.[+] On the condition that the $\Omega_j(\vec{b},\vec{e})$ are

---

[+] The terms "diffraction singlet" and "diffraction multiple" were originally used by Shtremel and Kaputkina[15,16] in connection with X-ray diffraction analysis of martensite reflections

sufficiently slowly varying functions in the vicinity of the directions $\vec{b} \,||\, \vec{h}_j$ the distribution $I_{pc}(g-h,\vec{e})$ can be approximated by the expression

$$I_{pc}(g-h,\vec{e}) = \sum_j W_j(\vec{g})\,\Omega_j(\vec{e})\,I_j(g-h_j) = \sum_j \omega_j I_j(g-h_j) \tag{5}$$

where $W_j(\vec{g}) = A_j(\vec{g})\,w_j(\rho)$ is the absorption-weighted lattice disorder statistics and $\Omega_j(\vec{e})$ denotes the mean value of $\Omega_j(\vec{b},\vec{e})$ with respect to the scattering lattice planes. The parameter $h = 2\sin\Theta_0/\lambda$ indicates the peak or center-of-gravity position of the observable reflection, and h_j is the corresponding quantity of the partial reflection $I_j(g-h_j)$.

As shown previously[5,6] in statistically homogeneous objects the statistics $W(\vec{g},\rho)$ becomes practically identical with the true lattice-disorder statistics $w(\rho)$, and for layer structures with small grains (t_i – layer position perpendicular to the sample surface) it is given by

$$W_j(\vec{g}) = w_j(\rho)\,\exp(-2\mu\,t_j/\sin\Theta)/\sum_j w_j\,\exp(-2\mu\,t_j/\sin\Theta) \tag{6}$$

In structures with very different grain sizes of the volume fractions $w_i(\rho)$ the relations between $W(\vec{g})$ and $w(\rho)$ cannot be described easily and an influence of microabsorption must be taken into account for both single and multiphase crystallite systems. The features of a diffraction multiplet, the meaning of which may be illustrated by Fig. 2, have already been considered in some detail in the previous work[1,4]. For this reason, here only some remarks concerning the parameters being of interest in X-ray profile analysis are given:

(i) Because of the relationship $I(0) = \sum_j \omega_j I_j(-x_j) = \sum_j \omega_j \alpha_j I_j(0)$ $= \sum_j \omega_j \alpha_j I_j(hk\ell)\,B_j^{-1}(hk\ell)$, where $x_j = \Delta g_j = h - h_j$ is the peak shift of the partial reflection j with respect to the peak position of the observable reflection and $I_j(hk\ell)$ the integrated intensity of this partial reflection, the integral breadth $B(hk\ell)$ of a diffraction multiplet is defined by

$$B^{-1}(hk\ell) = I(0)/I(x)dx = \sum_j \tilde{\omega}_j \alpha_j B_j^{-1}(hk\ell) \tag{7}$$

with

$$\tilde{\omega}_j = W_j\Omega_j(\overline{e})\,I_j(hk\ell)/\sum_j W_j\Omega_j(\overline{e})\,I_j(hk\ell).$$

The formula indicates that the line broadening of the resulting diffraction peak is determined by the weighted average of the reciprocal line breadths $B_j^{-1}(hk\ell)$ of the multiplet components. Therefore, the central part of the intensity distribution $I_{pc}(g-h,\vec{e})$ is preferently related to the crystallite fractions with higher lattice perfection, while the tails of the multiplets are determined by the crystallite fractions with higher defect content.

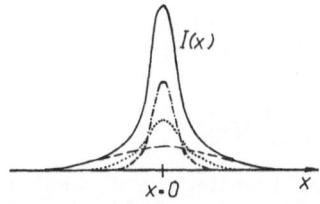

Fig. 2. On the definition of diffraction multiplets (case $x_j = 0$).

Moreover, peak shifts $x_j \neq 0$ of the partial reflections as caused by inhomogeneous internal stress or chemical inhomogeneity of the crystallite structures can be an additional source of significant line broadening.

(ii) The Fourier transform of the intensity distribution $I_{pc}(g-h,\vec{e})$ $I(x)$ is defined by

$$\Phi(t) = (1/2\pi)^{1/2} \int I(x) \exp(ixt)\ dx = \sum_j \omega_j\ \Phi_j(t)\ \exp(-ix_j t) \tag{8}$$

This expression leads to the following equation for the normalized complex Fourier coefficients ($t = n\pi/x_o = \pi L$) of the multiplet profile:

$$C(L) = \sum_j \tilde{\omega}_j\ C_j(L)\ \exp(-i\pi L\ x_j) \tag{9a}$$

The real cosine-coefficients of the diffraction profile are then given by

$$A(L) = \sum_j \tilde{\omega}_j\ A(L)\ \cos\pi L\ x_j) \tag{9b}$$

From the equations it follows that the line broadening as given by the line width and the Fourier coefficients (or the variance[5,6]) has different physical meaning. Let, for instance, the partial reflections be represented by Gauss profiles $\exp(-k_j^2\ x^2)$ and their line broadening caused by statistically distributed dislocations with the densities ρ_j. Then (7) leads to

$$B^{-1}(hk\ell) = \sum_j \tilde{\omega}_j\ k_j/\pi^{1/2} \sim\ <\rho^{-1/2}>$$

while from (8) the relationship

$$(2\pi)^{1/2}\ \Phi(t) = \sum_j \omega_j\ \exp(-t^2/k_j^2) = \sum_j \omega_j\ \exp(-B_j^2\ t^2/\pi)$$

and

$$\ln A(L) = \sum_j \tilde{\omega}_j\ B_o^2\ L^2\ \rho_j \sim\ <\rho>\ L^2 \quad\text{for}\quad B_o^2\ L^2\ <\rho>\ <<\ 1$$

can be derived. Accordingly, in real polycrystalline materials an estimation of the mean dislocation density $<\rho>$ should be carried out by means of the Fourier coefficients. On the other hand, in investigations of the line broadening due to a distribution of small crystallites (particles) causing Cauchy partial profiles $(1 + k_j^2\ x^2)^{-1}$ the following relationships are valid:

$$B^{-1}(hk\ell) = \sum_j \tilde{\omega}_j\ k_j/\pi \sim <T(hk\ell)>$$

and

$$A(L) = \sum_j \tilde{\omega}_j\ \exp(-B_o L/T_j) \approx 1 - B_o L\ <T^{-1}(hk\ell)> +\ \ldots$$

This indicates that in the examination of particle-size broadening the interpretation of line-breadths should be easier than that of the Fourier coefficients.

Experimental Results for Model Specimens

With regard to practical application the direct experimental verification of the formalism presented in Section 2 (Diffraction Peaks due to Real Polycrystals) seems necessary. This requires the examination of samples

130

with a well-defined lattice disorder statistics and the correct knowledge of the parameters of the diffraction singlets associated with its crystallite fractions. Of course, since the microscopical structures of inhomogeneous polycrystals occurring in practice is often complicated, model objects must be used for this purpose. A formally satisfactory procedure for the preparation of statistically inhomogeneous powder specimens is the careful mixing of given volume fractions of as-deformed and differently annealed fillings. Some results obtained in this manner with binary mixtures of α-iron fillings are shown in Fig. 3. According to (7) and (9) the line breadths and the Fourier coefficients of the diffraction peaks are described by the simple equations

$$B^{-1}(hk\ell) = w_1 B_1^{-1} + (1 - w_1) B_2^{-1}$$

$$A(L) = w_1 A_1(L) + (1 - w_1) A_2(L)$$

in this case and excellently confirmed in the experimental investigations. A very good agreement between the relationships of Section 2 (already mentioned) and the experiment was also obtained in studies of model objects with more complicated lattice-disorder statistics, and at last the formalism was proved to be correct for massive model objects composed of differently treated copper and α-iron sheets[17].

PROBLEMS OF X-RAY PROFILE ANALYSIS

On the Statistical Reliability of X-Ray Diffraction Experiments

X-ray diffraction phenomena of real polycrystals are usually related to a scattering volume which is very small in comparison with the macroscopical dimensions of the investigated material. For this reason it is suitable, analogous to texture analysis[18] to distinguish the local sample statistics $\omega_1 = w_1 \Omega_1$ of a selected partial volume (single specimen) and the global sample statistics $\omega_o = w_o \Omega_o$ of the total volume of a polycrystalline object which determines its macroscopical properties and behaviour. According to Chapter 'The Structures of Real Polycrystals' especially in X-ray profile analysis the following situations must then be taken into account:

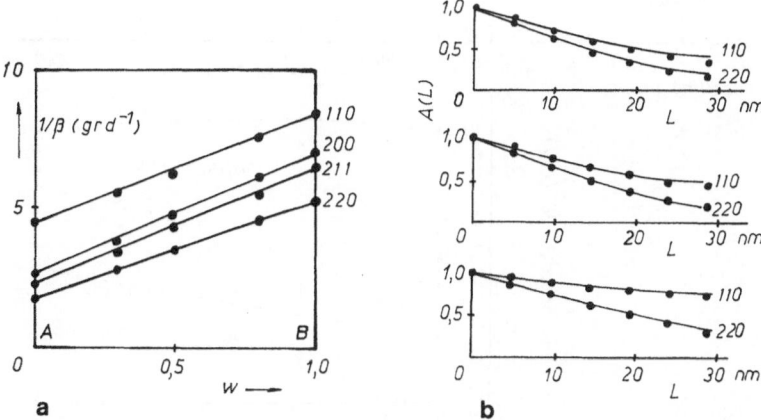

Fig. 3. Experimental verification of diffraction multiplets by means of α-Fe powders with a binary lattice-disorder statistics. The dots correspond to the experimental points. (a) integral breadths (b) Fourier coefficients (Mixing ratios A:B = 3:1, 1:1, 1:3).

(i) If the irradiated polycrystal is structurally homogeneous, the relation $w_1 \Omega_1 \approx w_o \Omega_o$ is valid. That means, the diffraction effects of a local object volume are representative for the total material and the examination of a single specimen is sufficient, in principle, for a statistically reliable structure characterization. The information content of the diffraction analysis is essentially limited by the uncertainties (e.g. the error due to the counting statistics, the instrumental line broadening, etc.) of the scattering experiment.

(ii) In structurally inhomogeneous polycrystals the microstructures of various partial volumes can extremely be different and $w_1 \Omega_1 \neq w_o \Omega_o$ must be considered as a rule. Accordingly, because of the connection $w_o \Omega_o \approx \langle w_1 \Omega_1 \rangle$ averaging is necessary over a sufficiently large number of carefully selected partial volumes (single specimens) for a statistically reliable structure characterization.

In order to illustrate the importance of the problems just mentioned, Fig. 4 informs on the results of line breadths measurements with martensitic specimens of low-alloy steel 55SiMn7 after various thermomechanical treatment (TMT). The scattering of the data, which essentially represent the physical line broadening of the reflections, is much greater than the error of the single measurement ($\Delta B/B < \pm 5\%$ as shown in[3]) and indicates considerable structural inhomogeneity of the material. The latter is caused by the fact that the microstructure of quenched steel 55SiMn7 is a mixture of lath (i.e. dislocated) and plate (i.e. twinned) martensite, the volume fractions of which are strongly influenced by local fluctuations of the microstructure and the chemical composition of the austenite before the γ-α phase transition. Fig. 4 demonstrates

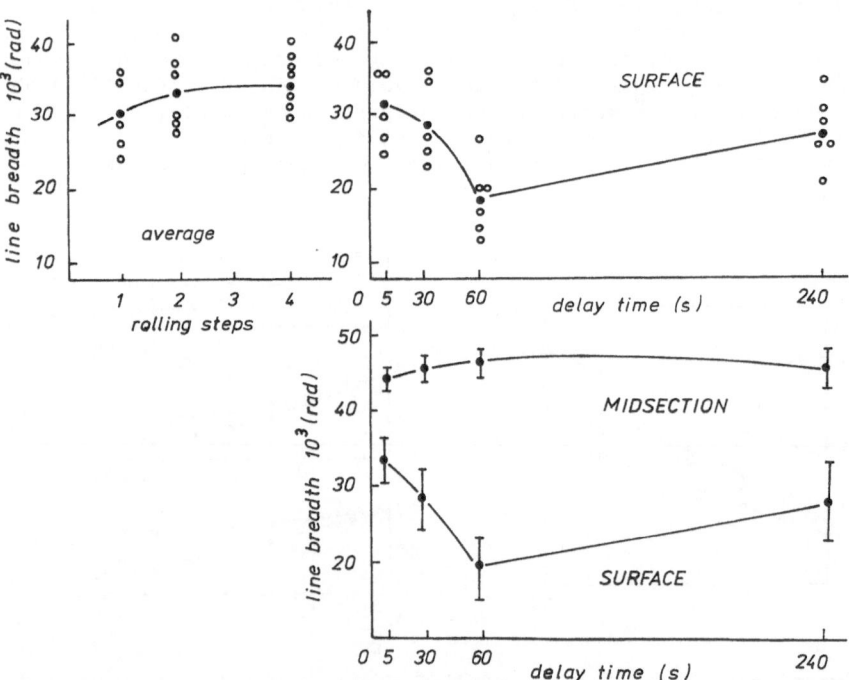

Fig. 4. Line-breadth scattering of the 211 reflection of structurally inhomogeneous martensites in thermomechanically treated low-alloy steel 55SiMn7 (Hot-rolling at 1123 K, $\varphi = 0.4$, $\dot{\varphi} = 10$ s^{-1}).

- The use of single specimens would be insufficient for the registration or could give rise to a misinterpretation of the dependence of the line broadening on the process parameters of the TMT. But already averaging of only a few single measurements, the statistical reliability of which can be estimated quantitatively by means of a confidence test if required[6], leads to a satisfactory valuation of this dependence (Fig. 4a,b).

- As illustrated by the differences of the line breadths measured at the surface and at the midsection of quenched steel 55SiMn7 (Fig. 4c), the information content and the reliability of line broadening studies with structurally inhomogeneous polycrystals is not only determined by the number but also by the special selection of the single specimens taken under investigation. This is generally important for a correct answer to the question, in which manner experimental results are representative for the sample material, and must in particular be taken into account in comparison of results obtained by different techniques of structure characterization or by various authors and, of course, in connection with the examination of structure-property correlations.

- Last but not least the investigations of the midsection structures (Fig. 4c) indicate that significant scattering of the results of single measurements may be a sufficient but cannot be a necessary criterion for structural inhomogeneity of a polycrystal, because its local and global sample statistics $\omega_j = w_j \Omega_j$ can be very similar. Moreover, because of the complicated meaning of the multiplet parameters the diffraction line broadening in structurally inhomogeneous objects can be very insensitive to (local) changes of the microstructure. Thus, from Fig. 4c it may be concluded, that there is no important change of the midsection structure of the martensite after different delay times of the hot-worked austenite before quenching, but actually the microstructure is significantly modified[2,3].

The Instrumental Line Broadening of Diffraction Multiplets

The observable line profile $F(x)$ of a diffraction multiplet is formed by convolution of the singlets $f_j(x-x_j)$ associated with the crystallite fractions w_j of the lattice-disorder statistics and the instrumental line profile $g(u)$ caused by the special conditions of the diffraction experiment as follows:

$$F(x) = \sum_j \tilde{\omega}_j \cdot \int g(u)\, f(x - x_j - u)\, du = \sum_j \tilde{\omega}_j\, F_j(x - x_j).$$

Accordingly, the Fourier transform $\Phi(t)$ of the distribution $F(x)$ is, independently of the shapes of the partial profiles $f_j(x - x_j)$, given by the equation ($\gamma(t)$ - Fourier transform of $g(u)$)

$$\Phi(t) = \sum_j \tilde{\omega}_j \cdot \gamma(t)\, \Phi_j(t) \cdot \exp(-ix_j t).$$

The correction of the instrumental line broadening of the Fourier coefficients (and also of the variance $\langle x^2 \rangle$ of the profile[5,6]) can therefore be performed as in the case of diffraction singlets. On the other hand, a simple correction of multiplet line breadths is not possible without further considerations, since

- the profile shape $F(x)$ of a multiplet is, in general, significantly different from the profile shapes $f_j(x-x_j)$ of its singlets and
- the quotient $y=b/B$ controlling the influence of the instrumental line broadening b has the complex meaning

$$y = \sum_j w_j \, \Omega_j \, y_j.$$

The formal application of a traditional correction formula as, for instance, the parabolic approximation[19,20]

$$\beta/B = 1 - (b/B)^2 \qquad (10)$$

can then lead to large systematic errors in the determination of the physical line broadening β . In order to demonstrate the effect, integral breadths B of diffraction multiplets were calculated numerically on the assumption that peak shifts are negligible, all partial profiles are Gaussian functions and only strain broadening is present.

The line breadths were formally corrected by either the equation (10) or by the well-known expression

$$(\beta/B)^2 = 1 - (b/B)^2 \qquad (11)$$

valid for Gaussian profiles. Results obtained in this manner for two reflection orders in the case of a binary lattice disorder statistics are presented in Fig. 5a. The upper diagram can qualitatively relate to diffraction experiments with partially recrystallized structures occurring in hot working of metals and alloys, for instance, while the situation of the lower one corresponds to the initial stages of cold work of polycrystals or to plastically deformed materials with complex (multicomponent) textures. A direct proof of the effect is given by Fig. 5b for the model objects discussed in Chapter dealing with Experimental Results for Model Specimens. From the data the following conclusions can be drawn:

(i) In general, it must be expected that formally determined physical line breadths of X-ray diffraction peaks are overestimated. The effect increases with increasing influence of the instrumental line broadening and can be very large particularly for low-angle reflections.

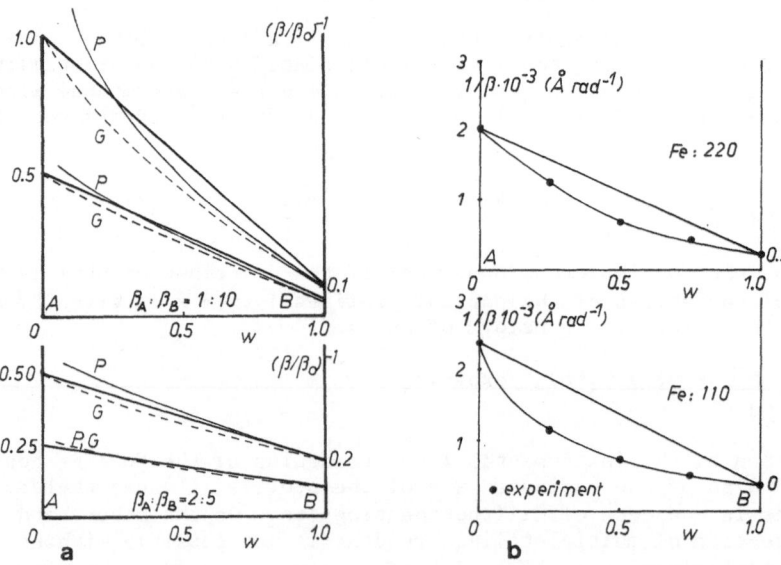

Fig. 5. Systematic errors in the correction of the instrumental broadening of diffraction-multiplet line widths connected with a binary lattice-disorder statistics
(a) model calculations (line breadths in arbitrary units)
(b) experimental verification by means of α-Fe powders

(ii) Since the ratio y=b/B does not increase proportional to the magnitude of the diffraction vector $\vec{g}$, the systematic error of the apparent physical line broadening can significantly be different for various reflection orders. Accordingly, in such cases the ratios of the line breadths of the reflection orders are modified and the classical interpretation of the diffraction line broadening leads to irregular conclusions concerning particularly so-called particle size effect.

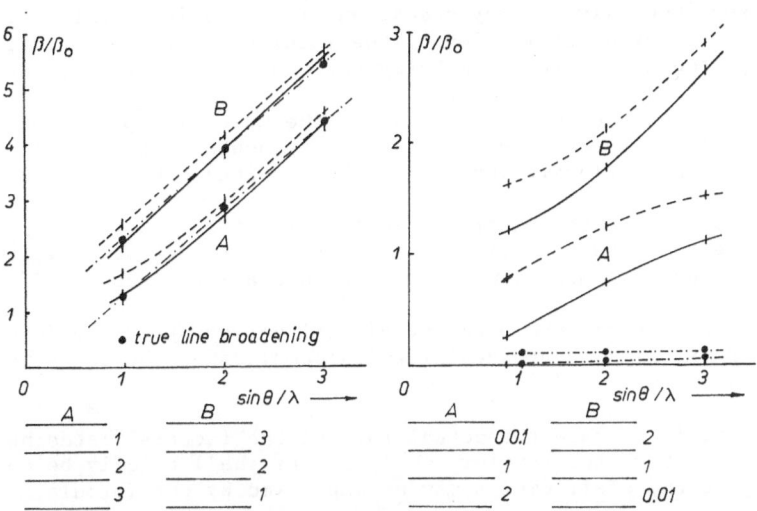

Fig. 6. Influence of a formal correction of the instrumental line broadening on the physical line breadths of diffraction multiplets due to single-phase layer structures with a ternary lattice-disorder statistics (model calculations).

In order to give an impression of the importance of (ii) in connection with the investigation of layer structures, line breadths of a system of three layers (crystallic fractions) were simulated on the same assumptions as used above (i.e. Gaussian profiles, no peak shifts, pure strain broadening) for three reflection orders and plotted as functions of $\sin\theta/\lambda$ (Fig. 6). Below the plots the structure models (lattice-disorder statistics) used are characterized in terms of line breadths $\beta_j = m_j \beta_o$ again. Fig. 6 shows that the relationship between the apparent physical line broadening and the magnitude of the diffraction vector $\vec{g}$ is not linear (as it should be the case for pure strain broadening). The shape of the curve depends on the structural perfection (line width or line width differences, respectively) and on the actual spatial arrangement of the structurally different layers (crystallite fractions). For this reason the usual evaluation of line breadths must lead to a non-zero line broadening $\beta(0)$ at $\sin\theta = 0$ in much cases, which for $\beta(0) < 0$ directly indicates an inaccurate correction of the instrumental line broadening and for $\beta(0) > 0$ simulates a really non-existing or considerably overestimated particle-size broadening.

The phenomenon just mentioned, a direct proof of which follows from Fig. 5b, for instance, is obviously able to explain physically unrealistic results of particle-size broadening as often observed in the practice. Therefore, in order to avoid it, the physical line widths of polycrystal reflections should be determined either by means of high-resolution X-ray diffractometry[21,22] or from Stokes-corrected Fourier coefficients. Of course, the effect can be neglected if the physical line broadening is large in comparison with the instrumental line breadth for all fractions of the lattice disorder statistics.

The Physical Line Broadening of Diffraction Multiplets

In the foregoing sections it was shown that by careful experimental work the physical line broadening of diffraction multiplets can correctly be determined. However, its interpretation must take into account the following problems:

(i) Structural inhomogeneity of a polycrystalline material should be detectable experimentally in many cases, but its detailed quantitative characterization (e.g. by determination of the lattice-disorder statistics) cannot be expected in general from X-ray (or neutron) diffraction analysis.

(ii) Because of the complex meaning of the multiplet parameters the identification and separation of different components of the physical line broadening (e.g. particle-size and strain broadening) can be very difficult.

(iii) According to the theory of information the number of separable line broadening components or singlet parameters, respectively, is very limited even in profile measurements with high statistic accuracy[23,24].

(iv) The structural inhomogeneity of real polycrystals can be a source of appearent physical line broadening which complicates the analysis of lattice disorder.

In order to illustrate the actual role of the factors listed here, the information content of the Fourier coefficients shall briefly be considered. For this purpose the coefficients may be expressed by the formula

$$A(L) = \sum_j \tilde{\omega}_j \, A_j(L) \cos\pi \, Lx_j = \sum_j \tilde{\omega}_j \, \exp[\ln A_j(L)] \cos\pi \, Lx_j$$

which for sufficiently small values L can be approximated by

$$A(L) = \sum_j \tilde{\omega}_j (1 + \ln A_j(L) + ...)(1 - \pi^2 \, L^2 \, x_j^2/2)$$

$$= 1 + \langle \ln A_j(L)\rangle - \langle\pi^2 \, L^2 \, x_j^2/2\rangle + ... \qquad (12)$$

$$= \exp[\langle \ln A_j(L)\rangle] \, \exp[-\langle\pi^2 \, L^2 \, x_j^2/2\rangle].$$

From the kinematical theory of X-ray or neutron scattering by distorted crystals it is known that each coefficient $A_j(L)$ represents a product

$$A_j(L) = A_j^1(L) \cdot A_j^2(L) \cdot A_j^3(L) \cdot ... \cdot A_j^\ell(L)$$

of Fourier coefficients $A_j^\ell(L)$ related to statistically independent sources of physical line broadening. For this reason equation (12) can be transformed into the expression

$$\ln A(L) \approx \sum_\ell \langle \ln A_j^\ell(L)\rangle - \pi^2 \, L^2 \, \langle x_j^2\rangle/2. \qquad (13)$$

On the assumptions of the traditional Warren-Averbach analysis[7,25] it leads to the well-known formula

$$-\ln A(L)/L = \langle T_j^{-1}\rangle + \{4\pi \, g^2 \, \langle\varepsilon_j^2\rangle + \pi^2\langle x_j^2\rangle/2\} \cdot L$$

$$= \langle T^{-1}\rangle + 4\pi \, g^2 \cdot \langle e^2\rangle \cdot L \qquad (14)$$

If the line broadening is caused by restrictedly randomly distributed dislocations, from the Krivoglaz-Wilkens theory (refs[9,10,26,27]) of X-ray

136

scattering by dislocated crystals the relationship

$$-\ln A(L)/L^2 = \langle B_o \rho_j \ln r_j/L + \pi x_j^2/2 \rangle$$

$$= \{B_o \langle \rho \ln r \rangle + \pi \langle x^2 \rangle /2\} - B_o \langle \rho \rangle \ln L \qquad (15)$$

is obtained. This important result indicates that the mean dislocation density $\langle \rho \rangle$ of a real polycrystal can be determined independently of the influence of peak shifts x_j on the total line broadening.

Generally the equations (13) to (15) demonstrate that the separation of different components of the physical line broadening is, in principle, possible without special knowledge of the structural inhomogeneity of a polycrystal. However, for a physically realistic interpretation of the averages $\langle \ln A^\ell(L) \rangle$ or $\langle e^2 \rangle$, $\langle \rho \rangle$, etc., at least some informations concerning the type of the sample statistics are needed. For this reason X-ray profile analysis of real materials should be combined with other procedures of microstructure characterization as texture analysis, stress measurements and, of course, microscopical methods.

From the equations (13) to (15) it becomes also clear that peak shifts x_j of the partial reflections must be taken into account in the analysis of the broadening of diffraction multiplets. They are caused by inhomogeneously distributed internal stress of the 1st and 2nd kind, respectively, or by lattice-parameter differences of the scattering grains and can be important even in such cases where line-broadening due to lattice defects is small. The phenomenon of stress-induced line broadening is illustrated by Fig. 7 for cold-worked ($\alpha+\beta$) brass and hot-worked ($\delta+\gamma$) stainless steel. In both cases one of the phases (β and γ, respectively) has a much higher flow stress than the other. For this reason, it does not take part practically in the initial stages of the plastic deformation but comes under high stress which usually is locally different[28,29]. Accordingly, the diffraction peak of the hard phase show a significant line broadening, although there is no important increase of the lattice defect content. However, because of the

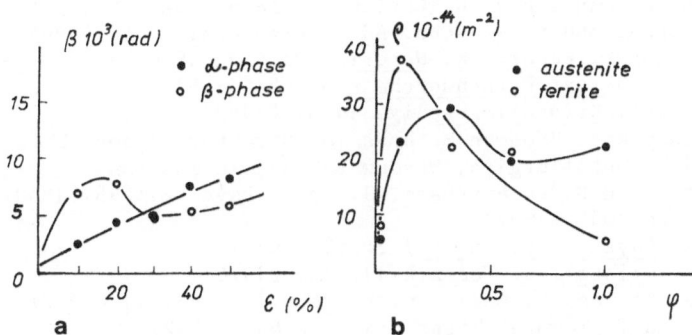

Fig. 7. Influence of inhomogeneously distributed internal stresses (peak shifts x_j) on the diffraction line broadening of

(a) cold-worked ($\alpha+\beta$) brass CuZn39Pb,
(b) hot-worked two-phase stainless steel X5CrNiTi26.6
The physical line broadening of (b) is formally described in terms of dislocation densities.

work hardening of the softer phase the internal stress of the harder component becomes more and more homogeneous with increasing degree of deformation and the line-broadening is reduced again. The result is a (rather unexpected) maximum of the physical line breadth of the hard phase. Simultaneously, the line-broadening of the softer phase is characterized by a parabolic increase (Fig. 7a: (α+ß) brass[30]) or by a behaviour which is typical for dynamic recrystallization (Fig. 7b: (δ+γ) stainless steel[31]).

REFERENCES

1. P. Klimanek, Crystal Res. & Technol. 18:K15 (1983).
2. P. Klimanek, G. Grosze, K.-E. Hensger, and U. Martin, Freiberger Forschungsheft B225:6 (1982).
3. G. Grosze, Thesis, Academy of Mining, Freiberg (1987).
4. P. Klimanek, "Annual Report 1986 on Nuclear Physics Activities and Applications", ZfK - Publ. 621, CINR Rossendorf (1987).
5. P. Klimanek, Freiberger Forschungsheft B265 (1988).
6. P. Klimanek, Proc. 5th Conference "Solid State Analytics", Karl-Marx-Stadt, GDR, 1987, Editor: TU Karl-Marx-Stadt (1988).
7. B. E. Warren, Progr. Metal Phys. 8:147 (1959).
8. A. Guinier, "X-Ray Diffraction", W. H. Freeman Comp., San Francisco (1963).
9. M. A. Krivoglaz, "Scattering Theory of X-Rays and Thermal Neutrons by Real Crystals", Nauka, Moscow (1967), in Russian.
10. M. A. Krivoglaz, "Diffraction Theory of X-Rays and Neutrons by Non-Ideal Crystals", Naukova Dumka, Kiev (1983), in Russian.
11. H. P. Klug and L. E. Alexander, "X-Ray Diffraction Procedures", John Wiley & Sons, New York (1954).
12. B. E. Warren, J.Appl.Phys. 31:2237 (1960).
13. H. J. Bunge, "Texture Analysis in Materials Science", Butterworth Publ. London (1982).
14. H. R. Wenk (editor), "Preferred Orientation in Deformed Metals and Rocks: An Introduction to Modern Texture Analysis", Academic Press, Orlando (1985).
15. M. A. Shtremel and L. M. Kaputkina, Kristallographiya 15:443 (1970).
16. M. A. Shtremel and L. M. Kaputkina, Fiz. Metallov Metalloved. 32:991 (1971).
17. A. Philipp, Thesis, Academy of Mining, Freiberg (1988).
18. H. J. Bunge, Z. Metallkde 75:97 (1984).
19. T. R. Anantharaman and J. W. Christian, Acta Cryst. 9:479 (1956).
20. C. N. J. Wagner and E. N. Aqua, Adv. X-Ray Anal. 7:46 (1964).
21. M. Wilkens and K. Eckert, Z. Naturf. 19a:459 (1964).
22. T. Ungar, Freiberger Forschungsheft B 265 (1988).
23. M. A. Shtremel, Kristallographiya 14:34 (1969).
24. Ya. D. Vishnyakov, "Modern Methods for Structure Research in Deformed Crystals", Metallurgiya, Moscow (1975), in Russian.
25. B. E. Warren and B. L. Averbach, J. Appl. Phys. 21:595 (1950) and 23:497 and 1059 (1952).
26. M. Wilkens, Phys. Stat. Sol.(a) 2:359 (1970).
27. M. Wilkens, Kristall u. Technik 11:1159 (1976).
28. H. Fischmeister and B. Karlsson, Z. Metallkde 68:311 (1977).
29. Y. Tomota and I. Tamura, Trans. ISIJ 22:665 (1982).
30. P. Klimanek, H. Richter, A. N. Ivanov, and Yu. A. Skakov, in preparation.
31. G. Küchhold, Thesis, Academy of Mining, Freiberg (1988).

TEXTURE INVESTIGATION OF NATURAL ROCK-SALT

BY NEUTRON DIFFRACTION

A. Ertel

Karl-Marx University, Dept. Chemistry
Institute for Crystallography
Talstr. 35, 7010 Leipzig, GDR

M. Betzl

Central Institute for Nuclear Research Rossendorf
POB 19, 8051 Dresden, GDR

H. Kaempf

Central Institute for Earth Physics
Telegrafenberg, 1500 Potsdam, GDR

The possibilities of texture studies using neutron diffraction on coarse-grained natural rock salt are investigated. The average grain diameter of our samples lies between 2 - 4 mm. For texture studies classical angle dispersive neutron diffraction have been applied, complete pole figures were measured using spherical sample method and also a combination of transmission and reflection geometry for rock salt slabs. Initial material is natural rock salt from Zielitz mine (GDR). Pole figures were measured for the (200), (220) and (222) reflections of NaCl, The crystallographic texture is weak, in general (100)-fibre textures were found.

INTRODUCTION

The crystallographic texture of minerals in rock contains important information, among others, on deformation history and on anisotropic mechanical properties of rocks[1]. Up to now, experimental values for plasticity of polycrystalline rocks differ in a wide range from one mineral aggregate to another. About the influence of mineral-texture is only rarely reported. The advatanges of neutron diffraction for texture investigation of geological materials were used to get characteristical information about mineral-texture in natural rock salt[2]. This contribution deals with interpretation of texture measurements on halit from Zielitz mine, Bleicherode mine and Werra region. The initial material for our studies was hanging wall salt of the stratigraphic layer $Na3_{\text{ß}}$ (650 m depth), Na2 and Na1.

139

EXPERIMENTAL

Neutron diffraction experiments were done at the research reactor RFN of the Central Institute for Nuclear Research Rossendorf, GDR. For texture measurements the classical angle-dispersive method has been applied. For geological materials the complete pole figure is needed because no sample symmetry can ad hoc be expected and made the experimental texture determination rather time-consuming.

Slabs ($90 \times 90 \times 8$ mm^3) and cubes ($20 \times 20 \times 20$ mm^3) for the neutron texture experiments and polished sections for reflection optical microscopy studies were prepared. The average grain diameter of the coarse-grained halite determined by microscopic studies is about 2 to 4 mm. Chemical analysis and X-ray powder diffraction patterns show halite is the dominant phase (more than 92 mass % NaCl) and they indicate that superpositions of Bragg-reflections by other minerals (sylvin, anhydrite and hematite in a small concentration up to 5 mass %) can be neglected.

RESULTS

Experimental (200) pole figures of halite are shown in Fig. 1. In the Zielitz and Bleicherode mines we found a (100) fibre texture and in the Werra region (Merkers) we found a (110) fibre texture. The (110) fibre texture can be interpreted as due to a ruptural geological deformation process. On the other hand, the (100) fibre texture of natural rock salt is caused by recrystallization processes.

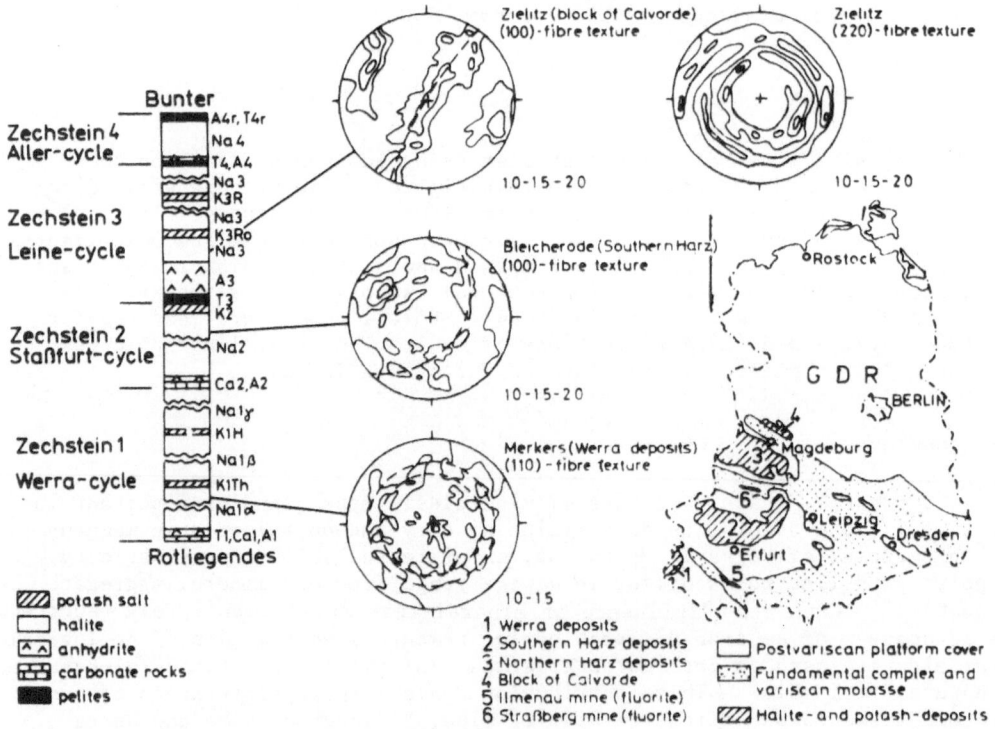

Fig. 1. Experimental pole figures of halite in natural rock salt from different stratigraphic layers.

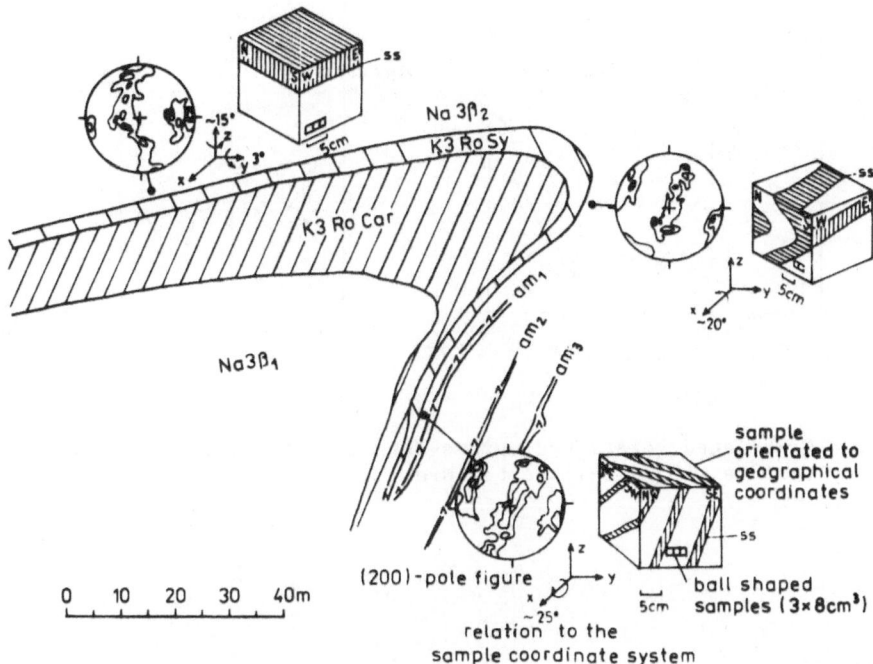

Fig. 2. Stratigraphic layer Na3$_\beta$ (Zielitz mine), halite texture on different
positions in a folded rock salt layer.

Fig. 2 shows a folded rock salt layer in the perturbation zone of the
Zielitz mine, more detailed investigation were done here. Samples were cut
and prepared in a definite geological and also in a reproducible geographical
coordinate system to give information about a global texture and its inhomogeneities over a geological folded perturbation zone. All investigated
samples from different points in the salt deposit showed a corresponding
weak (100) fibre texture. The (100) fibre texture of halite does not fit
the rock salt layering. Over a folded rock salt layer (Fig. 2) the halite
texture does not follow the geological fold geometry in the perturbation
zone. Our results show that the mineral-texture of halite in the Zielitz
mine is largely influenced by recrystallization processes during deformation
because a (110) fibre texture after experimental deformation reported by
Kern[3] could not be found in this perturbation zone.

For geological classification it becomes useful to compare the texture-
intensities (or sharpness) for a given pole figure type from samples of different geological conditions in a salt deposit. In Fig. 3 the pole densities
(multiples of random-orientation) are shown versus pole-distance for the
(100)- and (110)-fibre textures as calculated rotation diagrams (fibre axis
in the centre of pole figure). From our investigation we can observe a
weak dependence of texture-sharpness with increasing mining deepness in the
salt deposit of the Zielitz mine. We assume the main influence is the recent
deformation field for the increasing texture.

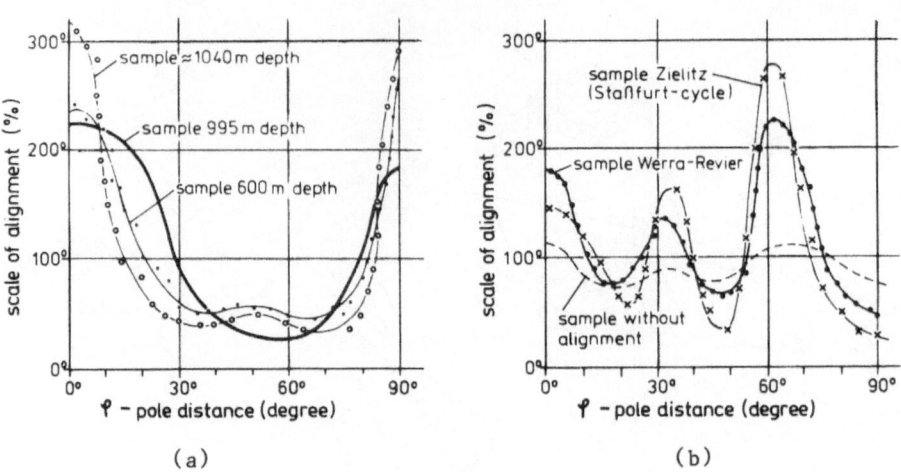

Fig. 3. Calculated rotation-diagrams for fibre textures: (a) 100-fibre texture, (b) 110-fibre texture.

REFERENCES

1. H. R. Wenk (Ed.): "Preferred Orientation in Deformed Metals and Rocks. An Introduction to Modern Texture Analysis". Orlando-Acad.-Press, Florida (1985).
2. A. Ertel, M. Betzl, P. Paufler, W. Voitus: Application of neutron diffraction methods to texture studies on natural rock salt, Cryst.Res. Technol. 22:209 (1987).
3. H. Kern, V. K. Gairola: Microstructure and texture in experimentally deformed single-layer rock salt, J.Struct.Geol. 6:201 (1984).

TEXTURE AND STRUCTURE OF ANISOTROPIC FeSi MATERIALS

AFTER HOT ROLLING

J. Švantner, V. Frič, D. Matisová, and A. Sólyom

Research and Testing Institute of VSZ Košice

Košice, Czechoslovakia

INTRODUCTION

In the research and development of oriented transformer steel efforts are aimed at decreasing core losses and increasing magnetic induction. Magnetic characteristics are directly connected with Goss texture, i.e. with the grain growth preference [110][001]. The aim of the present work was to study microstructure and texture changes in oriented anisotropic Fe-Si steels and to compare them with Sb microalloyed materials.

EXPERIMENTAL METHODS AND RESULTS

Samples 2.6 mm thick were used as hot-rolled sheet steel free of antimony and alloyed with 0.05 weight % Sb. Analyses upon microstructure and texture changes across the thickness of the examined steel were performed. The results in Figs 1,2 show that the microstructure across the thickness is heterogeneous. The surface layer of the material consists of large undeformed ferrite grains. The thickness of this layer varies in the range of 0.2 - 0.35 mm. In the middle part the structure consists of markedly elongated ferritic grains with a small proportion of fine recrystallized grains. Along the ferritic grain boundaries a small amount of fine pearlite has been identified. The metallographic analysis has not shown significant differences in the microstructure of the compared materials.

Texture measurements were performed on the surface of steel and then gradually after grind-off and etch-off of the affected layer in the inwards direction of the material. The analysis was carried out in six crystallographic planes: {110}, {200}, {211}, {310}, {222}, {321}. It results from the analysis that the plane 110 is of a maximum intensity in the surface zones in the compared series of samples. The intensities of other planes are low. In the direction towards the center the intensities of {200} and {222} planes increase and the reflection intensity of the {110} plane decreases. There exist more significant differences in the ready oriented transformer sheet as it is shown in Table 1.

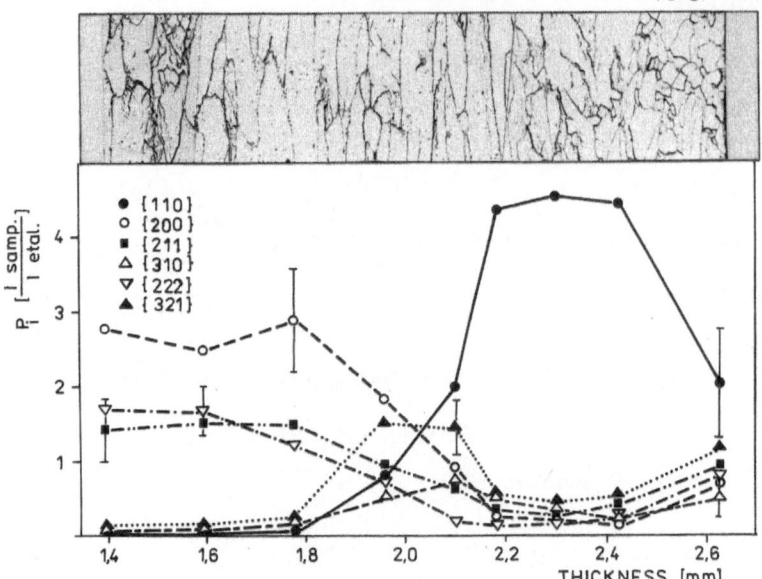

Fig. 1. Texture and microstructure across the sheet thickness of the antimony free material.

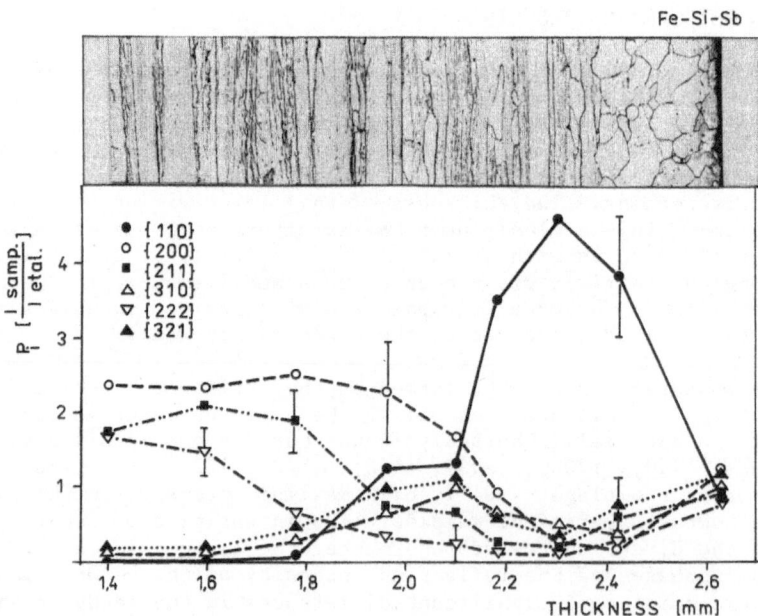

Fig. 2. Texture and microstructure across the sheet thickness of the Sb alloyed material.

Table 1. Intensities Diffracted from Different Planes in Ready-Oriented Transformer Sheet P_i [I_{sample}/I_{etalon}]

Plane	110	200	211	310	222	321
Fe-Si	5.13	0.25	0.27	–	0.18	0.17
Fe-Si-Sb	5.45	–	0.07	–	0.20	0.28

CONCLUSION

1. After hot-rolling there exists a difference in the microstructure and texture across the thickness of the steel sheet.

2. The addition of Sb has shown to be more significant in the ready oriented transformer sheet; larger share of {110} planes has been found there.

The connection between the volume proportion of the Goss texture and the magnetic parameters has been confirmed in a number of special studies[1-3]. This study gives an information about the positive effect of some surface-active elements (e.g. Sb) on the Goss texture formation[3]. The authors assume that antimony affects the surface energy of crystallographic grains by the equilibrium segregation and the absorption mechanism and thus influences the formation of texture constituents in the course of recrystallization.

REFERENCES

1. G. Y. Chin et al., "Ferromagnetic Materials", Vol. 2, North Holland Publ. Comp. (1980), p. 78.
2. A Sólyom, V. Frič, Kovové materiály 19:678 (1981), in Czech.
3. P. Marko, A. Sólyom, and V. Frič, J.Magn.Mat. 41:7 (1984).

REAL STRUCTURE OF Gd_3Co SINGLE CRYSTALS

A. Winiarski, E. Talik, A. Winiarska, J. Heimann, and J. Szade

Institute of Physics
Silesian University
Uniwersytecka 4, Katowice, Poland

Gd_3Co single crystals were grown by the Czochralski technique from a levitated sample in an argon atmosphere. The starting materials of high purity (Gd - 99.9 %, spectroscopically pure Co) were placed in a conic-shaped water cooled crucible. The crucible was placed in a quartz tube inside a RF induction coil. The temperature was controlled by a phototransistor. The crystals were pulled up from the levitated molten sample with the speed of about 10 mm per hour. The crystals were up to 50 mm long with a diameter of about 1.5 mm. The growth direction was close to the [110]. The crystals were studied by X-ray reflection topography and with the use of Auleytner's spectrometer with an oscillating film[1]. X-ray reflection topographies of "as-grown" crystals usually showed the crystals without mosaic spread. However, low-angle boundaries in some crystals were revealed. In this case, the crystals consisted of blocks elongated along the growth direction. The surface of growing crystals was not smooth but some roughnesses occurred. Therefore black spots and lines appeared in topographies of the "as-grown" crystals (Fig. 1).

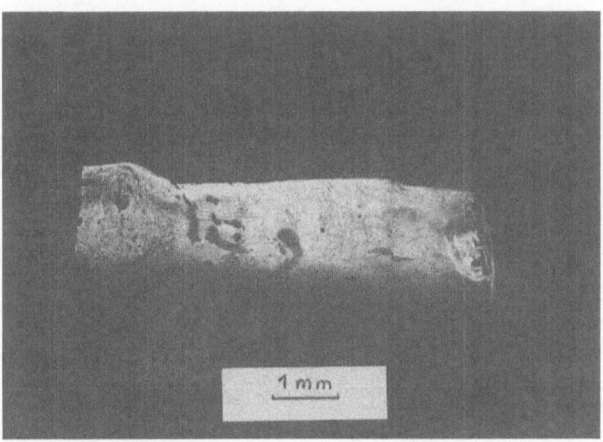

Fig. 1. X-ray reflection topography of "as-grown" Gd_3Co single crystal.
Fe-K_α, 1 3 4 reflection.

To determine the dislocation density the spectrometer with an oscillating film was used. The crystals were cut by a wire-saw and polished. To eliminate the effect of plastic deformation remaining on the surface after mechanical polishing the crystals were etched in a solution consisted of 1 ml HNO_3, 1 ml HCl and 100 ml C_2H_5OH and rinsed in ethyl alcohol. The half-widths of the intensity distribution functions were measured on films firstly in the case where the crystal and the film oscillated together, then when the crystal oscillated and the film remained stationary. The half widths of the intensity distribution functions of the primary beam traces were measured both for the film placed on the axis of the spectrometer and for the mid distance between the entrance slit and the crystal. The evaluated density of dislocations was about 5×10^5 cm^{-2}. No change of the dislocation density connected with change of the diameter of the growing crystals was noticed.

The work was supported by the Polish Academy of Sciences under Project CPBP 01.12.3.14.

REFERENCES

1. T. Bedyńska and J. Chmielewska, Investigation of dislocation density tensor components by the method of spectrometer with oscillating film, Acta Phys.Pol. 26:199 (1964).

HIGH TEMPERATURE ELASTIC DIFFUSE NEUTRON SCATTERING STUDY

OF THE DEFECT STRUCTURE IN $TiN_{0.82}$

Thierry Priem, Brigitte Beuneu, and Charles de Novion

CEA/IRDI/DMECN/DTech, Laboratoire des Solides Irradiés
Ecole Polytechnique
91128 Palaiseau Cédex, France

INTRODUCTION

Titanium mononitride TiN_x has the f.c.c. rocksalt crystal structure, with nitrogen vacancies accommodating the non-stoichiometry for $0.50 < x < 1.00$ (Ref. [1]) For $0.5 < x < 0.6$, a quadratic Ti_2N superstructure, where the nitrogen vacancies are long-range ordered, of space group $I4_1/amd$ and characterized by $(1\ 1/2\ 0)$ type superlattice reflections, occurs below 800 °C.[2] For larger nitrogen concentrations $(0.6 < x < 0.9)$, diffuse streaks were observed on electron diffraction patterns, and qualitatively interpreted in terms of short-range ordering (SRO) of nitrogen vacancies.[3] However, recent X-ray diffuse scattering measurements on the isomorphous compound $NbC_{0.72}$ showed that this diffuse intensity is dominated by static displacements of the metal atoms.[4] Therefore, because of their larger scattering amplitudes by light atoms, neutrons should be preferred to study the ordering contribution to the diffuse scattering. This is the object of the present paper. Preliminary results on a sample cooled at room temperature have been published recently.[5]

EXPERIMENTAL

The $TiN_{0.82}$ single crystal was prepared by a zone annealing technique in Aarhus University. The method of preparation and the characteristics of the sample are given by Priem et al.[5] The diffuse neutron scattering measurements were performed at three temperatures (700, 800 and 900 °C) on the two-axis spectrometer G4-4 (ref.6) at Laboratoire Léon-Brillouin*, CEN-Saclay, France, with an incident wavelength $\lambda = 2.56$ Å. The sample was held under secondary vacuum (10^{-6} torr). Time-of-flight analysis allowed to eliminate most of the inelastic scattering due to phonons and to obtain the elastic diffuse scattering cross-section in the $\{001\}$ and $\{1\bar{1}0\}$ reciprocal planes for $0.5 \leqq Q = 4\pi \sin\Theta/\lambda \leqq 4.5$ Å^{-1}. The data were calibrated to absolute units by comparison with a vanadium standard after background correction. They were corrected for effective absorption, total incoherent scattering and Debye-Waller factor ($B_{Ti} \simeq B_N \simeq 0.82$ Å^2 for T = 700 °C, 0.89 Å^2 for 800 °C and 0.97 Å^2 for 900 °C, ref. [7]).

---

*Laboratoire commun CEA-CNRS

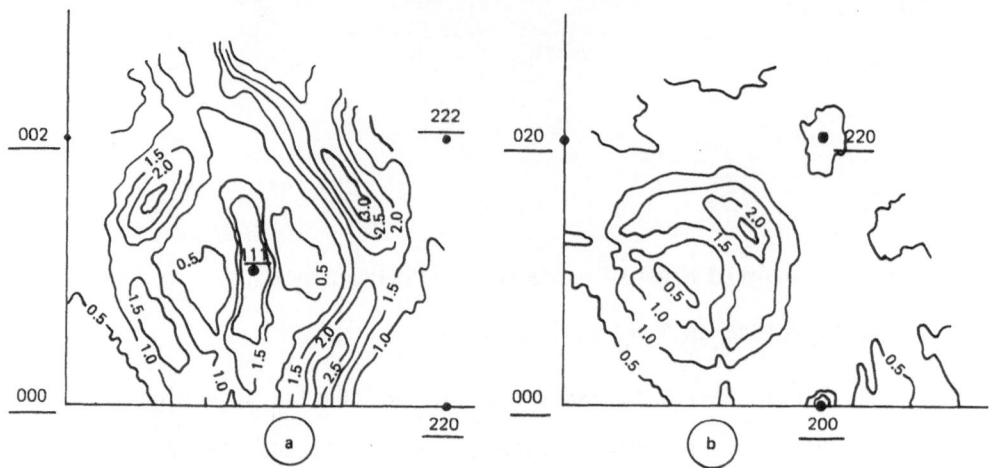

Fig. 1. Elastic diffuse neutron scattering differential cross-section of
TiN$_{0.82}$ (in Laue units: 1 Laue = 0.130 barns)
a: $\{1\bar{1}0\}$ and b: $\{001\}$ reciprocal lattice planes

RESULTS

The elastic diffuse differential cross-section $d\sigma/d\Omega$ of TiN$_{0.82}$ for
T = 800 °C is shown in Fig. 1: one observes streaks similar to those detected
by electron diffraction on quenched samples,[3] but the intensity is not periodic in the reciprocal space, showing the importance of the static displacement contribution. In the $\{001\}$ plane, the diffuse intensity is concentrated
along circles centered on the reciprocal nodes (110); no maxima are observed
at the Ti$_2$N superlattice reflection positions (1 1/2 0). In the $\{1\bar{1}0\}$ plane,
the diffuse intensity is maximum at the (1/2 1/2 1/2) type positions. For
T = 700 and 900 °C, the figures are similar, but the maxima decrease with
increasing temperature, i.e. as the atomic disorder increases.

As the static displacements (< 0.1 Å) are much smaller than the lattice
parameter a, the elastic diffuse cross-section can be expanded to the second
order in displacements. The formalism used, developed first for single sublattice compounds,[8] is described for the case of the rocksalt structure by
Priem et al.[5] In the above approximation, the diffuse scattered intensity
depends linearly of three sets of coefficients: $\alpha_{\iota mn}$, $\gamma_{\iota mn}$ and $\delta_{\iota mn}$, which
are respectively the Fourier transforms of the SRO contribution, the first
order and the second order displacement contributions. The $\alpha_{\iota mn}$ are the
Cowley-Warren short-range order coefficients $\alpha_{\iota mn} = 1 - P_{\iota mn}/(1 - x)$, where

$P_{\iota mn}$ is the conditional probability to find a vacancy in the lattice position
$\vec{R}_{\iota mn} = \iota\,\vec{a}/2 + m\,\vec{b}/2 + n\,\vec{c}/2$ (a, b, c vectors of the f.c.c. cell; ι, m, n integers), if a nitrogen is at the origin.

The experimental data were treated by the least squares fit method. The
second order displacement contribution was found small in the Q range studied.
The first $\alpha_{\iota mn}$ values are given in Table 1. The negative values of α_{011} and
α_{002} and the positive values of α_{112} indicate that vacancies avoid first and
second neighbour positions and prefer third neighbour positions. The $\gamma_{\iota mn}$
coefficients are consistent with an average radial repulsion of
4.3 10^{-2} Å by a vacancy of its Ti first neighbours.

Table 1. Short-range Order Coefficients $\alpha_{\ell\,mn}$ and Standard Deviations in $TiN_{0.82}$

	T = 700 $^\circ$C	T = 800 $^\circ$C	T = 900 $^\circ$C
α_{011}	− 0.101 (2)	− 0.096 (2)	− 0.090 (2)
α_{002}	− 0.114 (5)	− 0.102 (5)	− 0.096 (5)
α_{112}	+ 0.045 (2)	+ 0.040 (2)	+ 0.034 (2)
α_{022}	+ 0.018 (3)	+ 0.013 (3)	+ 0.013 (3)
α_{013}	+ 0.019 (1)	+ 0.016 (1)	+ 0.014 (1)
α_{222}	− 0.028 (1)	− 0.023 (1)	− 0.018 (1)

DISCUSSION

Preliminary measurements at heating and cooling between 700 and 900 °C gave reversible results, confirming that the nitrogen sublattice is in thermodynamical equilibrium within a few minutes at these temperatures. The short-range order coefficients $\alpha_{\ell mn}$ are related to the effective interatomic pair potentials $V_{\ell mn}$ responsible for ordering of nitrogen atoms and vacancies; in first approximation, the relation is given by the mean-field Krivoglaz-Clapp-Moss formula:[9]

$$\alpha(Q) = C \left[1 + 2x \, (1 - x) \, V(Q)/k_B T \right]^{-1},$$

where $\alpha(Q)$ and $V(Q)$ are the Fourier transforms of the $\alpha_{\ell mn}$ and $V_{\ell mn}$, k_B the Boltzmann constant, and C a constant related to the integrated diffuse intensity in the reciprocal lattice cell.

From a least squares fit with four interatomic potentials, the $V_{\ell mn}$ values were found practically temperature independent: $V_{011} = V_1 = 56.2$ meV, $V_{002} = V_2 = 39.8$ meV, $V_{112} = V_3 = 1.2$ meV, $V_{022} = V_4 = 2.4$ meV at 800 °C, whence $V_2/V_1 = 0.71$. More exact values were obtained from the $\alpha_{\ell mn}$ by Monte-Carlo simulations[10]: $V_1 = 81.7$, $V_2 = 60.5$, $V_3 = 2.4$ and $V_4 = 3.8$ MeV, $V_2/V_1 = 0.75$. The higher values of the Monte-Carlo $V_{\ell mn}$ are attributed to frustration effects, neglected in the mean-field theory.[10] However, the Monte-Carlo and mean-field ratios V_2/V_1, V_3/V_1 and V_4/V_1 are very similar.

Effective pair interaction potentials V_i for metalloid ordering in transition metal carbides and nitrides have been theoretically calculated from the band structure of the stoichiometric compounds, using the generalized perturbation method.[11] For nitrides of composition $MN_{0.83}$, qualitative agreement is found with the values deduced from neutron diffuse scattering: V_1 and V_2 are large and positive, V_3 and V_4 are very small and can be neglected in first approximation.[12] Figure 2 shows the phase stability of a binary f.c.c. alloy in terms of V_1 and V_2;[2] this scheme can be applied to the nitrogen-vacancy sublattice of TiN_x. The theoretical as well as experimental pair potentials found for $TiN_{0.82}$ fall in region III, characterized by maxima of the diffuse scattering in (1/2 1/2 1/2). It is interesting to remark that the superstructure Ti_2N falls in region II (1 1/2 0); the reasons of this discrepancy are not well understood[5]. For $TiN_{0.82}$, the long-range ordered structure is not known, but plot of $\alpha^{-1}(\vec{Q} = 1/2\ 1/2\ 1/2)$ versus 1/T as well as Monte-Carlo simulations suggest a critical ordering temperature in the range of 500–600 K (300 °C), where nitrogen mobility is practically frozen.

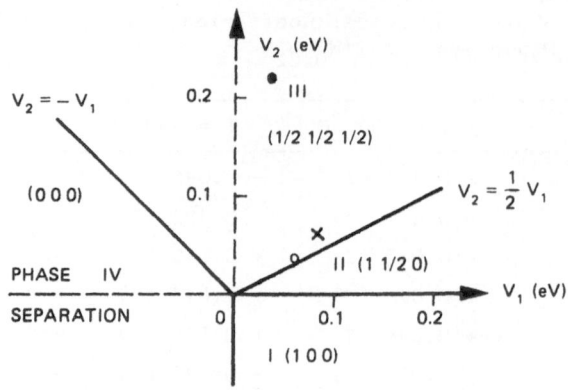

Fig. 2. Superstructure maxima of a f.c.c. binary alloy versus the first and second neighbour effective pair potentials.
o: Theoretical pair potentials calculated for TiN_{083}.[12]
o, x: Pair potentials deduced for $TiN_{0.82}$ from elastic diffuse neutron scattering data by the Clapp-Moss formula and Monte-Carlo simulation, respectively.

ACKNOWLEDGEMENTS

We wish to thank Dr A. N. Christensen for the $TiN_{0.82}$ single crystal, Dr F. Livet for the Monte-Carlo simulations and Dr G. Treglia for the pair potential calculations applied to the case of $MN_{0.83}$.

REFERENCES

1. L. E. Toth, "Transition Metal Carbides and Nitrides", Academic Press, London (1971).
2. G. Lobier and J. P. Marcon, Etude et structure d'une nouvelle phase du sous-nitrure de titane Ti_2N, C.R. Acad.Sc. Paris Serie C 268:1132 (1969).
3. J. Billingham, P. S. Bell, and M. H. Lewis, Vacancy short-range order in substoichiometric transition metal carbides and nitrides with the NaCl structure – I Electron diffraction study of the short-range ordered compounds, Acta Cryst.Sect. A 28:602 (1972).
4. K. Ohshima, J. Haraga, M. Morinaga, P. Georgopoulos, and J. B. Cohen, Distortion-induced scattering due to vacancies in $NbC_{0.72}$, Acta Cryst.Sect. A 44:167 (1988).
5. T. Priem, B. Beuneu, C. H. de Novion, R. Caudron, F. Solal, and A. N. Christensen, (1/2 1/2 1/2) versus (1 1/2 0) type ordering of nitrogen vacancies in TiN_x, Solid State Commun. 63:929 (1987).
6. R. Caudron and A. Finel, Technical Report ONERA no. 11/1221 M, 92322 Chatillon, France (1983).
7. C. R. Houska, Thermal expansion and atomic vibration amplitudes for TiC, TiN, ZrC, ZrN and pure tungsten, J.Phys.Chem:Solids 25:359 (1964).
8. B. Borie and C. J. Sparks, The interpretation of intensity distribution from disordered binary alloys, Acta Cryst.Sect. A 27:198 (1971).
9. P. C. Clapp and S. C. Moss, Correlation functions of disordered binary alloys. II, Phys.Rev. 171:754 (1968).
10. F. Livet, unpublished.
11. J. P. Landesman, G. Treglia, P. Turchi, and F. Ducastelle, J.Physique 46:1001 (1985).
12. G. Treglia, unpublished.

X-RAY DIFFRACTION STUDIES OF INTERDIFFUSION

IN SOLID SOLUTIONS OF $A^{III}B^{V}$ SEMICONDUCTORS

U. Voland and P. Deus

Sektion Physik, Bergakademie Freiberg
Cottastrasse 4, 9200 Freiberg, GDR

R. Černý and V. Valvoda

Dept of Semiconductor Physics, Charles University
Ke Karlovu 5, 121 16 Praha 2, Czechoslovakia

Interdiffusion in annealed powder blends of InAs-GaAs and InP-GaP was studied by Rudman's X-ray diffraction method. The experimental results were interpreted by means of a concentric spheres and a concentric cubes model. Temperature dependence of diffusion degree, diffusion coefficient and activation energy of interdiffusion were determined. The results were compared with corresponding parameters known for self-diffusion in these compounds.

INTRODUCTION

Solid solutions of $A^{III}B^{V}$ compounds are important materials for both opto- and microelectronics because they allow the combination of known properties of the $A^{III}B^{V}$ semiconductors. The diffusional homogenization of finely ground powder blends is a method of preparing such solid solutions below their melting point.

X-ray diffractometry is valuable to investigate the homogenization process in powder blends[1-3]. This was confirmed by measurements both on metal powder systems[2,4] and on crystal powder systems with NaCl-structure[5] or CaF_{2}-structure[5].

The method is based on the concentration dependence of the lattice parameter and the diffraction angle according to Bragg's law. The advantages of this technique are its capability to measure the concentration profiles in small sample volumes of the order of μm^{3} and its nondestructive character.

The aim of this paper is to show that X-ray diffractometry is useful for studying interdiffusion in powder blends of binary semiconductors with ZnS-structure as well. For the quantitative evaluation two simple geometric models[3,6] of powder blends are tested on the example of InAs-GaAs.

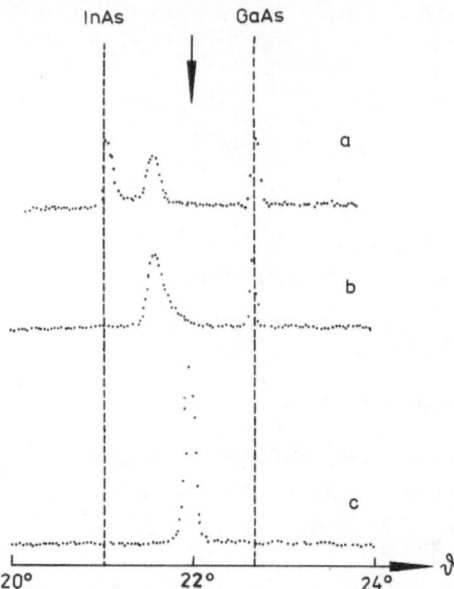

Fig. 1. 220 X-ray diffraction line profiles of InAs-GaAs powder blends
after diffusion anneal of (a) 8 h at 600 $^\circ$C, (b) 1 h at 650 $^\circ$C
and (c) 48 h at 800 $^\circ$C.
The peak position of the homogenized mixed crystal phase
$In_{0.43}Ga_{0.57}As$ is marked by the arrow.

The arsenides and phosphides of In and Ga have ZnS-structure and show
complete miscibility. Their lattice parameter differences are well suited
for researching on the diffusional state by the common powder diffractometer
technique (Fig. 1).

EXPERIMENTAL APPROACH

The starting powders of the Ga- and In-compounds were prepared from
polycrystalline ingots by grinding and sieving. Their grain size was chosen
with regard to the diffusion model (see next chapter) and to the microabsorption effects. For the Ga-compounds an average equivalent diameter of 11 μm
was used and for the In-compounds a grain size $\geq$ 5 μm.

The powder fractions were mixed in a molar fraction x = 0.43 and x = 0.5
for InAs and InP, respectively. In order to produce a close contact between
the grains, the mixed powders were cold-pressed at 188 MPa. The pressed
powder blends were sealed in quartz glass ampoules at vacuum of 4.10^{-3} Pa
and heated in a furnace controlled automatically to $\pm$ 2 $^\circ$C.

The diffusion state of the annealed specimen was studied with an X-ray
horizontal goniometer measuring the 220 reflection profile with CuK_α radiation. The α_2-component, Lorentz and polarization factor were corrected as
described by Valvoda et al.[4]. The dependence of the structure factor S(x)
on molar fraction x of In-compounds was considered as $S(x) = xf_{In} + (1-x)f_{Ga} + f_{As,P}$, where f are the corresponding atomic scattering factors.

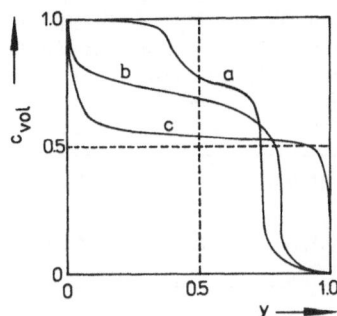

Fig. 2.   Concentration-penetration curves from the line profiles of Fig. 1.
c_{vol} is volume fraction of InAs and y is the penetration parameter.

METHOD OF PROFILE ANALYSIS

The analysis of X-ray diffraction line profile was carried out corresponding to Rudman's method. By means of concentration-penetration curves
(Fig. 2) the interdiffusion degree was calculated as

$$
H = \frac{\int_{0}^{c_M}(y_M - y)\, dc \; + \; \int_{c_M}^{1}(y - y_M)\, dc}{c_M\,(1 - y_M) + (1 + c_M)\, y_M} \; ,
$$

where the quantity of penetration y is defined in the same way by Valvoda
et al.[4], y_M is the position of the Matano interface and c_M is the related
volume fraction. The time and temperature dependence of interdiffusion
degree H_{exp} for InAs-GaAs is listed in Table 1.

The experimental values H_{exp} are interpreted by means of two models of
diffusional geometry assuming a concentration-independent interdiffusion
coefficient D:

1. The model of concentric spheres[3] in which the compounds with the higher
    melting temperature is represented by a spherical particle and is sur-
    rounded by a uniform shell formed from the smaller particles of the com-
    pound with the lower melting temperature;

2. the model of cubes[6] in which the higher melting compound is represented
    by cubes distributed within the matrix of the lower melting compound.

A comparison of the values of theoretical interdiffusion degree H_{th} calculated for both models as function of Dt is given in Table 2.

Table 1.   Interdiffusion Degree H_{exp} for InAs-GaAs Powder
           Blends Annealed 8 and 24 h at Temperatures   T
           between 550 $^{\circ}$C and 800 $^{\circ}$C

T, $^{\circ}$C	550	600	625	650	700	800
t/h						
8	0.035	0.26	0.695	0.82	0.88	0.90
24	0.04	0.51	0.849	0.86	0.89	0.91

Table 2. Interdiffusion Degree H_{th} = f (Dt) Calculated for Concentric Spheres and Concentric Cubes Models with Equi-volume Proportion InAs-GaAs

Dt	0.5	1.0	1.5	2.0	2.5	Model
H_{th}	0.32	0.61	0.81	0.86	0.96	concentric spheres
	0.25	0.48	0.65	0.75	0.82	concentric cubes

RESULTS AND DISCUSSION

From the movement of diffraction peaks in InAs-GaAs samples with a mean composition of molar fraction x = 0.43 (Fig. 1) and from the more rapid decreasing of the InAs peak we can conclulde that in the temperature range 550-625 °C Ga atoms are more easily diffusing compared with In atoms. For InP-GaP samples it was observed that the diffusion begins above 650 °C in analogous manner.

The temperature dependence of interdiffusion coefficient $D=D_o\exp(-Q/kT)$ obtained by means of H_{exp} and H_{th} is demonstrated for InAs-GaAs and InP-GaP powder blends in Fig. 3. For InAs-GaAs two regions of this dependence were observed. The linear part between 550 and 625 °C corresponds to a diffusion length lower than particle size. For this temperature region only the activation energy was estimated. At higher temperatures the homogenization proceeds so that the initial assumptions concerning the diffusion geometry do not hold.

The results for InAs-GaAs powder blends show that the concentric spheres and concentric cubes models are equally suitable for the determination of activation energy in a powder system with ZnS-structure under the condition that sample state and measurement conditions ensure the validity of these models.

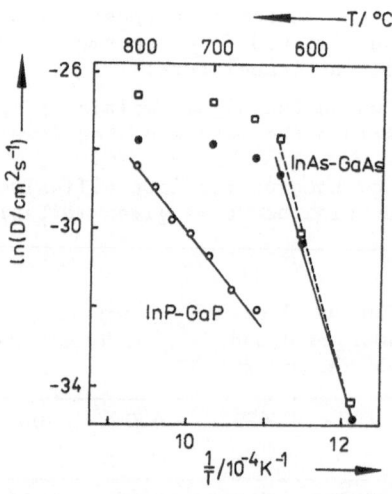

Fig. 3. lnD versus l/T for InAs-GaAs and InP-GaP powder blends. The values of D are averages of values obtained from 8 to 24 h annealed samples in the concentric spheres (o,o) and concentric cubes (□) model.

Table 3. Preexponential Term D_o and Experimental Activation Energy of Interdiffusion

Compound	$D_o/cm^2 s^{-1}$	Q/eV	$T/^\circ C$
InAs-GaAs	8.0×10^{17}	5.4	550 - 625
InP-GaP	2.8×10^{-3}	2.1	650 - 800

In Table 3 the values of activation energy Q and preexponential term D_o for InAs-GaAs and InP-GaP corresponding to the Arrhenius plot in Figure 3 are listed. From the comparison of these results with the activation energy of selfdiffusion in pure $A^{III}B^V$ compounds (e.g. given by Boltaks) it can be concluded that the dominant diffusion mechanism for interdiffusion in both powder systems under investigation is the same as for selfdiffusion in pure compounds.

REFERENCES

1. P. S. Rudman, An X-ray diffraction method for the determination of composition distribution in inhomogeneous binary solid solutions, Acta Cryst. 13:905 (1960).
2. E. J. Mittemeijer and R. Delhez, Determination of Compositional Variations by X-ray Diffraction Line Profile Analysis, National Bureau of Standards Special Publication 567, in: Proceedings of Symposium on Accuracy in Powder Diffraction held at NBS, Gaithersburg, MD, June 11-15, 1979, U.S. Government Printing Office, Washington, pp. 271-314 (1980).
3. B. Fisher and P. S. Rudman, X-ray diffraction study of interdiffusion in Cu-Ni powder compacts, J. Appl. Phys. 32:1604 (1961).
4. V. Valvoda, L. Dobiášová, and P. Karen, X-ray diffraction analysis of diffusional alloying of HfC and TaC, J. Mater. Sci. 20:3605 (1985).
5. R. Verma, Study of homogenization and cation interdiffusion in mixed UO_2-PuO_2 compacts by X-ray diffraction, J. Nucl. Mater. 120:65 (1984).
6. A. N. Raichenko, "Diffuzionnye raschety dlya poroshkovykh smesei", Naukova Dumka, Kiev (1969).
7. B. I. Boltaks, "Diffuzia i tochechnye defecty v poluprovodnikakh", Izdatelstvo "Nauka",Leningrad (1972).

IV. POLYCRYSTAL THIN LAYERS

KINEMATIC THEORY OF X-RAY AND NEUTRON SCATTERING

BY DEFECTS IN THIN FILMS AND SURFACE LAYERS

R. I. Barabash and M. A. Krivoglaz

Institute of Metal Physics, Academy of Sciences of the
Ukrainian SSR
Vernadsky 36, 252642 Kiev-142, USSR

The scattering by bounded defects in thin films and surface layers
has been considered by the fluctuation waves method. It is shown that the
finite crystal thickness and the relation of displacement fields near the
surface affect substantially the intensity distribution at the reciprocal
lattice points and lead to qualitative differences from the case of the
homogeneous distribution of defects in massive crystals. In the case of
free- or end-fixed thin films, the defects with small local distortions can
broaden the Bragg peaks or lead to a very strong diffuse scattering the intensity of which is inversely proportional to the fourth power of the distance to the reciprocal lattice point.

INTRODUCTION

The work deals with the scattering by bounded defects (point defects,
their clusters, dislocation loops of small radius, etc.) in thin crystal
films and crystal surface layers. The diffuse scattering intensity distribution $I(\vec{Q})$ ($\vec{Q}$ is the diffraction vector) has been studied in detail by such
defects in massive crystals. In the vicinity of the reciprocal lattice
points $\vec{G}$ of an "average crystal", $I(\vec{Q}) \propto q^{-2}$ increases sharply ($\vec{q} = \vec{Q} - \vec{G}$).
An investigation of $I(\vec{Q})$ is one of the most informative methods of the study
of defects.

The development of diffraction methods, in particular, the appearance
of high flux X-ray sources, e.g., synchrotron radiation, allows one to carry
out the study of scattering by small crystal volumes (the scattering by the
volume of ~ 200 μm was studied[1]) and makes the analysis of diffraction effects in films and surface layers of current concern.

At small thicknesses D of films or surface layers the distribution of
$I(\vec{Q})$ in the region $q \leq D^{-1}$ differs qualitatively from such distribution for
a massive crystal. The boundaries lead to the relaxation of defect displacement fields. As a result the shape of isodiffuse surfaces for defects
in surface layers and films on a substrate changes significantly though the
dependence $I(\vec{Q}) \propto q^{-2}$ is preserved. In free- or end-fixed films the dipoles
of forces produced by defects cause large flexural strains. These lead to
anomalous diffuse scattering $I \propto q_1^{-4}$ at small q_1 (q_1 is the projection of
$\vec{q}$ on the film plane) instead of usual law $I \propto q^{-2}$. In the case of large film
areas there arises the "infrared catastrophe" of static displacements, the
δ type Bragg peaks are suppressed, and the diffuse scattering intensity
is converted into broadened peaks.

It is convenient to determine the intensity $I(\vec{Q})$ by one of the variants of the static fluctuation waves method[2-4] considering the waves (the Fourier components) of defect concentration and the waves of static displacements engendered by the former. In bounded crystals, the displacement waves are not plane due to relaxation of displacements near the surface. The expansions in such waves were used in the Sneddon method for the solution of problems of the theory of elasticity.

Except for some special orientations of a very perfect crystal or some directions of scattering angles (in the vicinity of which the Laue-Bragg conditions for incident or scattering waves are fulfilled), the intensity $I(\vec{Q})$ can be determined by the kinematic scattering theory. The consideration given below will be carried out in the framework of this theory.[6] The dynamical theory of diffuse scattering by semi-infinite elastically isotropic imperfect crystals, necessary for the analysis of the above mentioned special directions, has been developed in[7,8].

GENERAL EXPRESSIONS FOR THE SCATTERING INTENSITY BY DEFECTS IN A FILM

On scattering by an ideal film of thickness D and longitudinal dimensions D', there appears a rodlike peak of a regular reflection in the reciprocal lattice space with the dimensions of $\sim 2\pi/D$ along the direction q_z (the axis z is perpendicular to the film plane) and of considerably less thicknesses $\sim 2\pi/D'$ along the directions $q_\perp = \{q_x, q_y\}$. Below in studies of scattering by imperfect crystals we shall consider the regions $q_\perp \gg 2\pi/D'$ of the above peak, where the scattering by defects shows up in the pure form. We restrict ourselves to the systems in which the thickness of a film or a near-surface defect layer is considerably less than the absorption length.

Let us consider the case of low concentration of defects, when one can neglect their correlation and restrict oneself to the first term in the expansion of $I(\vec{Q})$ in concentration. If the strength of bounded defects is not large or q is small, then one can preserve only the first terms in the expansion of $I(\vec{Q})$ in powers of displacements and similarly to the case of massive crystals (see, for example,[9])

$$I(\vec{Q}) = \sum_t \bar{c}(\vec{R}_t)\, e^{-2M} \left| f\vec{Q}\vec{U}_{\vec{q}t} - \varphi_{\vec{q}} \right|^2 , \tag{1}$$

$$\vec{U}_{\vec{q}t} = -i\sum_s \vec{u}_{st}\, e^{i\vec{q}\vec{R}_{st}} , \qquad \varphi_{\vec{q}} = \sum_s \varphi_{st}\, e^{i\vec{Q}\vec{u}_{st}} e^{i\vec{q}\vec{R}_{st}} ,$$

$$\vec{R}_{st} = \vec{R}_s - \vec{R}_t . \tag{2}$$

Here $\vec{R}_s$ and $\vec{R}_t$ are the radii vectors of the first points of the s-th cells and positions t in which the defect centres (lattice points or interstitial) can be found; $\bar{c}(\vec{R}_t)$ is the probability of locating a defect in the position t (in general $\bar{c}(\vec{R}_t)$ depends on the distance z_t to a crystal surface); f is the structure amplitude of an ideal crystal; $\exp(-2M)$ is the Debye-Waller factor resulting from static distortions (the Debye-Waller factor associated with thermal oscillations is included into f); φ_{st} and $\vec{u}_{st}$ are the changes of the structure amplitude of the first cell and the displacement of its first atom on introducing a defect into the position t. Near the surface $\vec{u}_{st}$ depends not only on $\vec{R}_{st}$, but also on z_t, but the dependence of φ_{st} on z_t in typical cases of small defect dimensions (as compared to D) can be neglected. Accordingly, $\vec{U}_{\vec{q}t}$ depends on z_t, and $\varphi_{\vec{q}}$ does not depend. If there are some types or orientations of defects α, then in (1) and (2)

$\bar{c}_\alpha (\vec{R}_t)$, $\vec{u}_{st\alpha}$ and $\varphi_{st\alpha}$ depend on α, and an additional summation with respect to α should be taken.

We introduce the fluctuation waves of defect concentration considering the expansion into a Fourier series of occupation numbers $c_t = 1, 0$ of the positions t by the defect centres,

$$c_t = c(\vec{r} = \vec{R}_t), \quad c(\vec{r}) = \sum_{\vec{k}} c_{\vec{k}} e^{-i\vec{k}\vec{r}}, \quad c_{-\vec{k}} = c_{\vec{k}}^*, \quad k_z = \frac{2\pi}{D} n. \quad (3)$$

Here $n = \ldots -1, 0, 1, 2 \ldots$, k_x, k_y take the quasi continuous values, the vectors $\vec{k}$ lie in the first cell of the reciprocal lattice. In a similar expansion for static displacements $\vec{u}_s = \vec{u}(\vec{R}_s)$,

$$\vec{u}(\vec{r}) = i \sum_{\vec{k}} A_{\vec{k}}(z) c_{\vec{k}} e^{-i \vec{k}_1 \vec{r}_1} ; \quad \vec{A}_{-\vec{k}}(z) = -\vec{A}_{\vec{k}}^*(z) \quad (4)$$

it is taken into account that the displacement waves are not plane waves ($\vec{A}_{\vec{k}}(z)$ depends on z), and unlike massive crystals, they are periodic only in the plane of layer XY ($\vec{k}_1$, $\vec{r}_1$) are the projections of $\vec{k}$, $\vec{r}$ on the plane XY), and their amplitudes are proportional to $c_{\vec{k}}$.

The displacements $\vec{u}_{st}$, generated by a single defect and their Fourier components $\vec{U}_{\vec{q}t}$ are determined by formulas (2) to (4) applied to the case when $c_t = 1$ only for one position t, i.e. when $c_{\vec{k}} = N^{-1} \exp(i\vec{k}\vec{R}_t)$. In this case $\vec{U}_{\vec{q}t}$, after summation over $\vec{R}_{s1}$ and $\vec{k}_1$ is expressed as a sum over k_z, containing the Fourier components of $\vec{A}_{\vec{k}}(z)$ function. Substituting this sum into (1), we find that $I(\vec{Q})$ for a layer with a constant concentration of defects $\bar{c}(\vec{R}_t) = c$ is determined by formula[6]

$$I(\vec{Q}) = N_d e^{-2M} \sum_{k_z} | f\vec{Q} \vec{A}_{\vec{q}_1 k_z}(q_z) - \varphi_{\vec{q}} g(q_z - k_z)|^2, \quad k_z = \frac{2\pi}{D} n, \quad (5)$$

where $N_d = cN$ is the total quantity of the defects in the layer

$$\vec{A}_{\vec{k}}(q_z) = D^{-1} \int_0^D \vec{A}_{\vec{k}}(z) e^{iq_z z} dz, \quad g(t) = D^{-1} \int_0^D dz\, e^{itz} = \frac{e^{itD}-1}{itD}. \quad (6)$$

As shown below, in the case of free films with large area $2M \gg 1$, the expression (5) is exponentially small (as well as the intensity of regular reflection peaks), and in determining $I(\vec{Q})$, one cannot expand the intensity of diffuse scattering in powers of displacements. More general consideration (see § 5, 19 in [9]) shows that the appearing bell-shaped intensity distribution is described by formulas

$$I(\vec{Q}) = |f|^2 \sum_{s,s'} e^{i\vec{q}\vec{\rho}} e^{-T}, \quad T = T' + i\,T'' = c \sum_t (1-e^{i\vec{Q}\vec{U}_{ss't}} + i\vec{Q}\vec{U}_{ss't}). \quad (7)$$

where $\vec{\rho} = \vec{R}_s - \vec{R}_{s'}$, $\vec{u}_{ss't} = \vec{u}_{st} - \vec{u}_{s't}$, $\vec{R}_s$ corresponds to an "average" crystal distorted by defects, and it is taken into account that in studies of peak region one can put $\varphi_{ts} = 0$.

The coefficients $\vec{A_k}(z)$ in (4) for small k are determined by equilibrium equation of the theory of elasticity

$$\frac{\partial \sigma_{ij}}{\partial \chi_j} = 0, \quad \sigma_{ij} = \sum_{l,m=x,y,z} c_{ijlm} \left[\frac{\partial u_l}{\partial \chi_m} - L_{lm} c(\vec{r})\right], \tag{8}$$

where $\hat{\sigma}$, $\hat{c}$, $\hat{L}$ are the tensors of stresses, elastic moduli and self-strain produced by defects, respectively. Substituting expressions (3) and (4) corresponding to k-th wave instead of $c(\vec{r})$ and $u(\vec{r})$ we obtain the system of three linear differential equations of the second order for determining $A_{\vec{k}j}(z)$. The boundary conditions for these equations in the case of films with free boundaries are obtained from the conditions $\sigma_{iz} = 0$ for z = 0,D, and in the case of films coupled with a substrate, the condition for z = D is reduced to the requirement of continuity of displacements.

In the general case of elastically anisotropic crystals and defects of arbitrary symmetry, the solution of these equations takes the form

$$A_{\vec{k}j}(z) = \sum_{\nu=1}^{6} B_{\vec{k}j\nu} \exp(-p_\nu z) + E_{\vec{k}j} \exp(-ik_z z), \tag{9}$$

where p_ν, $B_{\vec{k}j\nu}$ and $E_{\vec{k}j}$ are determined in[6].

In the case of elastically isotropic crystals and cubic symmetry of defects, the equations and boundary conditions $\sigma_{iz} = 0$ for $A_{\vec{k}x}(z)$, $A_{\vec{k}z}(z)$ (the axis x is chosen along $\vec{k}_1$) take a simpler form:

$$(1-2\sigma)A_{\vec{k}x}'' - ik_x A_{\vec{k}z}' - 2(1-\sigma)k_1^2 A_{\vec{k}x} = -2(1+\sigma)L_{xx}k_x \exp(-ik_z z),$$

$$2(1-\sigma)A_{\vec{k}z}'' - ik_x A_{\vec{k}x}' - (1-2\sigma)k_1^2 A_{\vec{k}z} = -2(1+\sigma)L_{xx}k_z \exp(-ik_z z), \tag{10}$$

$$iA_{\vec{k}x}' + k_x A_{\vec{k}z} = 0, \quad i(1-\sigma)A_{\vec{k}z}' + \sigma k_x A_{\vec{k}x} = (1+\sigma)L_{xx}, \quad z = 0,D \tag{11}$$

where $A_{\vec{k}j}' = dA_{\vec{k}j}/dz$; σ is Poisson's ratio. The characteristic equation for the system (10) has two pairs of multiple roots $p_1 = p_2 = k_1$, $p_3 = p_4 = -k_1$. Thus in the special case of elastically isotropic crystals the solution of equations for $A_{\vec{k}j}(z)$ is not described by formula (9) and takes the form

$$A_{\vec{k}j}(z) = B_{j1}z + B_{j2}) e^{-k_1 z} + (B_{j3}z + B_{j4}) e^{k_1 z} + E_j e^{-ik_z z}, \quad j = x,z. \tag{12}$$

Here

$$E_x = \frac{k_x}{k^2} a, \quad E_z = \frac{k_z}{k^2} a, \quad a = \frac{1+\sigma}{1-\sigma} L_{xx},$$

$$B_{x1} = \alpha B_{z1} = -\frac{k_x}{2k^2} (e^{k_1 D}-1)\left\{\frac{k_1}{shk_1 D+k_1 D} - \frac{i k_z}{shk_1 D-k_1 D}\right\} a, \quad \alpha = i\frac{k_x}{k_1},$$

$$B_{x2} = \frac{DB_{x1}}{e^{k_1 D}-1} + \frac{1}{2k^2} (e^{k_1 D}-1)\left\{\frac{(1-2)k_1}{shk_1 D+k_1 D} - \frac{2i(1-)k_z}{shk_1 D-k_1 D}\right\} \frac{a k_x}{k_1},$$

$$B_{x3} = -\alpha\, B_{z3} = -B_{x1}^{*}\, e^{-k_1 D}, \qquad B_{x4} = (B_{x2}^{*} + B_{x1}^{*} D)\, e^{-k_1 D},$$

$$k_1 B_{z2} = -(3-4\sigma)\alpha\, B_{x1} - \alpha\, k_1 B_{x2},$$

$$k_1 B_{z4} = -(3-4\sigma)\alpha\, B_{x3} + \alpha\, k_1 B_{x4}. \tag{13}$$

In the limit of a semi-infinite crystal $(D \to \infty)$ $B_{j3} = B_{j4} = 0$ and in formula (12) the second term disappears. In the depth of the crystal with $k_1 z \gg 1$ only the last term is preserved in (12) and expression (4) coincides with the result for the infinite crystal. In the case of thin free films with $k_1 D \ll 1$ and $k_z = 2\pi n/D \ne 0$

$$A_{\vec{k}z}(z) \approx \frac{6(1-\sigma)a}{\pi n D\, k_1^{2}} \propto \frac{1}{k_1^{2}} \tag{14}$$

instead of the dependence $A_{\vec{k}} \propto 1/k$ in massive crystals. Such strong divergence of $A_{\vec{k}z}$ is associated with a well-known pliability of plates to the long-wave flexural strains (the latter are produced by the waves of concentration with $k_z \ne 0$). In the case of films tightly coupled with a substrate, $A_{\vec{k}}(z)$ for $k_1 D \ll 1$, on the contrary, tends to a finite limit[6].

THE ANALYSIS OF SCATTERING IN FILMS

The intensity of diffuse scattering by cubic defects in elastically isotropic films is determined by formula (5), in which according to (6) and (12)

$$A_{\vec{q}_1 k_z j}(q_z) = E_j g(q_z - k_z) + D B_{j1} g_1 (q_z + i q_1) +$$

$$+ D B_{j3} g_1 (q_z - i q_1) + B_{j2} g (q_z + i q_1) + B_{j4} g (q_z - i q_1),$$

$$g_1(t) = \frac{i t\, e^{itD} + 1 - e^{itD}}{(itD)^{2}} \tag{15}$$

where the values $E_j \equiv E_j(\vec{k})$ and $B_j \equiv B_j(\vec{k})$ are taken for $\vec{k}_1 = \vec{q}_1 || \vec{x}$, $k_z = 2\pi n/D$.

In the limit of massive crystals $(q_1 D \gg 1)$ in (15) only the first term is substantial and with consideration for (13) expression (5) is reduced to a well-known formula of Huang[11]. In the range of $q_1 D \le 1$ the intensity distribution is substantially changed. It qualitatively differs in the cases of free and substrate – coupled films and substantially depends on the orientation of the reciprocal lattice vector $\vec{G}$ (i.e. $\vec{Q} \approx \vec{G}$) relative to the film plane. In the case of free films and $\vec{G}$ lying in the film plane (the scattering is connected only with the displacement vector and $A_{\vec{q}_1 k_z}(q_z)$ components parallel to this plane) for $q_1 D \ll 1$, $q_z D \ll 1$ according to (5), (13) and (15)

$$I = I_H + I_a + I', \quad I_H = (1-\sigma)^2 \, N_d \, e^{-2M} |f|^2 \, a^2 \, \frac{(\vec{Q}\vec{q}_1)^2}{q_1^4},$$

$$I_a = -2(1-\sigma)N_d \, e^{-2M} \, a \, \frac{\vec{Q}\vec{q}_1}{q_1^2} \, \mathrm{Re}\,(f\varphi_q^*),$$

$$I' = N_d \, e^{-2M} |\varphi_q|^2, \qquad\qquad (qD \ll 1, \, G_z = 0). \qquad (16)$$

Here I_H, I' and I_a describe respectively the scattering by distortions, by structure amplitude changes and the interferential effects. According to (16), in the vicinity of the reciprocal lattice points with $G_z = 0$ for $qD \ll 1$ as well as in massive crystals, I_H and I_a increase proportionally to q^{-2} and q^{-1}. However, as in massive crystals $I_H \propto (\vec{Q}\vec{q})^2 q^{-4}$ and isodiffuse surfaces are in the form of two touching spheres, in elastically iso-tropic films $I_H \propto q_1^{-2}$ is independent of q_z in the range of small q_1, $|q_z|$ and isodiffuse surfaces are in the form of two touching cylinders. In the range of $q_1 D \ll 1$, $|q_z| D \gg 1$ the intensity $I_H \propto q_1^{-2} q_z^{-2}$ decreases as q^{-4}, and isodiffuse curves in the plane $\vec{q}_1$, q_z are close to hyperbolas.

In the neighbourhood of the points with $G_z \neq 0$, I_H and I_a are mainly determined by the z-th projection of the vector $\vec{A}_{q_1 k_z}(q_z)$ which in free thin films is connected with flexural deformations and, as seen from (14), sharply increases for small q_1. As follows from (5), (13) and (15), in this case

$$I_H = 12(1-\sigma)^2 \, N_d \, e^{-2M} |f|^2 \, \frac{Q_z^2}{D^2 q_1^4} \, |g(q_z)|^2 a^2 \qquad (17)$$

$$(q_1 D \ll 1, \, G_z \neq 0, \, 2M \lesssim 1)$$

According to (17), for small $q_1 D \ll 1$ and $|q_z| D \ll 1$, I_H strongly de-pends on q_1 like q_1^{-4} and is independent of q_z. I_H rapidly decreases with increasing $|q_z|$ for $|q_z| D \gg 1$ (and $q_1 D \ll 1$) as $q_1^{-4} q_z^{-2}$.

The "infrared divergence" of the amplitudes of static displacements at small q leads not only to a sharp peculiarity of $I_H \propto q^{-4}$ at $q \to 0$, but also to a sharp increase of 2M exponent of $\exp(-2M)$ factor. From the formula for the estimation of 2M in the case of not very large local displacements around defects ($|\vec{Q}\vec{u}_{ts}| < 1$)[6]

$$2M \sim \frac{r_o \, D'^2}{D^3} \, ca^2 h_z^2 \qquad\qquad (G_z \neq 0) \qquad (18)$$

it is seen that 2M is proportional to the film area D'^2 (r_o is the inter-atomic distance, h_z is the Miller index). For example, for $ca^2 h_z^2 = 10^{-3}$,

$r_o/D \sim 10^{-2}$ and $D'^2/D^2 \sim (10^4 \div 10^8)$ according to (18) $2M \sim (10^{-1} \div 10^3)$, i.e. 2M can be both less or greater than unity.

Formulas (5) and (17) are true only in the case of a small film area, when $2M \leq 1$. For $2M \gg 1$ (in particular, in the $D' \to \infty$ limit), the δ-like peaks of regular reflections disappear and the diffuse scattering intensity $I(\vec{Q})$ is condensed into the bell-shaped peaks apprehended as broadened regular reflections. To determine the scattering intensity $I(\vec{Q})$ by free films in the neighbourhood of the reciprocal lattice points with $G_z \neq 0$, one needs to proceed not from formula (5) but from a more general formula (7). In this case a main contribution to expression (7) for $T \equiv T(\vec{R}_s, \vec{R}_{s'})$ is made by defects in the positions t withdrawn from the points s and s', for which $R_{st} \sim D' \gg D$, ρ and $u_{ss'tz} \approx \vec{\rho}_\perp \partial u_{stz}/\partial \vec{R}_{t\perp}$ are proportional to $\rho_\perp/R_{t\perp}$ ($\vec{\rho}_\perp$ and $\vec{R}_{t\perp}$ are the projections $\vec{\rho}$ and $\vec{R}_t$ on the plane XY). Here $|T''| \ll T'$, and the calculation of T' can be carried out as in the case of rectilinear dislocations[9,6]. In the actual region $D' \gg \rho_\perp \gg D$.

$$T' = \frac{1}{4\pi\ell} q_i^2 \rho_\perp^2 \ln\frac{D'}{\rho_\perp} , \qquad q_i^2 = \frac{6(1-\sigma)^2 c v Q_z^2 a^2 \ell}{D^3} , \ell = \ln(D' q_i) . \quad (19)$$

Substituting expression (19) into (7) and taking the summation with respect to s, s' (in the XY plane in the infinite limits, and along z-axis in the finite interval $0 < z < D$), we obtain a broadened distribution of $I(\vec{Q})$ near a reciprocal lattice point with $G_z \neq 0$:

$$I(\vec{Q}) = \frac{16\pi^2 N}{v} |f|^2 \frac{\sin^2(q_z D/2)}{Dq_z^2} \frac{1}{q_i^2} \exp\left[-\frac{\pi(q_x^2+q_y^2)}{q_i^2} \right] \quad (20)$$

This rod-shaped distribution has the integral width q_i and the Gaussian shape (in the central part) along the directions q_x and q_y, and much larger width $2\pi/D$ and the Laue shape along q_z. The ratio of the widths $q_i D/2\pi$ has the order $(c r_o/D)^{\frac{1}{2}} h_z a$, i.e. $\sim 10^{-2}$ for $r_o/D \sim 10^{-2}$, $\sqrt{c} h_z a \sim 10^{-1}$. In the region of distribution touls, for $q_\perp \gg q_i$, instead of the Gaussian function (20), $I(\vec{Q})$ is described by the power function of $q_\perp$. As in the case of $2M \ll 1$, it is determined by formula (17) in which $2M = 0$ is put.

The intensity distribution pattern is totally different if films are not free, but coupled with a subsstrate. In this case the anomalously large flexural deformations do not occur, 2M in the $D' \to \infty$ limit remains finite and the peaks of regular reflections are not broadened. The tight coupling of a thin film with a substrate substantially suppresses distortions in the film. As a consequence, the diffuse scattering intensity for small q strongly decreases. For example, in the limiting case of an absolutely rigid substrate for the reflections with $G_z = 0$, the intensities I_H and I_a do not increase like in massive crystals, but tend to zero ($I_H \propto q_\perp^2$ and $I_a \propto q_\perp$). For the reflection with $G_z \neq 0$, I_H and I_a tend to the finite limiting values.[6]

From the above results it follows that for small $q_\perp D$ the relaxation of distortions near the surface substantially changes the dependence of I both on the modulus and the direction of the vector $\vec{q}$. Thus, the analysis of defects, based on the study of angular dependences of I, must be substantially changed as compared to the case of massive crystals. On the other hand, using the data on the scattering by defects in films in the $q_\perp D \gg 1$ region, one can use the formulas obtained for massive crystals and carry out the analysis of defects based on these formulas.

The diffuse scattering intensity by defects of a surface layer can be determined with formula (1) by expressing the Fourier component of static displacements $\vec{U}_{\vec{q}t}$ through the static displacement wave amplitudes $\vec{A}_{\vec{k}}(z)$. In the simplest case of cubic defects in the elastically isotropic crystals, $Q\vec{U}_{\vec{q}t}$ is described by formula[7]

$$\vec{Q}\,\vec{U}_{\vec{q}t} = \frac{\vec{Q}\vec{q}}{q^2}\, a - \Phi\,(\vec{Q},\,\vec{q})\, a\, \exp(-q_\perp R_{tz} - iq_z R_{tz}),$$

$$\Phi(\vec{Q},\,\vec{q}) = \left[\frac{1}{q_\perp + iq_z} - \frac{3-4\sigma}{q_\perp - iq_z}\right]\left[\frac{\vec{Q}_\perp\vec{q}_\perp}{q_\perp} + i\,Q_z\right] + \frac{2q_\perp}{(q_\perp - iq_z)^2}\left[\frac{\vec{Q}_\perp\vec{q}_\perp}{q_\perp} - i\,Q_z\right]. \quad (21)$$

Substituting this expression into (1), we obtain

$$I_H = N_d\, e^{-2M}|f|^2\, a^2\, \left\{\frac{(\vec{Q}\,\vec{q})^2}{q^4} - 2\,\frac{(\vec{Q}\,\vec{q})}{q^2}\,\mathrm{Re}\big[c_L(q_\perp + iq_z)\Phi\big]\, c_L^{-1}(0) + \right.$$

$$\left. c_L(2q_\perp)\, c_L^{-1}(0)\, |\Phi|^2\right\},$$

$$I_a = -2N_d\, e^{-2M}a\, \mathrm{Re}\left\{f\varphi_{\vec{q}}^{*}\left[\frac{\vec{Q}\,\vec{q}}{q^2} - c_L(q_\perp + iq_z)\, c_L^{-1}(0)\Phi\right]\right\},$$

$$c_L(p) = \int_0^\infty e^{-pR_{tz}}\,\overline{c}(\vec{R}_t)d\,R_{tz}. \quad (22)$$

Formulas (21) and (22) determine the diffuse scattering intensity distribution near the reciprocal lattice points for the arbitrary $q_\perp D_e$, where D_e is the effective width of the layer with defects (i.e., the width of $c(\vec{R}_t)$ function). If $q_\perp D_e \gg 1$, then in formulas for I_H and I_a only the first terms are substantial and (22) coincide with the formulas for massive crystals. In the case of thin layers $c_L(q_\perp + iq_z) \approx c_L(2q_\perp) \approx c_L(0)$ and expressions (22) are simplified. The component I_H (being the main one for $q \to 0$); the formula for I_a is given in[6] and is described by the formula

$$I_H = N_d\, e^{-2M}|f|^2\, \frac{a^2}{q^4}\left\{\frac{(\vec{Q}_\perp\vec{q}_\perp)^2}{q_\perp^2}\,\left[4(1-\sigma)^2\, q_z^2 + (1-2\sigma)^2\, q_\perp^2\right] - \right.$$

$$\left. -2(3-4\sigma)(\vec{Q}_\perp\vec{q}_\perp)\, Q_z q_z + Q_z^2\,\left[(1-2\sigma)^2\, q_z^2 + 4(1-\sigma)^2\, q_\perp^2\right]\right\}. \quad (23)$$

According to (23) on scattering by the surface layer of defects, analogously to the case of a homogeneous distribution of defects in a crystal, $I_H \propto q^{-2}$. But the dependence of I_H on the orientation of $\vec{q}$ for small $q_\perp$ substantially changes. The isodiffuse curves $I_H(\vec{Q})$ = const are now oval in form as distinguished from the lemniscate-shaped curves for massive crystals, and the isodiffuse surfaces take the form of ellipsoids with the centres at the reciprocal lattice points as distinguished from the pair of spheres touching at these points. For example, for reflections with $\bar{G} \parallel \bar{X}$ or $\vec{G} \parallel \vec{Z}$, the relation of ellipsoid axes lengths perpendicular and parallel to $\vec{G}$ is $(1-2\sigma)(2-2\sigma)^{-1}$ (i.e. $\frac{1}{4}$ for $\sigma = 1/3$), and the isodiffuse surfaces are strongly elongated along $\vec{G}$.

Thus, the change of the shape of isodiffuse surfaces should be observed at $q_1 \sim D_e$. It can be used for the estimation of D_e.

REFERENCES

1. R. Bachmann, H. Kohler, H. Schultz, and H. P. Weber, Structure investigation of a 6 μm CaF_2 crystal with synchrotron radiation, Acta Cryst. A41:35 (1985).

2. T. J. Matsubare, Theory of diffuse scattering of X-rays by local lattice distortions, J. Phys. Soc. Japan 7:270 (1952).

3. H. Kanzaki, Point defects in four-centered cubic lattice. I. Distortion around defects, J. Phys. Chem. Solids 2:24 (1957).

4. M. A. Krivoglaz, Theory of diffuse scattering at X-ray and thermal neutrons by solid solutions. III Role of distortions, Zh. Eksp. Teor. Fiz. 34:204 (1958) (English translation: Sov. Phys. – JETP 34:139 (1958).

5. I. N. Sneddon, F. J. Lockett, On the steady-state thermoelastic problem for the half-space and the thick plate, Quarterly Appl. Math. 18:145 (1960).

6. R. I. Barabash, M. A. Krivoglaz, Static fluctuation waves and scattering of X-rays or thermal neutrons by defects in thin films and surface layers, Preprint of the Institute of Metal Physics, N 11, Kiev (1987),

7. V. B. Molodkin, S. I. Olikhovski, and M. E. Osinovski, Dynamical theory of diffuse scattering of X-rays and electrons in crystals with Coulomb-type defects, Metallofizika 5:3 (1983) (English translation Physics of Metal 5:1 (1984).

8. V. M. Kaganer, V. L. Indenbom, The analysis of diffuse scattering with the effects of dynamic diffraction, Metallofizika 8:25 (1986).

9. M. A. Krivoglaz, "Diffraction of X-rays and Neutrons by Imperfect Crystals", Naukova Dumka, Kiev (1983) (in Russian).

10. L. D. Landau, E. M. Lifshitz, "Theory of Elasticity", Pergamon Press, Oxford (1970).

11. K. Huang, X-ray reflections from dilute solid solutions, Proc. Royal Soc. 190:102 (1947).

METHODICAL ASPECTS OF X-RAY DIFFRACTION ANALYSIS

OF SURFACE TREATED MATERIALS

Heinrich Oettel

Department of Material Science
Mining Academy
Freiberg 9200, GDR

INTRODUCTION

In modern material engineering, surface treatment by thermal, thermo-
chemical or physical methods represents an important technology for improving
the fatigue behaviour, wear properties and corrosion resistance of materials.
Such technology is also of importance for the production of special functional compounds. Therefore in the last few years the interest in microstructure analysis of materials with modified surfaces has increased considerably. Using X-ray diffraction analysis it is necessary to take into
account, that frequently the structure parameters are changed within the
mean penetration depth T of the X-rays. For that reason the phase concentration, lattice parameters, residual stresses, line breadth or pole density,
estimated in usual manner, represent absorption weighted effective parameters, different from the true surface parameters. This effect is significant if the condition $T \geq m_y/(dy/dx)$ is valid (m_y - error limit of the experimental determination of y, x - surface distance). In case of BRAGG-BREN-
TANO and SEEMANN-BOHLIN diffractometry the penetration depth can be calculated according to

$$T = \{\mu[\sin^{-1}(\vartheta + \alpha) + \sin^{-1}(\vartheta - \alpha)]\}^{-1}, \tag{1}$$

where ϑ is the glancing angle, α is the angle between surface normal and
diffraction vector within the diffraction plane and μ is the linear absorption coefficient.

The effective parameters y_{eff} can be estimated by integration over all
contributions of infinitesimal layers of the whole specimen with the thickness x_0 (see Figure 1)[1]. For instance, in case of varying phase concentration $c(x)$ perpendicular to the specimen surface this integration yields

$$c_{eff} = \int_0^{x_0} c(x).e^{-x/T} dx / \int_0^{x_0} e^{-x/T} dx. \tag{2}$$

The conventional formalism, applied in quantitative phase analysis, results
in c_{eff} instead of $c(0)$. However, c_{eff} can be quite different from $c(0)$.
It often leads to the misinterpretation of exprimental results. For example,
Figure 2a shows the results of the phase analysis of a nitrided steel, carried out with copper radiation with a small penetration depth and cobalt
radiation with a large penetration depth. The differences amount up to
50 %!

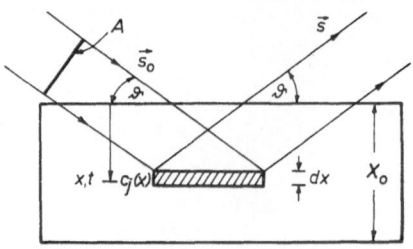

Fig. 1. Diffraction of an infinitesimal layer in the depth x

Definitions of the same kind can be obtained for d-values $d(x)$ or lattice parameters $a(x)^2$, residual stresses $\sigma(x)^{3,4}$, pole densities $\Omega(x)$ and reciprocal line breadths $B^{-1}(x)^{5,6}$. In general, the definition of the corresponding effective parameters is given by

$$y_{eff} = \int_0^{x_o} y(x) \cdot e^{-x/T} \, dx \Big/ \int_0^{x_o} e^{-x/T} \, dx \; . \tag{3}$$

The aim of a complete diffraction analysis of surface modified specimens consists of a determination of the true parameters $y(x)$ as a function of surface distance x, calculated from the measured effective parameters y_{eff}. For these two possibilities can be offered:

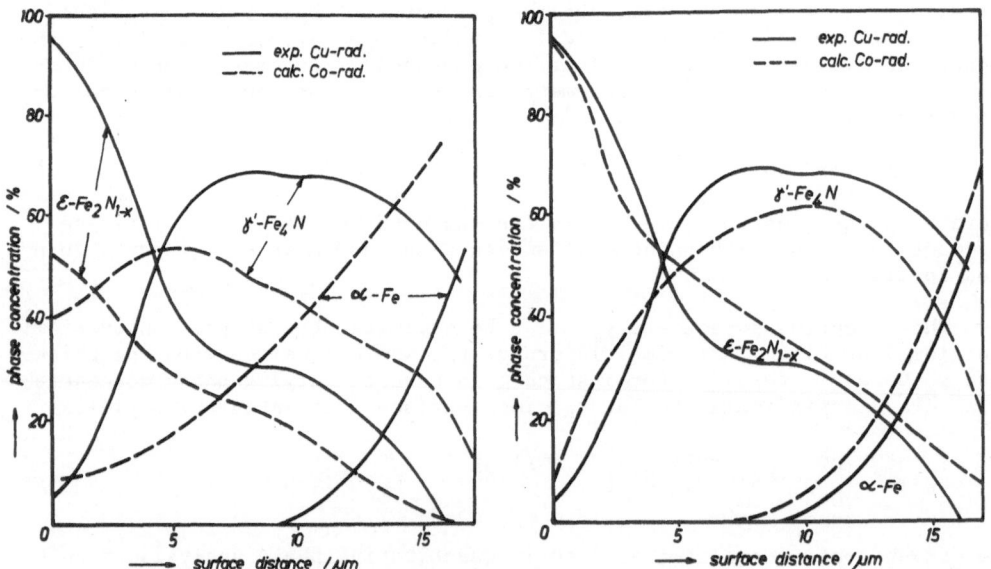

Fig. 2. Phase analysis of a nitride layer
(a) comparison of experimental results obtained by Cu-radiation (full lines) and Co-radiation (broken lines)
(b) comparison of experimental results of Cu-radiation (full lines) and corrected results of Co-radiation (broken lines).

172

Variant I: variation of the penetration depth T by changing the wavelength (absorption coefficient) and/or changing the angle α (see formula 1)

$$y_{eff}(T) = \int_0^{x_o} y(x).e^{-x/T} \, dx / \int_0^{x_o} e^{-x/T} \, dx \qquad (4a)$$

Variant II: estimation of effective parameters after step by step removal of layers with the thickness z

$$y_{eff}(z) = \int_z^{x_o} y(x).e^{-x/T} \, dx / \int_z^{x_o} e^{-x/T} \, dx \; . \qquad (4b)$$

In principle, the experimental determination of either $y_{eff}(T)$ or $y_{eff}(z)$ allows the calculation of the true parameter function $f(x)$. In the next section these two possibilities will be discussed in more detail.

INTERPRETATION OF EFFECTIVE PARAMETERS

Let us start with a short consideration of well defined layer structures, characterized by box-type distributions of the phase concentrations ($c_j = 1$ inside and $c_j = 0$ outside of the layer j). In this case each layer j with thickness x_j produces its own interference with the intensity

$$I_j = K.I_o.G_j.V_j(1-e^{-x_j/T}) \exp[-2 \sum_{k=1}^{j-1} \mu_k x_k / \sin \vartheta_j], \qquad (5)$$

where K is a constant, I_o the primary beam intensity, G_j is the theoretical intensity factor, A is the cross section of the primary beam and V_j is the effective irradiated volume of an infinite thick sample, consisting of the phase j (e.g. in case of $\alpha = 0, V_j = A/2 \mu_j$).

As shown by Hejdová et al.[7,8] on the base of this relation it is possible to derive the thickness x_j of all the layers. Supposing a phase composition independent on the surface distance inside of the layer both the layer thickness and the phase concentrations can be estimated [9,10].

Normally surface modified specimens cannot be described by such simple box-type phase distributions. The c(x) are continuous functions of the surface distance x, which is true also for lattice parameters, stress, line breadth and pole density. A more general treatment of the problem was given by the author[1], approximating y(x) by Fourier series respectively by power series. Doing this the integration according to relation (3) is a feasible one. In the case of the Fourier series expansion according to

$$y(x) = A_o + \sum_n [A_n \cos(2\pi \, nx/x_o) + B_n \sin(2\pi \, nx/x_o)] \qquad (6)$$

the effective parameter in dependence on T (variant I) is given by

$$y_{eff} = A_o - T^{-1} \sum_n A_n^* \qquad (7)$$

with

$$A_n^* = -(A_n \, p_n + B_n \, q_n)$$

173

and

$$p_n = (T^{-1} + 4\pi^2 T x_o^{-2}); \qquad q_n = (x_o/(2\pi nT^2) + 2\pi n/x_o)^{-1}.$$

Using the layer removal technique (variant II) we find the relation

$$y_{eff}(z) = A_o + \frac{\sum_n A_n^*(e^{-x_o/T} - e^{-z/T}\cos(2\pi n\frac{z}{x_o})) - e^{-z/T}\sum_n B_n^* \sin(2\pi n\frac{z}{x_o})}{T(e^{-z/T} - e^{-x_o/T})} \qquad (8)$$

$$B_n^* = A_n q_n - B_n p_n.$$

The estimation of a sufficient number of $y_{eff}(T)$ respectively $y_{eff}(z)$ enables the calculation of the A_n, B_n by means of a linear equation system.

The expansion of $y(x)$ in a power serie (see also Zevin et al.[11])

$$y(x) = \sum_m a_m x^m \qquad (9)$$

allows a similar procedure. The integration yields in the case of T-variation and with the abbreviation $t_o = x_o/T$

$$y_{eff}(T) = \frac{\sum_m a_m T^m e^{-t_o} (t_o^m + mt_o^{m-1} + m(m-1)t_o^{m-2} + \ldots + m!) - \sum_m a_m T^m m!}{e^{-t_o} - 1} \qquad (10)$$

and in the case of layer removal technique with $t = x/T$

$$y_{eff}(z) = \frac{[\sum_m a_m T^m e^{-t}(t^m + mt^{m-1} + m(m-1) t^{m-2} + \ldots + m!)]_z^{x_o}}{e^{-t_o} - e^{-z/T}} . \qquad (11)$$

The coefficients a_m of the power series (9) follow from measurements of different $y_{eff}(T)$ or $y_{eff}(z)$ by solution of appropriate equation systems again.

Two examples will demonstrate the applicability of the relations (10) and (11). At first, the Co-radiation results of the quantitative nitride phase analysis, shown in Figure 1a, were corrected with the help of relation (11). After this we find a satisfactory coincidence within the error limits between the corrected phase concentrations and the results of the measurement with Cu-radiation (see Fig. 2b), which yields nearly the true values due to the small penetration depth of about 1 µm only.

Secondly, nickel specimens were investigated after surface deformation by fine grinding with two different abrasive papers (grain size number 400 and 600). Line breadth measurements were performed at the 311-interference using Cu-Kα₁-radiation and a thin film diffractometer with SEEMANN-BOHLIN focusing geometry(Fa Huber, FRG). With this diffractometer the angle of incidence of the primary beam relative to the specimen surface could be changed between 2 and 10°, corresponding to a variation of the penetration depth between about 0.75 and 3.5 µm. Figure 3 summarizes the results. The curves with empty marks represent the measured effective breadth ß(T),

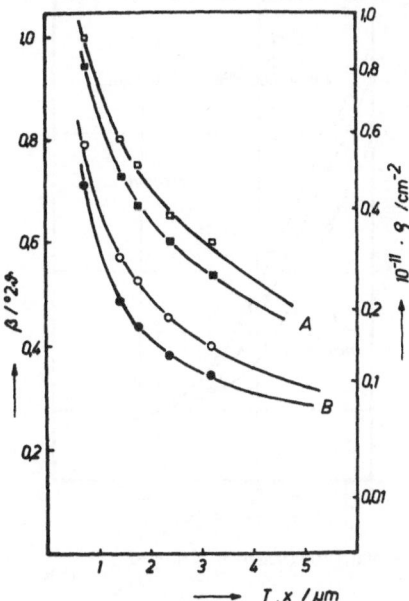

Fig. 3. Line breadths and dislocation densities of fine grinded Ni
A – grain size number 400, B – grain size number 600, empty marks:
effective values, full marks: corrected values.

corrected for instrumental broadening, the curves with full marks the true
breadth $ß(x)$, recalculated from $ß(T)$ according to relation (10). The dislocation density (for evaluation see ref. 12,13,14) near the surface reaches
about 10^{11} cm^{-2} and drops down to 10^{10} cm^{-2} within a range of 3 respectively
6 μm. Using ordinary BRAGG–BRENTANO diffractometry, the penetration depth
is $\approx$ 12 μm and therefore the effective line breadth of both specimens is
nearly the same ($\simeq$ 0.2° 2ϑ), not reflecting clearly the difference in dislocation densities near the surface.

If we approximate the function $y(x)$ linearly by

$$y(x) = y(0) + a.x ,$$ (12)

the simple relation

$$y_{eff}(0,T) = y(0) + a.T. \Phi(\frac{x_o}{T})$$ (13)

with

$$\Phi(\frac{x_o}{T}) = [1 - e^{-x_o/T} (\frac{x_o}{T} + 1)] / (1 - e^{-x_o/T})$$

can be obtained. Measurements with two different penetration depths T_1 and
T_2 now allow the estimation of

$$a = (y_{eff}(0,T_2) - y_{eff}(0,T_1))/(T_2 \Phi(\frac{x_o}{T_2}) - T_1 \Phi(\frac{x_o}{T_1}))$$ (14a)

and

$$y(0) = y_{eff}(0,T_1) - a.T_1. \Phi(\frac{x_o}{T_1}) .$$ (14b)

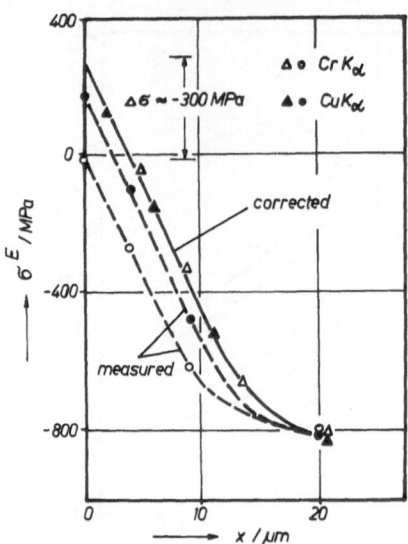

Fig. 4. Residual stress measurements in ε-Fe$_2$N$_{1-x}$ of nitrided 20 MnCr5

For example, measurements of residual stresses in ε-Fe$_2$N$_{1-x}$ of the compound layer of nitrided steel 20 MnCr5 (x_o = 25 µm) show + 175 MPa in case of Cu-radiation (T = 1.7 µm) and -20 MPa in case of Cr-radiation (T = 5.1 µm). From this we derive a = -57 MPa/µm and $\sigma(0)$ = +270 MPa (see also Figure 4). It is obvious that under these conditions the true surface stress can be determined directly neither by Cu- as by Cr-radiation. Possibly the dis-crepancies in results of residual stress measurements in nitride layers, published in literature (see [15]), are caused by neglecting this effect.

Furthermore, in connection with relation (11) and (12) we find

$$y_{eff}(z) = y(z) + a(z).T.\Phi\left(\frac{x_o-z}{T}\right) = y\left(z + T.\Phi\left(\frac{x_o-z}{T}\right)\right). \qquad (15)$$

Equation (15) means that the effective parameter $y_{eff}(z)$, measured at the removed layer thickness z, corresponds to the true parameter at $z + T.\Phi\left(\frac{x_o-z}{T}\right)$. Therefore a simple shift of the z-axis by $T.\Phi\left(\frac{x_o-z}{T}\right)$ yields a first approximation for y(x) (Figure 5a). In Figure 4 this procedure was applied to residual stress measurements in the ε-phase mentioned above [16]. After transformation the curves obtained with Cu- and Cr-radiation coincide very well, giving us the true stresses in a good approximation. Note that in this example also a correction is included concerning apparent stresses due to stress gradients (see section 3).

Another correction procedure is based on the analysis of the slope y'(z) and $y'_{eff}(z)$. According to (15) the relation

$$a = y'(z) = y'_{eff}(z) - a.T.d\Phi\left(\frac{x_o-z}{T}\right)/dz , \quad \text{respectively}$$

$$a = y'_{eff}(z).\left(1 + T.d\Phi\left(\frac{x_o-z}{T}\right)/dz\right)^{-1}$$

is valid. Finally we obtain

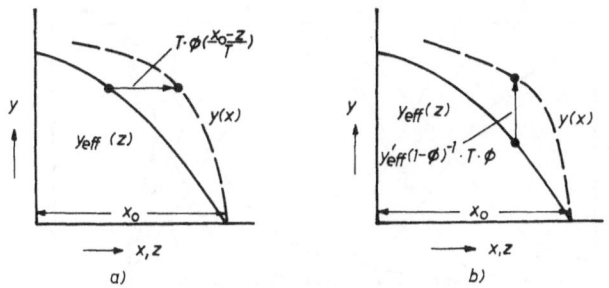

Fig. 5. Linear correction procedures
(a) shift correction, (b) slope correction

$$y(z) = y_{eff}(z) - y'_{eff}(z).(1-\Phi'(\frac{x_o-z}{T}))^{-1}.T.\Phi(\frac{x_o-z}{T}) , \qquad (16)$$

where

$$\Phi'(\frac{x_o-z}{T}) = d\Phi(\frac{x_o-z}{T})/d(\frac{x_o-z}{T}) .$$

Because Φ and Φ' are known (see Figure 6), the determination of $y'_{eff}(z)$ enables the estimation of $y(z)$. In Figure 7a the results of such a slope correction of computer-simulated effective phase concentration profiles are shown. The true concentration distributions for ϵ-Fe$_2$N$_{1-x}$, γ'-Fe$_4$N and α-Fe were chosen trapezoidally. For comparison Figure 7b demonstrates the results of a correction procedure according to relation (11). It is clear that the slope correction yields similar success like the more complicated polynom correction.

PROFILE SIMULATIONS

The considerations of section 2 suppose two main simplifications: first-ly, only one structure parameter is changed with the surface distance and secondly, the line position is given by the centre of gravity. In practice, e.g. the experimental determination of phase concentrations or line breadths is influenced also by variable pole densities. Also residual stress measure-ments can be affected by variable lattice parameters due to solid solution

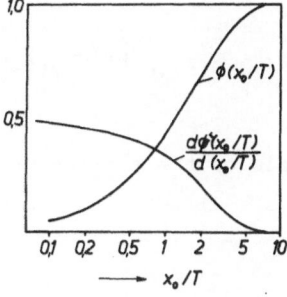

Fig. 6. Correction functions $\Phi(\frac{x_o}{T})$ and $\Phi'(\frac{x_o}{T})$.

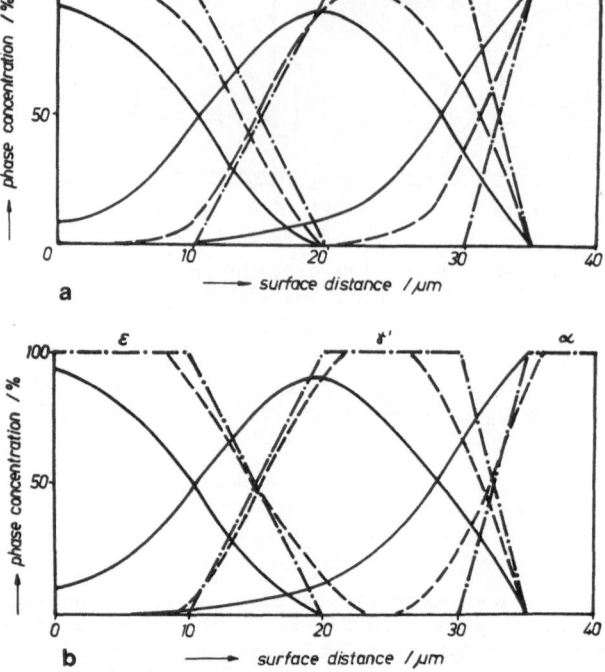

Fig. 7.   Computer-simulated effective phase concentrations in a nitride
          layer and their corrections
          (a) slope correction; (b) polynom correction (full lines: effective
          concentrations; broken lines: corrected concentrations)

effects, by variable phase concentration, pole density and line breadth.
Furthermore, line position measurements are carried out by determination of
the line maximum instead of the centre of gravity. Therefore, it is necessary to calculate the complete intensity distribution diffracted by a specimen, including all microstructure parameters, which are dependent upon surface distance, such as lattice parameters $a(x)$, residual stresses $\sigma(x)$, line
breadth $B(x)$, relative pole densities $\Omega(x)$, phase concentrations $c(x)$, the
mean linear absorption factor $\mu(x)$ and the theoretical intensity factor $G(x)$.
($G(x)$ represents the product of structure factor, Lorentz-polarization factor,
multiplicity factor and the reciprocal elementary cell volume squared.) As
pointed out in [ ], each layer in the depth x and with the thickness dx produces
an infinitesimal profile at the position $\vartheta_o + \Theta(x)$ and the breadth $B(x)$ (see
Figure 8). After integration we find

$$I(\vartheta) = \frac{K.I_o.A}{\sin(\vartheta\pm\alpha)\cos\psi} \int_0^x \frac{G(x)}{B(x)}.c(x).\Omega(x).e^{-\overline{\mu x}/\eta}.\varphi(\vartheta-[\vartheta_o + \Theta(x)])\ dx \qquad (17)$$

where ψ is the inclination angle of the surface normal to the diffraction
plane,

$$\overline{\mu x} = \int_0^x \mu(v)\ dv\ ,$$

$$\eta = \cos\psi\ [\sin^{-1}(\vartheta + \alpha) + \sin^{-1}(\vartheta - \alpha)]^{-1}\ ,$$

178

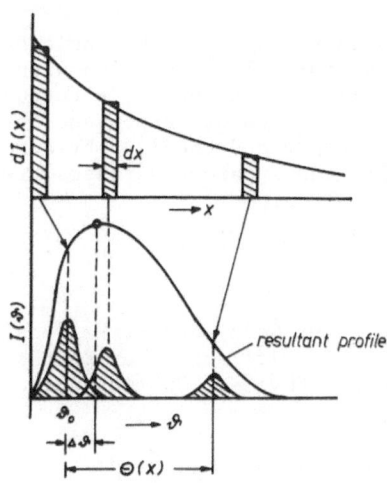

Fig. 8.   Contributions of infinitesimal profiles to the resultant profile

and $\varphi(\vartheta-[\vartheta_o + \Theta(x)])$ denotes the normalized distribution profile of an infinitesimal layer. It can be of Lorentzian, Gaussian or other type. The parameter $\Theta(x)$ describes the shift of the infinitesimal profile relative to the glancing angle ϑ_o at x = 0. In case of solid solutions with concentration gradients $\Theta(x)$ is determined by Vegard's rule.

$$\Theta(x) = -\frac{1}{a}\frac{da}{d\zeta}(\zeta(0) - \zeta(x))\tan\vartheta_o\frac{180}{\pi} ,$$

where ζ is a concentration of the solid solution component. If $\psi = \alpha = 0$, two-dimensional residual stresses produce a shift of

$$\Theta(x) = \frac{\gamma}{E}[\sigma_1(0) + \sigma_2(0) - \sigma_1(x) - \sigma_2(x)]\tan\vartheta\frac{180}{\pi} ,$$

where E, γ are the Youngs modulus and Poisson constant, respectively. The numerical evaluation of the equation (17) allows an accurate description of the interference profile of surface modified materials and a profitable simulation of the combined influence of different microstructure parameters on the profile. For example, the last Figure shows the results of the computer-simulation of apparent stresses in compound layers of nitrided steels, which are characterized by steep stress gradients near the surface.

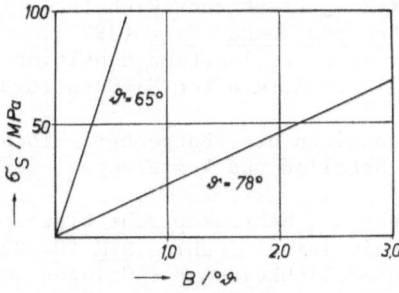

Fig. 9.   Computer-simulated apparent stresses as a function of line breadth.

Residual stresses are measured by changing the angle between the surface normal and the diffraction vector. In connection with this the penetration depth and consequently the effective stress is altered. This effect results in so-called apparent stresses, which are dependent on the stress gradient itself and also the glancing angle and the line breadth. With increasing line breadth and decreasing   the apparent stresses are growing up considerably.

ACKNOWLEDGEMENT

The author is indebted to Dr Černý and Dr Valvoda for supporting the investigation with the thin film diffractometer in the X-ray Laboratory of the Department of Semiconductor Physics, Charles University, Prague.

REFERENCES

1. H. Oettel, Methodik der röntgenographischen Analyse von beschichteten Werkstoffen, Freiberger Forschungshefte, in press.
2. R. Prümmer and H. W. Pfeiffer-Vollmar, Einfluss eines Konzentrationsgradienten bei röntgenographischen Spannungsmessungen, Z.f.Werkstofftechnik 12:282 (1981).
3. H. Krause and H.-H. Jühe, Untersuchungen über die Auswirkungen der Eindringtiefe und die Wahl der Strahlung bei der röntgenographischen Spannungsmessung, Forschungsberichte des Landes Nordrhein-Westfalen Nr. 3026, Westdeutscher Verlag (1981).
4. T. Hanabusa, H. Fujiwara, and K. Nishioka, X-Ray Residual Stress Analysis by the Weighted-Averaging Method for the Layer Having Steep Stress Gradient, in: HTM - Beiheft "Eigenspannungen und Lastspannungen", Carl Hanser-Verlag, München 204 (1982).
5. H. Oettel and H. Helbig, Röntgenographische Analyse des Erholungsverhaltens des nichtrekristallisierten Gefügebestandteils während der primären Rekristallisation, Neue Hütte 21:571 (1976).
6. P. Klimanek, X-ray scattering by real polycrystalline materials, Cryst. Res. Technol. 18:K15 (1983).
7. H. Hejdová and M. Čermák, A new X-ray diffraction method for thin film thickness estimation, Phys. Stat. Sol.(a) 72:K95 (1982).
8. J. Neumann and H. Hejdová, Zpřesnění metody stanovení tloušťky tenkých vrstev rentgenovou difrakcí, Čes. Čas. Fyz.(A) 35:603 (1985).
9. H. Oettel, I. Haase, and H. Wolter, X-Ray Phase Analysis of Multicomponent Systems according to Fiala, in: Proceedings of the Conference on Applied Crystallography, Kozubnik, Poland (1978).
10. G. Lange and H. Oettel, Quantitative röntgenographische Phasenanalyse an kleinsten Probenmengen, Neue Hütte 31:259 (1986).
11. L. S. Zevin, P. Rozenak, and D. Eliezer, Quantitative X-ray phase analysis of surface layers, J. Appl. Cryst. 17:18 (1984).
12. M. A. Krivoglaz, Diffrakcija rentgenovskikh lucej i neitronov v neidealnykh kristallakh, Naukova Dumka, Kiev (1983).
13. H. Oettel, X-ray analysis of dislocation densities and resistivity changes of plastically deformed Fcc Ni-Co-alloys, Phys. Stat. Sol.(a) 6:265 (1971).
14. H. Oettel, Über Möglichkeiten der röntgenographischen Versetzungsdichtebestimmung an kfz. Metallen und Legierungen, Z. Exp. Techn. Physik 21:99 (1973).
15. H. Oettel and B. Ehrentraut, Makroskopische Eigenspannungen in der Verbindungsschicht gasnitrierter Stähle, HTM 40:183 (1985).
16. T. Schubert, Variationsmöglichkeiten des Gefüges der Verbindungsschicht beim kontrollierten Gasnitrieren ausgewählter unlegierter und niedrig legierter Stähle, Dissertation Bergakademie Freiberg (1986).

STRUCTURE INVESTIGATION OF HARD COATINGS

BY TOTAL PATTERN ANALYSIS

Václav Valvoda

Charles University, Faculty of Mathematics and Physics

Ke Karlovu 5, 121 16 Praha 2, Czechoslovakia

INTRODUCTION

   X-ray diffraction is used in structure investigation of hard coatings as a complementary method to the complicated techniques of electron microscopy and electron diffraction. However, the routine X-ray methods are in many cases insufficient for a complete characterization of microstructure of thin films and unusual results are often said to be confusing. The main reason for these confusions is the fact that we are facing a new type of polycrystalline microstructure which is unknown in bulk materials. The only way how to understand the structure of these coatings is to use methods which are as complex as possible and which are accurate enough. The method of total pattern analysis (TPA), based on an analytical approximation of the whole diffraction spectrum, can fulfil these demands if it is accompanied by other methods for direct investigation of stresses and texture in the layers. Application of these methods for TiN coatings is presented here as an example.

   TiN coatings are intensively studied[1] during last few years for their great industrial importance. They are applied as wear resistant coatings on cutting tools, anticorrosive and antiabrasive coatings, selectively transmitting coatings on architectural glass, diffusion barriers in integrated circuits and in jewellery industry. Microstructure of these coatings strongly depends on deposition conditions and their properties can be accordingly modified. TiN coatings deposited by magnetron sputtering were investigated in our laboratory and new aspects of these coatings, namely an occurrence of large strains and stresses, were observed. Further, a dependence of microstructure of the coatings on their thickness was found together with a new type of inhomogeneity which is dependent on grain orientation.

Deposition Conditions

   TiN coatings of thickness ranging from 1 µm to 12.5 µm were reactively sputtered onto polished substrates from 12 % Cr steel at substrate temperatures 50 $^\circ$C or 150 $^\circ$C. The sputtering was carried out in an atmosphere of argon and nitrogen at a total pressure of 0.3 Pa in a planar magnetron with a circular Ti-target. The deposition rate was in the range from 0.1 to 0.5 µm/min. Composition of the coatings was controlled by changing the N_2/Ar flow rate ratio and thickness by time of deposition.

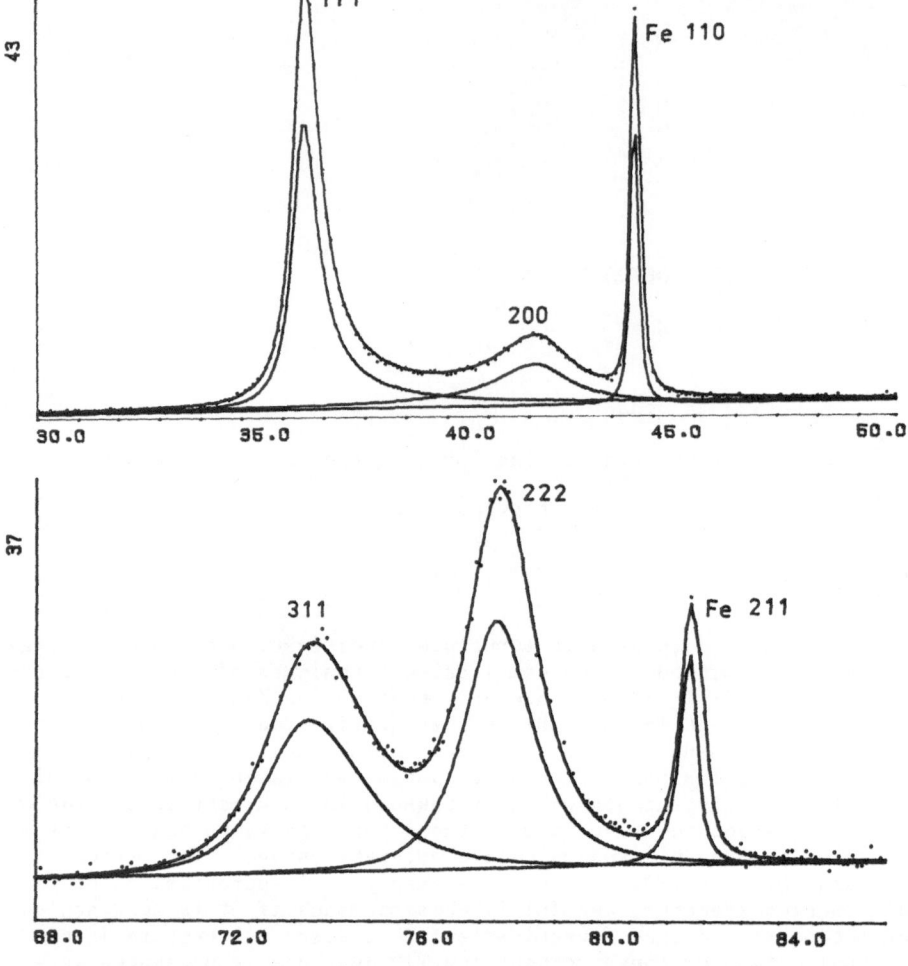

Fig. 1. Total pattern analysis of TiN coatings. Profiles of $K\alpha_1$ components
of individual reflections are also shown.

EXPERIMENTAL

The coatings were investigated by X-ray diffraction using the Bragg-
Brentano goniometer and monochromatized CuKα radiation. The angular position,
integral breadth and integrated intensity of the Kα_1 components of individual reflections were determined by fitting analytical functions[2] to the
measured diffraction profiles (see Fig. 1). Most measurements were carried
out with the flat specimen in symmetrical reflection position and thus only
diffraction from crystallographic planes parallel to the specimen surface
was registered. An asymmetric reflection position was used for the study
of residual stress and preferred orientation only. Line broadening was
analysed using the Williamson-Hall plots

$$B = \frac{1}{D} + \frac{4\ e}{\lambda} \sin \Theta,$$

where $\underline{B}$ is the line broadening expressed in reciprocal space units ($B =$
$= \text{ß}\cos \Theta/\lambda$, ß is line broadening in radians), $\underline{D}$ is the domain size and
$\underline{e}$ strain ($e = \Delta d/d$).

Microhardness of the coatings was determined by the Vickers indenter with a load of 20 gf. The chemical composition of the coatings was determined by the electron probe microanalysis (EPMA).

RESULTS

Phase composition. Two crystallographic phases were observed in the coatings investigated: hcp solid solution of nitrogen in α-titanium, α-Ti(N), or cubic titanium nitride, δ-TiN$_x$, in dependence on the nitrogen content. In some cases only the reflections 00ℓ of the hexagonal phase or the reflections hhh of the cubic phase were measurable in symmetrical reflection position owing to the strong preferred orientation of these planes. Diffraction pattern of α-Fe from the substrate was superimposed on the diffraction pattern from the coatings in all cases. The ε-Ti$_2$N phase was never clearly seen but a small amount of this phase cannot be ruled out in the coatings deposited under conditions approaching the phase transition from α to δ-phase (see Fig. 2).

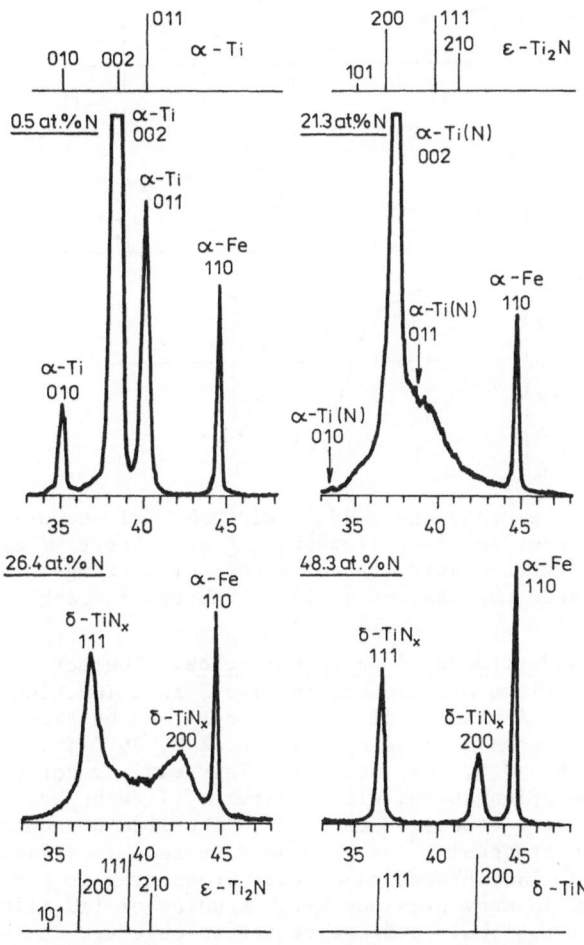

Fig. 2. ε-Ti$_2$N phase may be present in some coatings in small amounts. The layer containing 26.4 at.% N is the hardest one in given set.

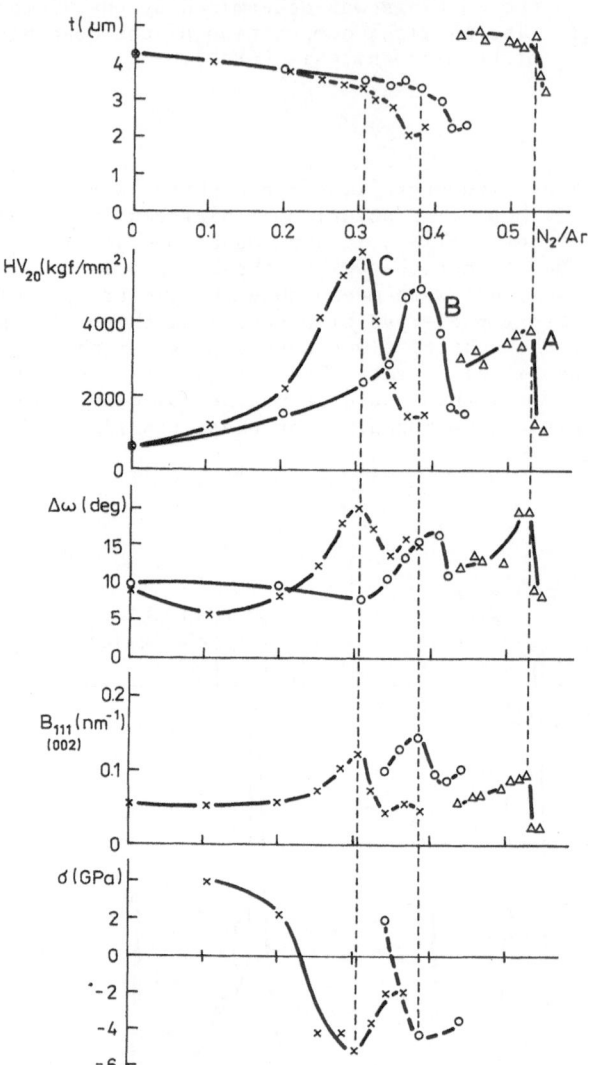

Fig. 3. Thickness $\underline{t}$, microhardness HV_{20}, width $\Delta\omega$ of the orientation distribution function, line broadening $\underline{B}$ and stress σ as a function of the gas flow rate ratio for three sets of coatings. Vertical dotted lines indicate the maximum of microhradness in each set.

<u>Microhardness and structural characteristics</u>. The results obtained from three sets of the coatings are plotted in Fig. 3 as a function of the N_2/Ar gas flow rate ratio[4]. A maximum of microhardness can be seen in all three sets occurring at the compositions of 36.8 at. % N, 36.0 at. % N and 38.0 at. % N for the sets A, B and C, respectively. This maximum correlates with a minimum of preferred orientation, with maximum of line broadening and with a maximum of compressive stress[3] in each set. The line broadening is mainly caused by the strain broadening[3], as can be also seen from Fig. 4 (in contrast to unjustified interpretation of this broadening in terms of domain size broadening only in many previous X-ray studies of TiN films). The values of strain and of residual compressive stress observed in the hardest coatings are unusually high as compared, for example, with deformed metals. This is also the reason for large broadening of the peaks and for the high values of lattice parameters determined from the interplanar spacings of

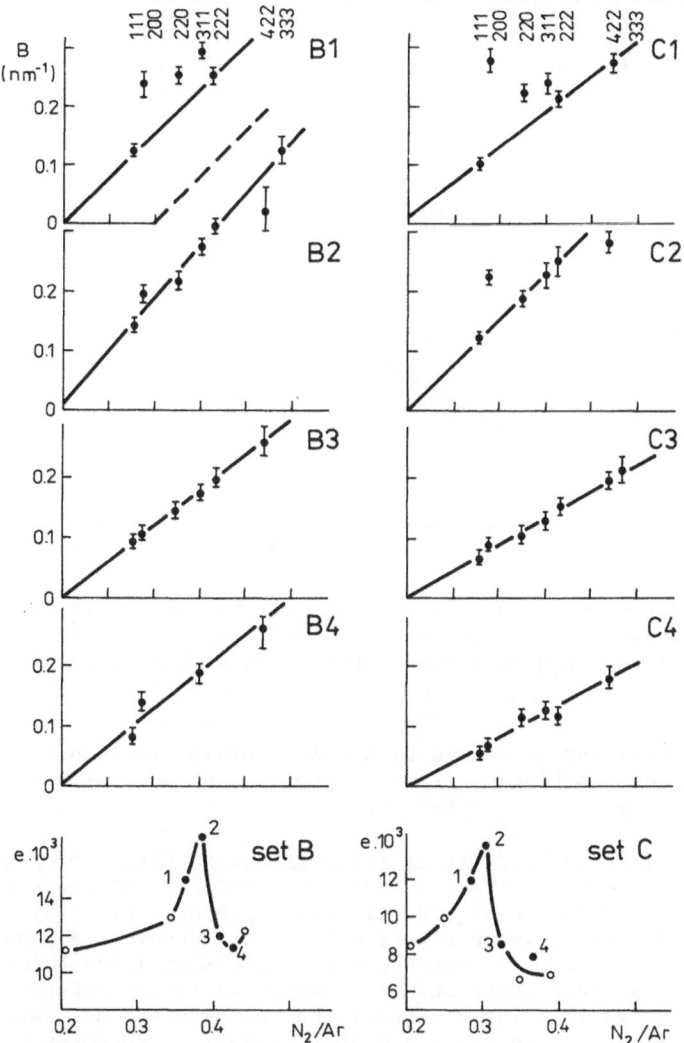

Fig. 4. Williamson-Hall plots of line broadening for selected samples of
the sets B and C and strain $\underline{e}$ (bottom) as a function of the gas
flow rate ratio.

planes parallel to the substrate surface (see Fig. 5). On the other hand,
the line broadening may be also partly caused by small domain size but determination of this contribution is difficult (for example, the domain size of
30 nm makes only 14 % of the total broadening of the 222 reflection which
belongs to the medium-broadened reflections). And also, the large values of
lattice parameters cannot be completely explained by the compressive stresses
present (Fig. 5).

It was also observed that the total intensity of all reflections together reaches minimum in the hardest coatings. This effect is illustrated
by Fig. 6 in which are presented intensities corrected for different thickness of the coatings, for different nitrogen content (by relating them to
bulk values of randomly oriented specimens with the same composition) and
partly also for preferred orientation (by summing the intensities of all
observable reflections together). Occurrence of such a minimum in the cor-

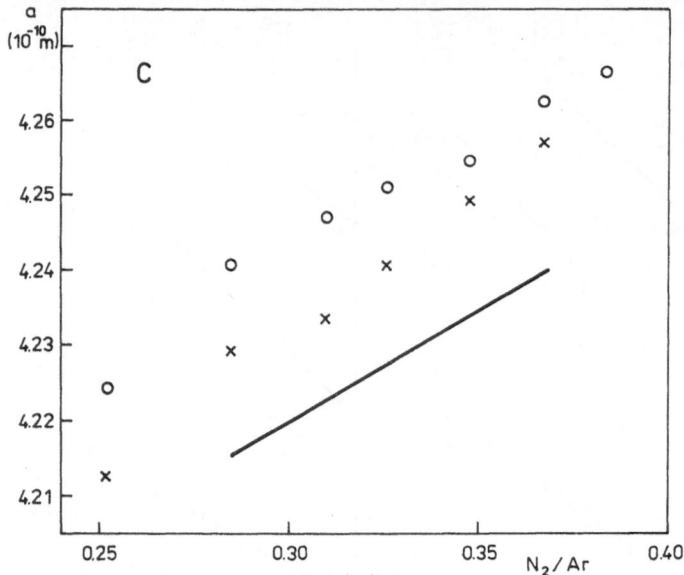

Fig. 5. Lattice parameters of the coatings of the set C. Circles represent
the observed values, crosses - the observed values corrected for residual stress and full line corresponds to bulk values of Nagakura
et al.[5]

rected total intensities is caused by a small volume fraction of crystalline
phase in the hardest coatings. It means that these coatings are also the
crystallographically most distorted ones.

Dependence of microstructure of TiN coatings on their thickness was also
investigated in two sets of the coatings containing on the average 34.5 at.%N
(set D) and 50.3 at.% N (set E). No systematic changes of strain and microhardness with the layer thickness were observed[6]. However, a general tendency of lattice parameters to decrease with increasing layer thickness was
found together with their systematic dependence on the orientation of grains
in the coatings (Fig. 7). These results and the observed decrease of stress
with the layer thickness are explained by a gradual improvement of the struc-

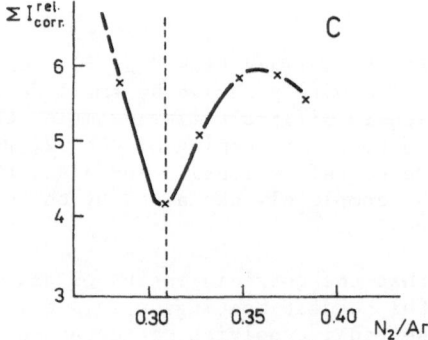

Fig. 6. Degree of crystallinity in the cubic-phase coatings of the set C
expressed by the sum of the corrected relative intensities of reflections 111, 200, 220, 311, 222 and 422.

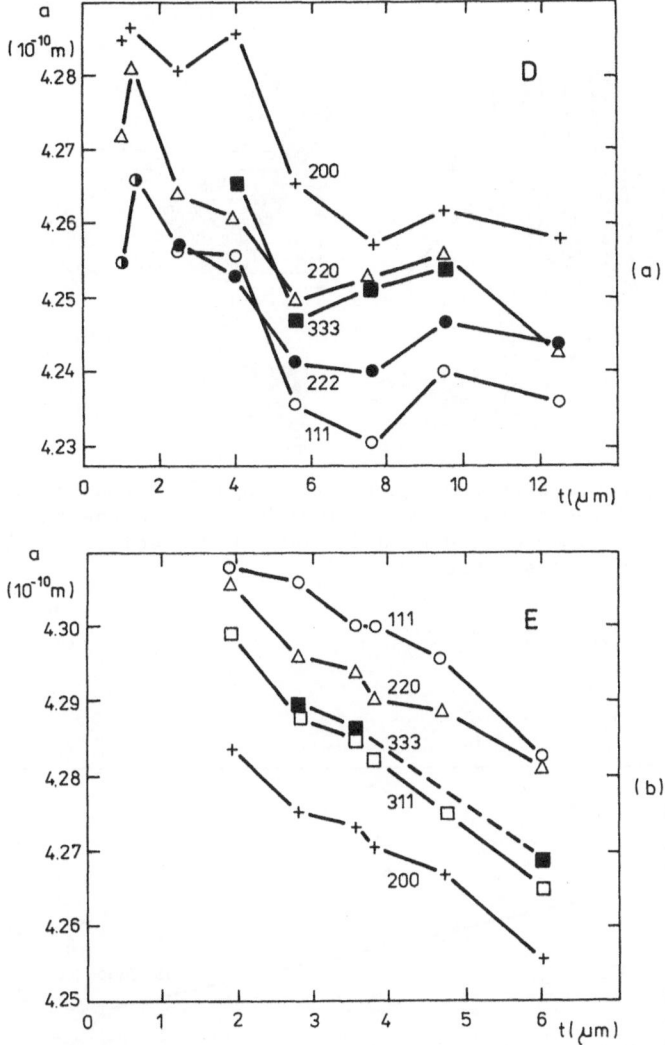

Fig. 7. Lattice parameters as a function of the layer thickness for the set D (a) and the set E (b) containing 34.5 and $50_0 3$ at. % N respectively. The bulk values are equal to 4.22×10^{-10} m and 4.24×10^{-10} m for the set D and E, respectively.

ture of the coatings with increasing thickness and by a heterogeneous nature of the structure. It can be assumed that the growth of grains oriented with their (111) or (100) planes parallel to the substrate surface may be different (Fig. 8); the (111) planes are alternatively occupied by Ti or N atoms whereas each of the (100) planes is occupied by both kinds of atoms arranged in rectangular, mutually interpenetrating nets. Defects occurring in both stacking sequences can be expected to be different and these differences can result in their different interplanar spacings (compared by means of lattice parameters calculated from them).

Annealing of the coatings at temperatures ranging from 500 °C to 800 °C for 3 hours in vacuum (10^{-3} Pa) resulted in a decrease of the lattice parameter values (Fig. 9) and in a decrease of line broadening (Fig. 10). The influence of annealing at 700 °C for 3 hours in vacuum on microhardness and some structural characteristics is compared for two selected samples in Tab. 1.

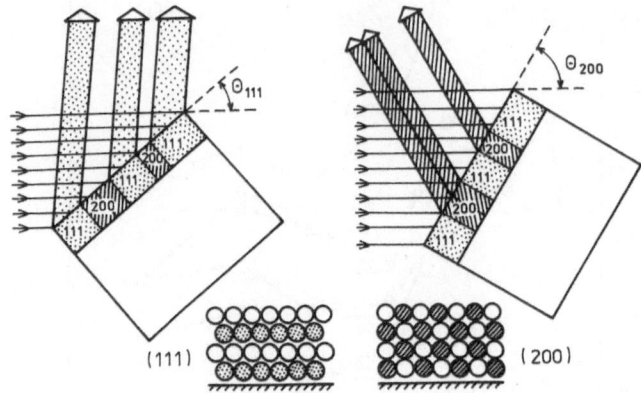

Fig. 8.   Heterogeneous structure of the coatings is dependent on crystallographic orientation of grains with respect to the substrate surface and may be caused by different growth defects in both lattice
plane sets illustrated.

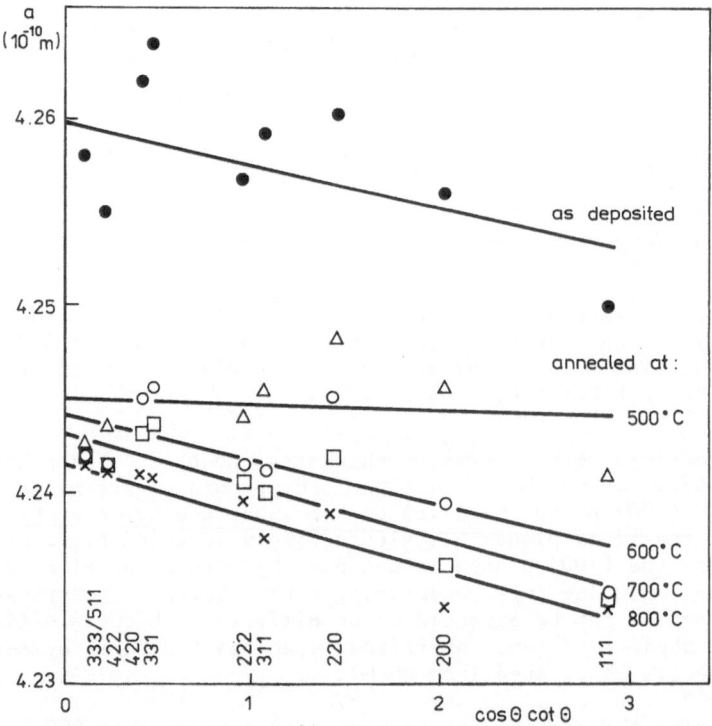

Fig. 9.   Lattice parameters of as-deposited and annealed coatings plotted
as a function of $\cos\Theta \cdot \cot\Theta$.

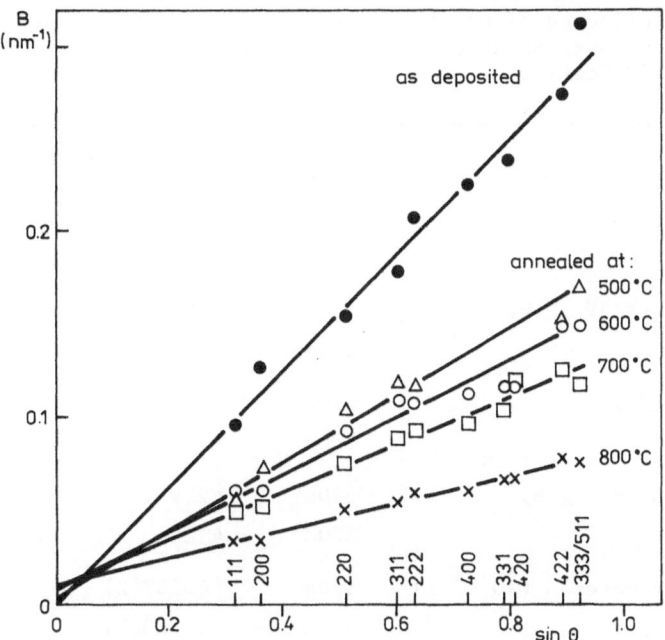

Fig. 10. Williamson–Hall plots of line broadening for as-deposited and annealed coatings.

DISCUSSION AND CONCLUSIONS

It was found that the hardest coatings are the most disordered ones. That is in agreement with the generally accepted idea of strengthening influence of structural defects. The growth defects and grain interactions are also mostly responsible for the extremely large values of compressive stress found in the hardest coatings because the deposition temperatures were certainly too low to explain these stresses by different thermal contraction of the film and the substrate during cooling. These compressive stresses are partly responsible for the large values of lattice parameters, for the large values of microstrain and for the extremely large values of microhardness observed. All these quantities returned back to nearly bulk TiN values when the coatings were annealed and the structure of films recovered.

Two types of inhomogeneities were found in the layers. The first one is represented by gradients in the lattice spacings and in microstrain ($\Delta d/d$) in some layers, the second one is given by the dependence of the lattice parameters on a crystallographic orientation of the grains with respect to the substrate surface (Figs 7 and 8). These inhomogeneities may lead to confusions when only an incomplete set of diffraction data is available. However, it is not clear yet how these inhomogeneities are connected with deposition conditions (see e.g. the different behaviour of samples of the set D and E in Fig. 7).

In any case, the unusual diffraction features observed in polycrystalline thin films must be regarded as a reflection of specific microstructure of these materials and our knowledge based on structure investigation of bulk samples can be extrapolated only carefully.

189

Table 1. Influence of Annealing at 700 °C for 3 Hours in Vacuum on Some Structural and Mechanical Characteristics of TiN Coatings

	Specimen	As deposited	Annealed
Thickness t (μm)	530a	2.7	–
	530d	5.0	–
Microhardness HV_{20} (kgf/mm^2)	530a	4900	3100
	530d	4900	2100
Stress σ (GPa)	530a	–1.25	–0.40
	530d	–1.62	–0.27
Composition (at. % N)	530a	42.9	45.5
	530d	41.5	39.7
Lattice parameter a (10^{-10} m)	530a	4.257(2)	4.2442(5)
	530d	4.260(2)	4.2428(8)
Strain $e \times 10^3$	530a	10.8	4.7
	530d	12.0	5.0

REFERENCES

1. J. E. Sundgren, Structure and properties of TiN coatings, Thin Solid Films 128:21 (1985).
2. V. Valvoda, R. Černý, R. Kužel, Jr., and L. Dobiášová, Method of complex X-ray diffraction analysis of TiN coatings, Cryst. Res. Technol. 22: 1301 (1987).
3. V. Valvoda and J. Musil, X-ray analysis of strain in titanium nitride layers, Thin Solid Films 149:49 (1987).
4. V. Valvoda, R. Kužel, Jr., R. Černý, and J. Musil, Structure of TiN coatings deposited at relatively high rates and low temperatures by magnetron sputtering, Thin Solid Films 156:53 (1988).
5. S. Nagakura, T. Kusunoki, F. Kazkimoto, and Y. Hirotsu, Lattice parameter of the non-stoichiometric compound TiN$_x$, J. Appl. Cryst. 8:65 (1975).
6. V. Valvoda, R. Černý, R. Kužel, Jr., J. Musil, and V. Poulek, Dependence ofmicrostructure of TiN coatings on their thickness, Thin Solid Films 158:225 (1988).

THE DETERMINATION OF LATTICE PARAMETERS AND STRAINS

IN STRESSED THIN FILMS USING X-RAY DIFFRACTION: EXTENSIONS

E. L. Haase

Kernforschungszentrum Karlsruhe, Institut für Nukleare
Festkörperphysik, P.O.B. 3640
D-7500 Karlsruhe, Federal Republic of Germany

A method is presented that allows the simultaneous determination of the lattice parameters and the strains of stressed thin films. The method employs either a thin film Guinier camera or a diffractometer both with Seemann-Bohlin focussing. The generalization of the method for non-cubic system and samples having a Poisson ratio unequal to 1/3 is given. Applications are given for silver and gold films and several phases in the Nb-Ge system. A simplification of the method is given when only the lattice parameter is desired. Departures from linearity are discussed.

INTRODUCTION

The recent years have brought a tremendous growth of the field of thin films. Until recently[1] the people concerned were unaware that a special method has to be applied to determine the lattice parameter a_o of thin films, which are normally in a state of stress. As a result of this, there exist hundreds of papers in the literature which quote false values of a_o. The treatment in ref.[1] is restricted to films with cubic symmetry. It is the aim of the present paper to provide a more general treatment, extending the method to non-cubic systems and to the case where the Poisson number deviates from 1/3. The method of ref.[1] can be simplified when only the lattice parameter is wanted. In any case it is recommended to read ref.[1] for a better understanding of the present paper.

The method yields simultaneously the lattice parameters and strains for films of arbitrary symmetry and specified extinction rules. It employs either a Guinier camera or a diffractometer with Seemann-Bohlin focussing. Ref.[2] gives a good description of the Seemann-Bohlin geometry and method. It also describes a method of strain determination of thin films when a powder sample with the same lattice parameter is available for cubic systems. This papere is also good background reading. The various aspects of stresses have been reviewed in ref.[3], including the full tensorial treatment.

The present method is quite simple. It employs a Fortran IV indexing program developed by Schneider[4] and modified by the author to take care of the effects of stresses (available upon request).

In the present work a Huber system 600 thin film instrument was employed (Huber Diffraktionstechnik, D-8211 Rimsting, FRG). All of the work described was carried out using X-ray films; the accuracy can be considerably improved by using the available diffractometer.

Fig. 1 shows the diffraction situation and geometry schematically for a film under tensile stress on an exaggerated scale. For the left "grain" the diffraction planes are parallel to the surface. One can consider the rectangles to represent strain distorted unit cells. Then the lattice plane spacing d' is smaller than the unstrained value d. γ is fixed angle of grazing incidence, Θ the diffraction angle for forward scattering. It is immediately obvious that for the Bragg-Brentano geometry, where the incident scattering angle is always equal to the outgoing angle (dashed lines), diffraction occurs only from lattice planes parallel to the surface, and for stressed films one will obtain false lattice parameter values. By tilting the sample, one gets scattering from planes not parallel to the film surface and this is employed largely in the $\sin^2 \psi$ method of stress analysis[5,6].

In the right side of Fig. 1 the Seemann-Bohlin geometry is shown for a "grain" with lattice plane spacings d' larger than d and more nearly parallel to the normal to the film (long dashed line). Φ is the angle between the normal to the film and the normal to the diffraction plane (dot-dashed line). From Fig. 1 it is obvious that $\Phi = \Theta - \gamma$. Hence as one scans Φ from zero (left side) to values somewhat below $90°$ (d'-d)/d varies from negative values via zero, at a value depending on ν (formula 6) to positive values. When d'=d the effect of strains on the lattice plane spacing vanishes.

As the film surface is free, the strain σ_3 normal to the film is taken to be zero. The two stresses in the plane of the film σ_1 and σ_2 are taken to be equal and independent of direction. To avoid the full tensorial treatment the approximation is made that the elastic properties of the polycrystalline film in any direction are given by an average value of Young's modulus E and Poisson's ratio ν. Then the strain components are

$$\varepsilon_1 = \varepsilon_2 = \frac{(1-\nu)\,\sigma_1}{E} \tag{1}$$

$$\varepsilon_3 = -\frac{2\nu\,\sigma_1}{E} \tag{2}$$

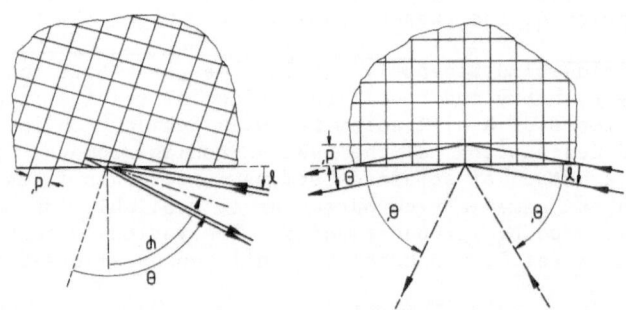

Fig. 1. Schematic representation of the scattering geometry for a film under tensile stress. The grid lines may be taken to represent distorted unit cells.

By combining equations (1) and (2) one obtains

$$\varepsilon_3 = - \frac{2\nu}{1-\nu} \cdot \varepsilon_1.$$ (3)

One can show that by rotating the axes of the strain tensor that

$$\varepsilon(\Phi) = \varepsilon_1 \sin^2\Phi + \varepsilon_3 \cos^2\Phi = \varepsilon_1(\sin^2\Phi - \frac{2\nu}{1-\nu} \cos^2\Phi)$$ (4)

$$= - \varepsilon_1 \cdot E(\Phi)$$

For the special case that $\nu = 1/3$ this simplifies to

$$\varepsilon(\Phi) = - \varepsilon_1 \cos 2\Phi,$$ (5)

which vanishes for $\Phi = 45°$. In general, by setting equation (4) to zero, the effect of stresses vanishes at Φ_o, where

$$\Phi_o = \text{arc tan} \sqrt{2\nu/(1-\nu)}$$ (6)

As Φ becomes smaller than Φ_o the measured lattice plane spacings become smaller than their unstressed values (for $\varepsilon_1 > 0$) such that $\Delta d/d$ approaches $-\varepsilon_1$ as Φ becomes zero.

$$\frac{\Delta a}{a}\bigg|_{\Phi=0} = \frac{\Delta d}{d}\bigg|_{\Phi=0} = \frac{d_{obs}-d_c}{d_c}\bigg|_{\Phi=0} = -\varepsilon_1,$$ (7)

where d_{obs} are the experimental lattice plane spacings calculated from the observed 2Θ values and d_c are the calculated unstressed lattice plane spacings from the last refinement of the data. On the other hand, as Θ becomes larger than Φ_o, $\Delta d/d$ increases from zero (for $\varepsilon_1 > 0$).

In practice, a straight line is fitted to the $(d_{obs}-d_c)/d_c$ data points as a function of $E(\Phi)$. For the strain determination a problem arises in that for $\nu \neq 1/3$, $E(\Phi=0)-E(\Phi_o)$ is different from $E(\Phi_o)-E_o(\Phi=90°)$. To minimize the error it is proposed to take $\varepsilon_1 = (E(\Phi=90°)-E(\Phi=0°)/2$. Following the convention ε_1 is taken to be positive for tensile stress.

At this point in a few additional iterations of step of the program the d_{obs} values are corrected to minimize the slope of the fit subject to the condition that the fitted line passes through zero at Φ_o. In this process the fitted lattice parameter values are modified minimally to allow for the fact that, due to statistical fluctuations in the input data, the fit of the zeroth iteration may not have the value zero at Φ_o. The lattice parameters thus obtained are then the final values.

The line focus correction in analogy to the Hadding correction is discussed in detail in ref.[1]. It leads to a line shift

$$\Delta\Theta = 2F \cot (2\Theta),$$ (8)

$$F = (1+x)(h/R)^2.$$ (9)

x is a constant that has to be determined empirically, h is the beam spot or the counter slit height and R is the radius of the focussing circle. h was somewhat arbitrarily been taken to be 5 mm. The $\Delta\Theta$ correction has to be applied both to the calibration lines and the lines of the substance to be

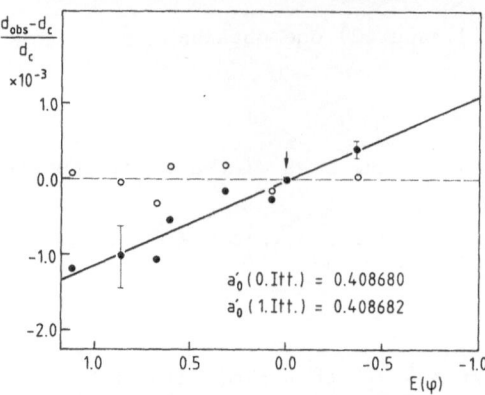

Fig. 2. Plot of $(d_{obs}-d_c)$ as a function of $E(\Phi)$. Solid line and solid dots are the result of the last refinement before correcting for stresses. Open circles and dashed line are the results after correction for stresses in the second step of the program. The lines should and do intersect at $(d_{obs}-d_c)/d_c$ and $E(\Phi) = 0$.

measured. As calibration standard 1 μ Ni powder was used. As substance to be measured 1 μ Ag powder was employed. After several iterations x was found to be −0.10 by requiring the "thermal stress" of the silver powder to be zero. To improve the accuracy of x, it is desirable to run several powders and take an average value.

Fig. 2 gives an example of this procedure for a 500 nm thin silver film evaporated at RT and then annealed for one hour at 300 °C. The solid circles are the $(d_{obs}-d_c)/d_c$ values of the zeroth iteration and the solid line is the corresponding fit. The open circles give the corrected values after the first iteration with the dashed line being the corresponding forced fit. The lattice parameter values after the zeroth and the first iteration are also given in the figure. They differ by much less than the a_0 error of about 0.0001 nm.

A problem arises when the Poisson ratio is not known. In this case one usually takes $\nu = 1/3$, and then one would have to take the lattice parameter values of the zeroth iteration of the second step. A better solution appears to be to adjust ν until $(d_{obs}-d_c)/d_c$ crosses zero at $E(\Phi)=0$ and take the lattice parameters after the iterations converge after a few times. This would then constitute also a rough determination of ν.

If one only wants an approximate value of the lattice parameters, especially for cubic systems, the amount of labor can be reduced considerably by taking only a few, in the extreme case two, lines to carry out above procedure. The lines should lie near Φ_0. For refractory compounds like NbN, where $\nu \simeq 0.20$, the (400) line lies near the position where the effect of stresses vanishes. Having once carried out the full procedure, the difference in the position of the nearby (311) gold calibration line and the (400) line can be used to yield an approximate value of a_0 without using the program, just by calculating Δa_0 from the measured $\Delta\Theta$.

RESULTS AND DISCUSSION

Fig. 3 shows $(d_{obs}-d_c)/d_c$ as a function of $E(\Phi) = \cos 2\Phi$, which equals $E(\Phi)$ for $\nu = 1/3$, for a set of 500 nm silver films evaporated at room temperature (RT) and annealed at the temperatures indicated. As ν for silver is

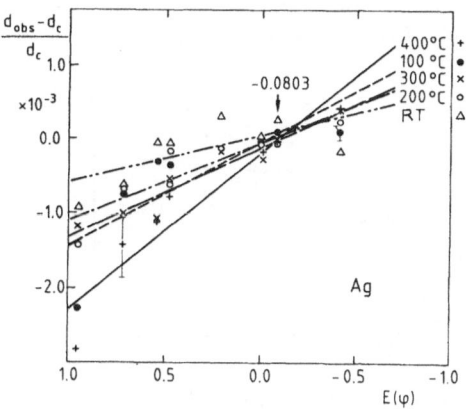

Fig. 3. Plot of $(d_{obs}-d_c)/d_c$ as a function of cos 2Φ for a set of silver films under tensile stress. The lines should intersect at cos 2Φ = -0.0803.

0.37 rather than 1/3 the effect of stress vanishes and the lines intersect within experimental error at cos 2Φ = -0.0803. As always in the following, the lines are the fit to the experimental points. The error bars are given by Δd/d = ΔΘ.cotΘ with ΔΘ = 0.015° as a representative value for the evaluation of the X-ray films. Note that the error bars decrease to the right.

Fig. 4 shows the same data now as a function of E(Φ) (equation 4). Now E(Φ) starts at 1.175 for Φ=0 and ν=0.37. This is the general treatment for ν≠1/3 and now the lines should and one sees that indeed they do intersect within experimental error at $(d_{obs}-d_c)/d_c$=0 and E(Φ)=0. The rising slope means that the films are under tensile stress.

Fig. 5 shows a plot of two data sets of the hexagonal C40 $NbGe_2$ phase. The solid circles stem from sample Nb2GeTl while the open circles are from sample 223 VI. As the author was unable to find a value for the Poisson ratio, ν was taken to be 1/3. This is taken for all of the following data sets except the gold data. For the two data sets shown in Fig. 5 this assumption is quite good as the $(d_{obs}-d_c)/d_c$ lines cross zero near E(Φ) = 0.

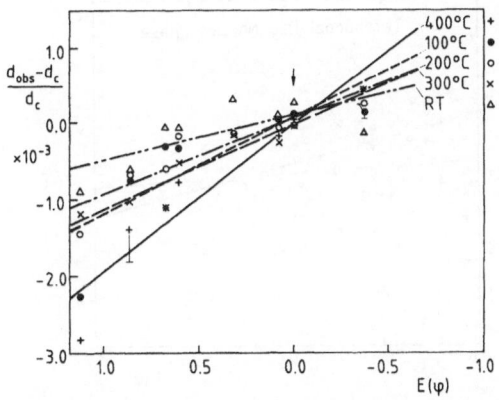

Fig. 4. Plot of $(d_{obs}-d_c)/d_c$ as a function of E(Φ) for the same data as shown in Fig. 3. Now the lines should interesect at E(Φ) = 0.

195

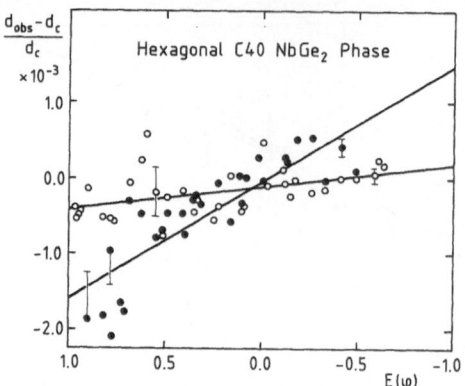

Fig. 5. Plot of $(d_{obs}-d_c)/d_c$ as a function of $E(\Phi)$ for two samples having
the hexagonal C40 phase for the Poisson ratio $\nu = 1/3$ under
tensile stress.

With a $(d_{obs}-d_c)/d_c\big|_{\Phi=0}$ value of about $\varepsilon_1=1.6\times10^{-3}$ sample Nb2GeT1 is under
moderate tensile strain.

Fig. 6 shows the data and the fitted lines for sample 166VI (open
circles) and 222V (crosses) having the hexagonal $D8_m$ Nb_5Ge_3 phase. Both
samples are under moderate compressive stress. Two representative error
bars are shown. Again ν has been taken to be 1/3, but in this case the assumption is not well justified as the lines cross zero near $E(\Phi) = -0.2$.
In fact, a rough empirical determination of ν could be carried out by increasing ν until the lines cross at $E(\Phi) = 0$.

Fig. 7 shows the data and their fits for the films 224III (open circles,
solid line) and 222VII (crosses) for the hexagonal $D8_8$ Nb_3Ge_2 phase. Both
samples are under moderate tensile stress. ν has been taken to be 1/3, which
is well justified as the lines cross 0 near $E(\Phi) = 0$.

Fig. 8 shows the data points for the bulk sample NbGe2A2 having the
C40 $NbGe_2$ phase. The points show a pronounced departure from linearity.
The rather meaningless fitted line is also shown. The sample was annealed

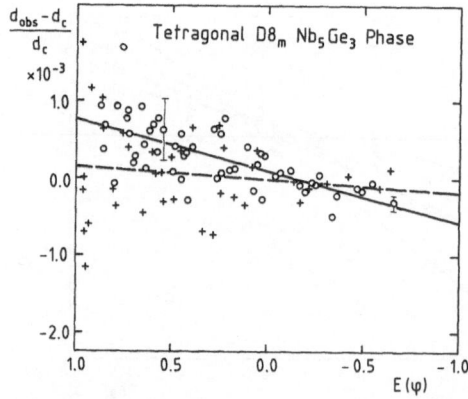

Fig. 6. Plot of $(d_{obs}-d_c)/d_c$ aas a function of $E(\Phi)$ for two films having
the $D8_m$ phase. Both samples are under moderate compressive stress.

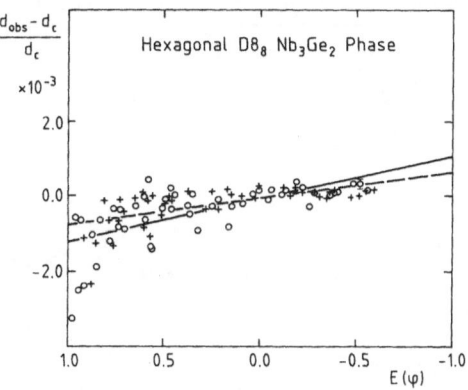

Fig. 7. Plot of $(d_{obs}-d_c)/d_c$ against $E(\Phi)$ for two films having the $D8_8$ phase. The assumption that $\nu = 1/3$ is well justified.

twice at temperatures around 800 °C. During the annealing presumably oxygen and perhaps carbon diffused into the surface and caused the stress. Such a departure has occasionally been observed, but rarely so pronounced. Note the magnitude of the stress. Until the cause of the non-linearity is understood, no stress analysis is possible. It appears that the annealing is causing the non-linearity. Note that the points near $\Phi = 0$ in Fig. 7 also show an onset of non-linearity.

Fig. 9 shows the data and fits for three gold films evaporated at RT, two of which were then annealed at 100 and 500 °C. While the RT and 100 °C film show only a small departure from linearity and can still be fitted with straight lines, the 500 °C film shows a pronounced non-linearity associated with the annealing. Departures from linearity have also been found for the $\sin^2\psi$ method[6]. A stress analysis is not possible until the cause is understood. However, the lattice parameter of the 500 °C sample at $E(\Phi) = 0$ agrees quite closely with the values of the other two samples which show powder values[1]. Both the RT and 500 °C sample show texture with a half width of about 9.5°. The linewidths of the 500 °C film decreased only by a 13 % during annealing.

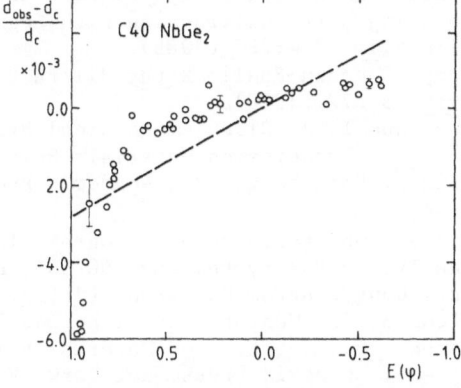

Fig. 8. Plot of $(d_{obs}-d_c)/d_c$ for a C40 film. Note the pronounced departure from linearity and the magnitude of the stress.

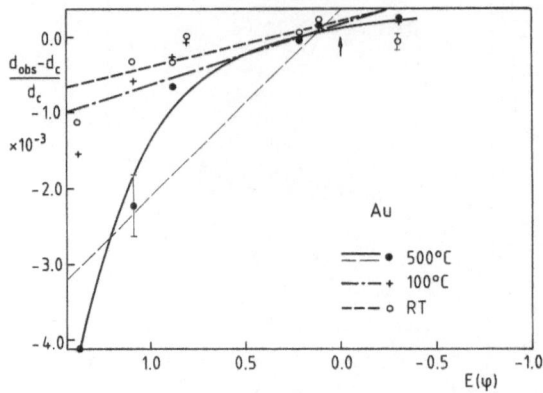

Fig. 9. Plot of $(d_{obs}-d_c)/d_c$ for three gold films, two of which were annealed at the temperature indicated. While the RT and the 100 °C films can be fitted rather well linearly, the 500 °C films show a pronounced departure from linearity.

CONCLUSIONS

In ref. [1] a simple method has been developed that yields both the lattice parameter and the strains of thin films for cubic substances assuming $\nu = 1/3$. The method requires Seemann-Bohlin focussing and uses either a Guinier camera or a diffractometer. In the present work this method has been extended to films with non-cubic symmetry and to the case that $\nu \neq 1/3$. The method uses a Fortran IV program. A simplification of the method is discussed when only the lattice parameter is needed. The method has been applied to a number of systems. The problem of non-linearity needs further theoretical clarification.

ACKNOWLEDGEMENT

The author wants to thank V. Jung for the texture and linewidth measurement of the Au films.

REFERENCES

1. E. L. Haase, The determination of lattice parameters and strains in stressed thin films using X-ray diffraction with Seemann-Bohlin focussing, Thin Solid Films 124:283 (1985).
2. R. Feder and B. S. Berry, Seemann-Bohlin X-ray diffractometer for thin films, J. Appl. Cryst. 3:372 (1970).
3. M. Murakami, T.-S. Kuan, and I. A. Blech, Mechanical Properties of Thin Films on Substrates, in: "Treatise on Materials Science and Technology", K. N. Tu and R. Rosenberg, eds, Academic Press, New York, Vol. 24 (1982), p. 163.
4. W. Schneider, Ein Indizier- und Verfeinerungsprogramm für Pulverbeugungsaufnahmen in Fortran IV für das System IBM 360/365, Externer Berich 6/69-4, IMF, Kernforschungszentrum Karlsruhe (1969).
5. M. R. James and J. B. Cohen, The Measurement of Residual Stress by X-Ray Diffraction Techniques, in: "Treatise on Material Science and Technology", H. Herman, ed., Academic Press, New York, Vol. 19A (1980), o. 1.
6. V. M. Hauk, X-Ray Methods for Measuring Residual Stress, in: "Residual Stress and Stress Relaxation", E. Kula and V. Weiss, eds, Plenum, New York (1982), p. 117.

USE AND PERSPECTIVE OF X-RAY DIFFRACTION IN SCIENCE

AND TECHNOLOGY OF CERAMIC COATINGS

R. Delhez, Th. H. de Keijser, and E. J. Mittemeijer

Laboratory of Metallurgy
Rotterdamseweg 137, 2628 AL Delft, The Netherlands

ABSTRACT

X-ray diffraction provides a major tool to investigate the relation between the properties of a coating/substrate composite and its microstructure. Stresses, compositions, stress-depth profiles and composition-depth profiles can be determined. Further, microstructural parameters as crystallite size and microstrains can be obtained. The role of the penetration of X-rays deserves special attention, in particular when analysing thin coatings.

In various CVD TiC coating/substrate systems the strain (and stress) can be predicted quantitatively from thermal shrinkage. In some of these systems stress relaxes partially by strain-accommodating processes either in the layer and/or in the substrate.

In commercial CVD coatings foreign elements can be present either in solid solution or as a second phase and these can significantly influence the microstructure and (mechanical) properties.

In the surface (compound) layer of nitrided steels both stress and composition depth profiles occur. The stress profile is caused by the thermal shrinkage and by the composition-depth profile. Partial strain accommodation occurs due to the presence of pores.

INTRODUCTION

The main characteristics of an X-ray diffraction line profile are its intensity, its position and its shape (width). From the occurrence of line(s) of a phase its presence is deduced. The (integrated) intensity of a line is a measure for the quantity and for the degree of preferred orientation (texture) of the phase concerned. From the peak position of line profiles stresses and compositions can be determined, and from the peak shape microstructural parameters as crystallite size and microstrains can be derived.

The interpretation of these characteristics for thick, homogeneous, polycrystalline specimens is well known and can be found in textbooks and review papers[1,2]. If thin coatings are analysed and/or stress and composition gradients occur, complications arise. Some of these are treated in the next section.

The power of X-ray diffraction for the analysis of ceramic coatings is illustrated by case studies of Chemically Vapor Deposited (CVD) TiC coatings on various substrates and of nitride coatings produced by nitriding iron and steels.

CHARACTERISTICS OF X-RAY DIFFRACTION LINE PROFILES

The complications met when analyzing surface layers by X-ray diffraction are not always fully appreciated. In a recent review paper[3] a treatment of some of these difficulties has been given. Here the basic issues will be indicated briefly.

Obviously the depth of penetration of the X-rays plays an important role in the analysis of surface layers, the more because gradients of stresses, composition and microstructure often occur.

Measures for X-ray Penetration

When quantifying the penetration of X-rays into a specimen, the basis of any treatment is formed by a description of the decrease of the intensity by a layer of thickness z, according to the absorption factor, A, which reads

$$A = \exp[-\mu k z] , \tag{1}$$

where μ denotes the (effective) linear absorption coefficient and k is a trigonometric function determined by the diffraction geometry (see Table 1). Then it follows that the so-called integrated intensity dP, i.e. the number of counts observed from a layer of thickness dz at depth z below a surface is given by

$$dP = C.A.dz = C\exp[-\mu k z]dz , \tag{2}$$

where C depends on the diffraction geometry. Hence, the total integrated power, P(t), from a surface layer of thickness t is given by

$$P(t) = C\{1-\exp[-\mu k t]\}/\mu k = P_\infty \{1-\exp[-\mu k t]\} . \tag{3}$$

On this basis four different measures for penetration can be defined[3]:
- The <u>penetration depth</u>, ζ_r, is the depth where the ratio $dP(z=\zeta_r)/dP(z=0)$ equals an arbitrarily chosen fraction r.
- The <u>information depth</u>, ζ_i, is the intensity-weighted average depth; for a layer of thickness t it follows from

$$\zeta_i(t) = \int_o^t zAdz / \int_o^t Adz.$$

- The <u>contributing thickness</u>, τ_R, of a diffracting surface layer of thickness t is the thickness for which the ratio $P(z=\tau_R)/P(z=t)$ equals an arbitrarily chosen fraction R.
- The <u>equivalent thickness</u>, τ_{eq}, is the thickness of a hypothetical non-absorbing layer giving the same amount of diffracted intensity as the actual absorbing layer of thickness t.

The formulae for these quantities are given in Table 1. The parameter most frequently used in the literature to account for absorption, $1/\mu k$, is not cited there, because it can be interpreted in various ways. From Table 1 it follows that for an infinitely thick layer:

Table 1. Measures for Penetration in Powder Diffractometers where ω or ψ Tilt is Possible

Depth Probed

ζ_r = penetration depth

$= \dfrac{1}{\mu k} \ln \dfrac{1}{r}$

$(0 < r < 1)$

$\zeta_i(t)$ = information depth

$= \dfrac{1}{\mu k} - \dfrac{t \exp[-\mu k t]}{1 - \exp[-\mu k t]}$

Thickness Probed

$\tau_R(t)$ = contributing thickness

$= \dfrac{1}{\mu k} \ln\left\{ \dfrac{1}{(1-R) + R \exp[-\mu k t]} \right\}$

$(0 < R < 1)$

$\tau_{eq}(t)$ = equivalent thickness

$= \dfrac{1}{\mu k} - \dfrac{\exp[-\mu k t]}{\mu k}$

μ = linear absorption coefficient
t = coating thickness
k = $2 \sin \Theta \cos\omega / (\sin^2\Theta - \sin^2\omega)$ for ω tilt performed about the $\Theta/2\Theta$ axis
k = $2/\sin\Theta\cos\psi$ for ψ tilt performed about an axis which lies in the specimen surface and is perpendicular to the $\Theta/2\Theta$ axis.

$$\zeta_i(t=\infty) = \tau_{eq}(t=\infty) = \dfrac{1}{\mu k} \tag{4a}$$

and, in the case that $r = 1/e$ and $R = (e-1)/e$:

$$\zeta_{r=1/e} = \zeta_i(t=\infty) = \tau_{R=(e-1)/e}(t=\infty) = \tau_{eq}(t=\infty) = 1/\mu k \tag{4b}$$

In general, for the characterization of absorption effects a depth or thickness parameter should be selected dependent on the kind of information to be extracted from the diffraction experiment; for example a stress value should be related to a depth. Obviously, for the analysis of surface layers a measure for penetration dependent on t should be adopted and considering the arbitrary choice of r or R as a disadvantage, the measures ζ_i and τ_{eq} are to be preferred. It should be noted, that these measures also have a well defined meaning for infinitely thick specimens.

The information desired determines the radiation (i.e. wavelength) to be applied. If lattice-spacing gradients occur in a surface layer, the depth analysed in a diffraction experiment should be as small as possible, implying utilization of radiation which is absorbed relatively strongly; in this case the spacing-depth profiles can be determined by successive layer removal (e.g. by polishing). On the other hand, if a substrate beneath a coating should be analysed, this may be achieved without removal of the coating by applying radiation which is absorbed relatively weakly in the coating.

Intensities

Integrated intensities and peak intensities observed for thin coatings are smaller than those for infinitely thick specimens of the same material, as is immediately seen from eq.(3). This simple equation can be applied to normalize data from thin coatings, e.g. for the quantitative comparison of the degrees of preferred orientation of the crystallites in various coatings of different thicknesses.

Peak Position

Peak positions are used to identify phases, to determine the composition of solid solutions and to measure macrostress. If composition and/or macrostress change significantly within the depth, t, probed by the X-rays, the peak position observed yields the lattice spacing, $<d>$, averaged over t:

$$<d> = \int_o^t d(z)\exp[-\mu kz]dz \;/\; \int_o^t \exp[-\mu kz]dz \qquad (5)$$

It can be shown that $<d>$ is a very good approximation for the lattice spacing at depth $z = \zeta_i(t)$. A more detailed discussion of eq. (5) is given elsewhere.[3]

A lattice-spacing profile can be traced by X-ray diffraction analyses after successive sublayer removals, if the measure used for X-ray penetration is significantly smaller than the extent of the lattice-spacing profile. After each sublayer removal some average lattice spacing is extracted from the diffraction data and assigned to some location (see above and ref. [3]) in the layer. The resulting spacing-depth profile can be transformed into a concentration-depth profile if the relation between lattice parameter and composition is known (in many cases a linear equation will satisfy). For practical situations the following remarks are made: (i) successive sublayer removals could optimally be made in steps of the order of the depth analysed in the diffraction experiment; (ii) a correction for line shift by residual macrostress is necessary usually.

Surface-concentration estimates by X-ray diffraction analysis can be more accurate than those obtained by electron-microprobe analysis of specimen cross-sections, not only because the very surface layer has the largest weight in the diffracted X-rays recorded but also because the presence of porosity and/or second-phase particles in the surface layer leads to too small X-ray intensities generated in the electron-microprobe and thus to a seemingly too small concentration, as has been shown for a nitride layer on steel.[4]

Profile Breadth

Profile breadth originates from the imperfection of the microstructure of the specimen, from the instrument and from the X-ray spectrum used. Methods exist to extract the contribution from the imperfection of the microstructure.[1,2] It is common practice to interpret this contribution (structural line breadth) in terms of lattice distortions and of smallness of size of the diffracting crystallites. Lattice distortions can be induced, for example, by concentration variations and misfitting second-phase particles.

DETERMINATION OF MACROSTRESSES

A pragmatic division of the types of stresses/strains which can be found in crystalline material involves the distinction between

- macrostrain/stress, which is homogeneous over macroscopic distances and is related with a change of the average lattice spacing (associated with line shift),
- microstrain/stress, which changes over distances of the order of one lattice spacing and is related with variations of the lattice spacing with respect to the average value (associated with line broadening).

202

Many properties of surface layers are highly dependent on their state of stress. In common practice one usually tends to ignore the role of microstrains and one restricts the stress analysis to the determination of the (residual) macrostress. It can be shown that for an understanding of material behaviour often the combined effects of both macro- and microstrains should be considered.[3] In this section only the determination of macrostress is elucidated; for the determination of microstrain, see previous section.

Macrostresses can be determined using the so-called $\sin^2\psi$ method. The X-ray diffraction method has the potential to determine the elastic (macro)-strain, $\varepsilon_{\phi\psi}$, by utilizing the lattice spacing as an internal strain-gauge:

$$\varepsilon_{\varphi,\psi} = (d_{\varphi,\psi} - d_o)/d_o , \tag{6}$$

where d_o is the strain-free reference spacing and where Φ and ψ denote the specimen rotation angle and the specimen tilt angle, respectively (see Fig. 1).

For coatings a state of plane stress, characterized by two principal stresses σ_1 and σ_2, can usually be assumed. Then:

$$d_{\varphi,\psi} = d_{\psi=0} + m_\varphi \sin^2\psi , \tag{7a}$$

where

$$d_{\psi=0} = d_o + S_1 d_o (\sigma_1 + \sigma_2) \tag{7b}$$

and

$$m_\Phi = 1/2 \; S_2 d_o \sigma_\Phi \tag{7c}$$

with

$$\sigma_\Phi = \sigma_1 \cos^2\Phi + \sigma_2 \sin^2\Phi \tag{7d}$$

The so-called X-ray elastic constants S_1 and $1/2 \; S_2$ can be measured by external loading of a test specimen during X-ray diffraction analysis or may be calculated from single-crystal elastic constants using a grain-interaction model (Voigt-Reuss mean or Kröner[5]). By plotting $d_{\phi\psi}$ versus $\sin^2\psi$ a straight line with slope m_Φ is obtained from which σ_Φ can be calculated (eq. (7c)). In general $d_{\psi=0}$ will not deviate more than 1 % from d_o and an accurate value for σ_Φ is obtained by replacing d_o by the experimental $d_{\psi=0}$ in the expression for m_Φ.

The reference spacing is identical to the spacing measured in the strain-free direction $\psi_o(\Phi)$. If the ratio of S_1 and $1/2 \; S_2$ is known and - in general - two values of σ_Φ for Φ values $\pm$ 90° apart are available ($\sigma_\Phi + \sigma_{\Phi+90°} = \sigma_1 + \sigma_2$; see eq. (7d)), $\psi_o(\Phi)$ can be calculated according to[3]

$$\sin^2\psi_o(\Phi) = - \frac{S_1}{1/2S_2} \frac{\sigma_1 + \sigma_2}{\sigma_\Phi} . \tag{8}$$

In surface layers it holds for many practical cases $\sigma_1 = \sigma_2 = \sigma_\Phi = \sigma_{//}$ and $\sin^2\psi_o(\Phi)$ is independent of $\sigma_{//}$ and Φ and equal to $-S_1/(1/2 \; S_2)$.

The simultaneous determination of d_o and $S_1/(1/2 \; S_2)$ is possible if for a certain phase a number of d_ψ-$\sin^2\psi$ plots with different slopes, m_Φ, are

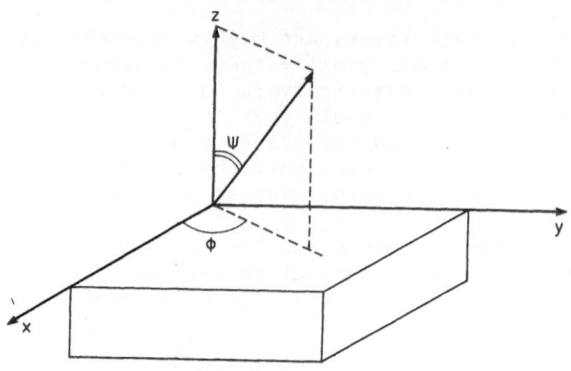

Fig. 1.   Definitions of angles Φ and ψ as used in eqs (6-8).

available (for constant Φ). As follows from eqs 7a, b for $\psi = \psi_o$, a plot of $d_{\psi=0}$ vs. m_Φ yields a straight line with a slope equal to $\sin^2\psi_o(\Phi)$ and a $d_{\psi=0}$ at $m_\Phi = 0$ equal to d_o. Of course d_o and $\sin^2\psi_o(\Phi)$ should be constants for the experiments considered.

Two common origins of internal stresses in surface layers are (i) the differences in thermal expansion/shrink between phases present in the layer/ substrate composite and (ii) the presence of **concentration profiles**. If, for example, after cooling from the temperature where the surface layer was produced, the lateral equilibrium dimensions of surface layer and substrate do not match, a lateral linear misfit occurs and a state of stress can develop. Owing to the presence of a concentration profile, the lateral linear misfit can change as a function of depth below the surface. Then a stress profile develops which is closely bound up with the concentration profile as has been found for the nitride layer on nitrided steels (see the second Case Study).

Compositional Variations and Macrostress Simultaneously

Until now it was tacitly assumed that within the diffracting volume the specimen is homogeneous with respect to composition (constant d_o value) and that no stress gradients occur perpendicular to the surface.

With X-ray stress measurements the depth analysed decreases as the specimen tilt angle increases (cf. Table 1). If composition- (and stress-) depth profiles occur, this variation in effective depth will lead to affected $d_{\Phi\psi}$ $-\sin^2\psi$ data. Inconsiderate fitting of straight lines to these data will lead to erroneous stress values. Even in the absence of stresses a composition-depth profile will lead to a non-zero slope for the straight line fitted to the $d_{\Phi\psi}$ $-\sin^2\psi$ data and thus an apparent (ghost) stress is detected. In general however, by selection of X-ray radiation which is strongly absorbed, the effect can be minimized and in some cases even be made insignificant.[6]

To determine a concentration-depth profile, by lattice-spacing measurements often the method of successive layer removals (e.g. by polishing) is applied. In the presence of depth-independent stresses, a concentration--depth profile can be still traced unambiguously by determination of the strain-free spacing, d_o, according to eq. (8), after each sublayer removal. Thereafter a correction for the effect of penetration of the X-rays has to be performed (see above). Moreover, each sublayer removal induces a redistribution of stresses within the remaining specimen to maintain mechanical equilibrium. The stress at the occurring surface then will deviate from the original one at that location.[7] The original stress profile can be reconstructed from the measured one.[7] For relatively high stress levels near the surface of the specimen the correction can be ignored if only some tens of μm's are removed from specimens thicker than a few mm's. The correction is always significant at depths where the stress reverses its sign.

If $\sigma_1 = \sigma_2 = \sigma_\phi = \sigma_{//}$, the strain-free direction ψ_o does not depend on $\sigma_{//}$. Then, even if a $\sigma_{//}$-depth profile occurs, correct d_o values are obtained at ψ_o as indicated below eq. (8), provided the X-ray elastic constants do not depend on composition (and depth).

Hence, in the presence of curvature in the $d_\psi - \sin^2\psi$ plot imposed by either a concentration profile in conjunction with depth-independent stresses or a concentration profile in conjunction with a $\sigma_{//}$ profile, the determination of the strain-free lattice spacing remains possible according to eq. (8).

CASE STUDIES

Macrostresses and Microstructure of TiC Coatings on Mo, W, Fe and Fe-C

In TiC coatings produced by Chemical Vapor Deposition (CVD) high internal stresses occur. These stresses influence the adherence of the layer and its resistance to cracking, fatigue, wear and corrosion. X-ray diffraction enables the measurement of such stresses present in coating and substrate without separating coating and substrate.

If it is assumed that no stresses develop during the deposition, then a quantitative prediction of the stress $\sigma_{//}$ (for this type of coating $\sigma_1 = \sigma_2 = \sigma_\phi = \sigma_{//}$) can be obtained from the difference in shrink of TiC layer and substrate during cooling from the deposition temperature to room temperature. Because the TiC coating is so much thinner than the substrate, the difference in shrink between layer and substrate will manifest itself predominantly[8,9] in the layer. The strain parallel to the surface as calculated[8,9] from the difference in shrink between layer and substrate, $\varepsilon_{//}^{th}$, will be compared with the value $\varepsilon_{//}$ obtained from the stress measured, $\sigma_{//}$, according to $\varepsilon_{//} = \sigma_{//}(1-\nu)/E$, where ν is the Poisson ratio and E is Young's modulus (note $\varepsilon_{//} \neq \varepsilon_{\psi=90°}$ because of elastic anisotropy).

Table 2. Strains Calculated from Thermal Shrinkages, $\varepsilon_{//}^{th}$, and Experimental Strains, $\varepsilon_{//}$, and Stresses, $\sigma_{//}$, in TiC Layers on Various Substrates

Layer	Substrate	Calculated $\varepsilon_{//}^{th}$ (%)	Experimental $\varepsilon_{//}$ (%)	Experimental $\sigma_{//}$ (MPa)
TiC	Mo	+0.16	+0.15	+820
TiC	W	+0.23	+0.21	+1130
TiC	Fe	-0.26	-0.32	-1715
TiC	Fe-1.0C	-0.65	-0.38	-2130

In Table 2 the results for TiC on Mo, W, Fe and Fe-1 wt% C are given. For TiC on Mo and on W the values of $\varepsilon_{//}^{th}$ and $\varepsilon_{//}$ agree very well; for TiC on Fe and in particular on Fe-1 wt% C a discrepancy occurs. This discrepancy can be explained for Fe from the phase transition γ Fe (f.c.c.) $\rightarrow$ α Fe (b.c.c.) at 1183 K (the CVD temperature was 1203 K) and for Fe-1 wt% C from the phase decomposition of austenite into ferrite + cementite at the eutectoid temperature of $\approx$ 996 K. During these transformation processes the "mobility" in the substrate is high and therefore strains can be accommodated. After the phase transformations recovery and recrystallization can contribute to a further stress relaxation. The strain measured at room temperature in the TiC coating on Fe indicates that relaxation effectively continues to about 760 K. For the TiC coating on Fe-1 wt% C such relaxation processes are hindered by the presence of carbon in the substrate: the strain measured in the coating indicates that relaxation effectively stops at about 840 K. Mo and W show neither phase transformations nor recovery or recrystallization in the temperature range concerned: the values of $\varepsilon^{th}_{//}$ and $\varepsilon_{//}$ do agree.

It is not generally realized that elements like Cr and Fe are usually present in TiC coatings chemically vapor deposited in an industrial reactor system. Cr and Fe may influence coating properties and/or the rate of formation of the coating. By comparing measured lattice parameters of TiC (after correction for the stress present) with lattice parameters calculated (using the known influence of Cr and Fe on the lattice parameter) it has been found that the Cr present in the TiC coatings is in solid solution, whereas the Fe is not.[10] This is consistent with the larger solid solubility in TiC of Cr than of Fe and with the presence of Fe-rich precipitates (see below).

X-ray diffractograms indicated that in the coating precipitates occur, which consist of austenite (in coatings with $\approx$ 3 at.% Cr) or of austenite and ferrite (for lower Cr contents). The fact that mainly austenite is present, indicates that the Fe particles were already present at the CVD temperature and did not form by a precipitation process during cooling after CVD. Normally one would expect that austenite transforms to ferrite on cooling to room temperature. However, the presence of Cr may have hindered the transformation, as is also known for steels alloyed with Cr.

The integral breadth observed for the region near the free surface is independent of the coating thickness. This cannot be explained if selective growth of crystallites occurs during layer deposition: then crystallite size should increase and lattice distortion should decrease with coating thickness, resulting in a decrease of the line broadening with coating thickness.

Line-broadening analysis showed that the crystallite size is small and constant throughout the coating. For the region near the free surface in addition an appreciable microstrain was observed, which can be explained as due to the strain fields surrounding the Fe precipitates which are much smaller in the surface region than near the coating/substrate interface.[10]

The presence of macrostress alone can already evoke a spectrum of lattice spacings as a consequence of elastic/plastic grain interaction, which leads to line broadening. For the present case this is not likely to occur since $\sigma_1 = \sigma_2$.[11] Indeed, no significant differences in integral breadths were observed before and after detaching the TiC coatings from their substrates, although the internal stress thereby reduces from about -3000 MPa to approximately zero. This agrees with the interpretation above where independent origins of macro- and microstrains have been distinguished.

Macrostress and Composition Profiles in γ'-Fe$_4$N Coating on Fe

Nitriding of steel is applied frequently (also) because the nitride layers formed have high resistance to wear and corrosion.

If pure Fe is nitrided in an NH_3/H_2 gas mixture a coating of γ'-Fe$_4$N, a non-stoichiometric compound, is formed. By X-ray diffraction analyses after successive sublayer removals the strain parallel to the surface as well

as the strain-free lattice spacing in the γ'-Fe$_4$N layer can be determined
as a function of the distance to the original free surface (Fig. 2a).[12]
Although the relation between the lattice parameter of γ'-Fe$_4$N and its nitrogen content is not precisely known, the lattice-spacing profile can be taken
as isomorphous with the nitrogen-concentration profile. (Because the nitrogen-concentration range for γ'-Fe$_4$N is less than ≈ 0.2 wt% N, it follows
that in this case nearly no method offers a better depth resolution than X-ray
diffraction.)

The interpretation of the stress profile is as follows. During cooling
from the nitriding temperature to room temperature the nitride shrinks less
than the iron substrate and a compressive stress in the layer ($\sigma_{//}$ and $\varepsilon_{//} < 0$)
will arise. The strain in the layer at the layer/substrate interface is
ascribed to this (thermal) effect (see discussion on TiC coatings).

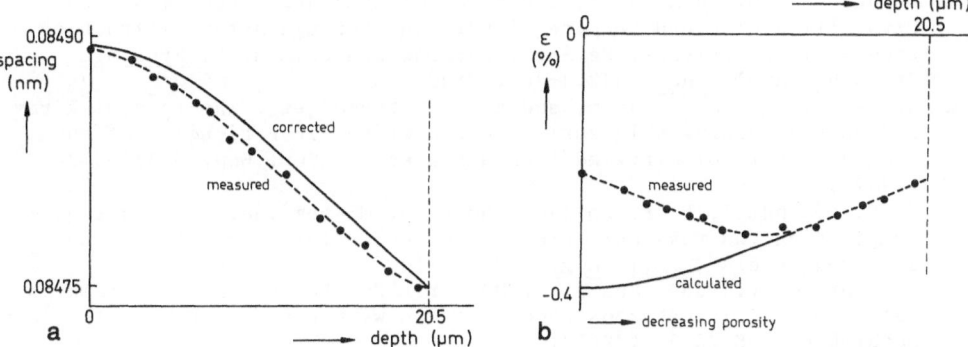

Fig. 2.: Results of X-ray diffraction analyses (after successive sublayer
removals) of a γ'-Fe$_4$N layer on nitrided steel (after ref. [12]).

(a) The as-measured and the penetration corrected strain-free lattice spacing as a function of distance to the free surface.
The corrected profile is isomorphous with the composition-depth
profile.
(b) The as-measured and the calculated (compressive) strain as a
function of distance to the free surface. The strain is calculated from the composition profile (Fig. 2a) and the difference
in thermal shrink between layer and substrate. Clearly relaxation has occurred. This is due to pore caused by N$_2$ gas originating from the decomposition of γ'-Fe$_4$N.

Because of the presence of a nitrogen-concentration gradient a compress-
ive-stress profile will develop in the layer, because the lateral dimension
tends to increase from the layer/substrate interface to the outer surface.
From this stress profile the corresponding strain profile can be calculated.
The latter can be added to the thermal strain mentioned. Fig. 2b shows the
measured and the thus predicted strain profiles.

It follows that near the free surface the absolute value of the strain measured is smaller than the one predicted. This is ascribed to pore formation, originating from N_2 formation in the unstable γ'-Fe_4N (an aging phenomenon). Because the nitrogen concentration near the outer surface is highest, and because that region of the layer is the oldest part, the pores will be present there mainly. By coalescence of pores channels develop in open connection with the gas atmosphere and, as a consequence, part of the thermal strain $\varepsilon_{/\!/}$ is accommodated during cooling.

ACKNOWLEDGEMENT

We are highly indebted to our co-workers Ir. W. G. Sloof and Ir. M. A. J. Somers, who contributed in an essential way.

REFERENCES

1. H. P. Klug and L. E. Alexander, "X-ray Diffraction Procedures for Poly-crystalline and Amorphous Materials, 2nd Ed., John Wiley & Sons, New York (1974).

2. R. Delhez, Th. H. de Keijser, and E. J. Mittemeijer, "Determination of Crystallite Size and Lattice Distortions through X-ray Diffraction Line Profile Analysis. Recipes, Methods and Comments", Fresenius Zeitschr. Anal. Chem. 312:1-16 (1982).

3. R. Delhez, Th. H. de Keijser, and E. J. Mittemeijer, "The role of X-ray diffraction analysis in surface engineering; investigation of the microstructure of nitrided iron and steels", Surf.Eng. 3:331-342 (1987).

4. H. C. F. Rozendaal, P. F. Colijn, and E. J. Mittemeijer, "Morphology, composition and residual stress of compound layers of nitrocarburized iron and steels", Surf. Eng. 1:30-42 (1985).

5. F. Bollenrath, V. Hauk, and E. H. Müller, "Zur Berechnung der vielkris-tallinen Elastizitätskonstanten aus den Werten der Einkristalle", Zeit. Metallkunde 58:76-82 (1967).

6. H. C. F. Rozendaal, "The Influence of Nitriding on the Microstructure and Stress State of Iron and Steel", Ph. D. thesis, Delft (1985).

7. M. G. Moore and W. P. Evans, "Mathematical corrections for stress in removed layers in XRD residual stress measurements", SAE Trans. 66: 340-345 (1958).

8. W. G. Sloof, H. J. M. Rijpkema, R. Delhez, Th. H. de Keijser, and E. J. Mittemeijer, "Prediction and measurement of compressive and tensile stresses in coatings of TiC and TiN chemically vapour deposited on various substrates", Surf. Eng. 3:59-63 (1987).

9. W. G. Sloof, R. Delhez, Th. H. de Keijser, and E. J. Mittemeijer, "Development and partial relaxation of internal stresses in thin TiC layers chemically vapour deposited on Fe-C substrates", J. Mat. Sci. 22:1701-1706 (1987).

10. W. G. Sloof, R. Delhez, Th. H. de Keijser, D. Schalkoord, P. P. J. Ramaekers, and G. F. Bastin, "Chemical constitution and microstructure of TiC_x coatings chemically vapour deposited on Fe-C substrates; effects of Fe and Cr", J.Mat.Sci. 23:1660-1672 (1988).

11. M. Hendriks, "X-ray Diffraction Study of Polycrystalline Silicon Layers", Ph. D. thesis, Delft (1985).

12. M. A. J. Somers and E. J. Mittemeijer, in preparation.

APPLICATION OF X-RAY DIFFRACTION TECHNIQUES

TO STUDY HIGHLY DISPERSED SUPPORTED METALS

Jerzy Pielaszek

Institute of Physical Chemistry
Polish Academy of Sciences
Warsaw, Poland

INTRODUCTION

In practice, the most important case of highly dispersed metals on poly-crystalline or quasi-crystalline substrates are supported catalysts. Of special interest are those with low metal loading (not exceeding few wt.%) and with small particle size of the metal phase (less than 100 Å). Their diffraction patterns are a superposition of a weak broad diffraction profile from the metal which is very often obscured by a predominant diffraction profile from the support. It is assumed that the contribution of the intensity from the metal-support interphase boundary to the total diffracted intensity is negligible. One of the objectives in catalyst studies is to follow structural changes induced into its active phase by the catalytic conditions. For the supported metal catalysts it means to observe changes which the diffraction profiles originating from the metal have undergone. The aim is then to obtain the diffraction profile originating from the metal phase alone which can be subsequently handled using conventional procedures of profile analysis (see e.g. [1]).

Most of the X-ray diffraction experiments with catalysts are currently being performed using conventional diffractometry and sealed-off X-ray tubes. It should be, however, stressed that modern powerful X-ray sources as rotating anodes used with conventional diffractometers greatly facilitate the work and the use of synchrotron radiation or storage ring opens new possibilities of recording diffraction spectra with different wavelengths [2].

EXPERIMENTAL

The following catalysts already investigated in this laboratory were studied: 0.69 wt.% Pd/SiO_2 [3], 0.76 wt.% Pd/SiO_2 [4], 1 wt.% $Pd-\gamma Al_2O_3$, 3 wt.% $Pd/\gamma-Al_2O_3$ [5]. Pure SiO_2 and 0.1 wt.% $Pd/\gamma-Al_2O_3$ were used as standards for the procedure for support background subtraction. The measurements were performed in an X-ray diffraction camera-gradientless catalytic reactor of the type described elsewhere [6]. The catalysts were deposited on slices of porous glass and mounted in the sample holder of the camera. The total weight of the catalyst sample was about 20 mg, i.e. for the sample area of about 1.5 cm^2 it gives, depending on the metal loading, the concentration of the metal from about 0.1 mg/cm^2 to about 0.4 mg/cm^2.

The measurements were done on a conventional X-ray Rigaku-Denki diffractometer using CuK$_\alpha$, Ni filtered radiation from a 1.5 kW tube. The estimation
of an optimum counting time at each point[7] for a counting rate of 100 c/s
(close to typical for our case at the maximum of the peak), the peak to background ratio of 0.1 and assumed relative error of 5 % gives the value of
about 370 s. This is unacceptable for any kinetic study of a catalyst. As
a compromise between the need of a reasonably good statistics and a total
time of the measurements the counting time was 100 s. The recordings were
at 2Θ steps of 0.05 and occasionally for some kinetic studies at 0.1. Well
annealed palladium fillings were used for instrumental correction. The spectra were handled using C-128 microcomputer.

BACKGROUND SUBTRACTION

When a support of a catalyst is amorphous and metal loading is high
(of the order of 10 wt.%) the diffraction profile from the metal can be
truncated by subtracting linear background defined either on the diffraction
pattern at points far away from the diffraction peak from the metal or by
measuring independently the scattering from the support and the matching
(normalizing) it with the observed intensity from the catalyst. These methods were used mainly for catalysts with silica gel as a support. For the
catalysts with crystalline support and high metal loading the exact spectrum
from the metallic phase can be obtained by employing a Fourier unfolding
technique[8]. It was successfully used for studies of zeolite-supported
catalysts[9].

A special interest lies however in catalysts with quasi-amorphous supports, low metal loading and high dispersion which produce diffraction profile with very broad low intensity peaks from the metallic phase. In such
a case the support background subtraction solves several problems. First,
the long tails of the metal diffraction profile makes difficult proper background level estimation and secondly the overall counting statistical error
at given diffraction angle influences substantially the resulting diffraction
profile from the metal phase making necessary the use of proper smoothing
procedures.

Schematically, the problem of the support background separation is represented in Fig. 1. Both catalyst and support diffraction profiles are
registered in a chosen range of diffraction angles $2\Theta_A$-$2\Theta_B$ (lower part of

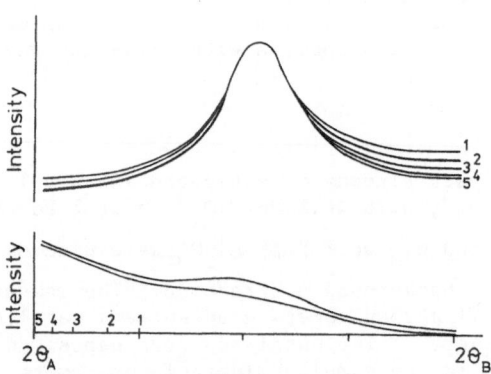

Fig. 1. Effect of the background level choice on the support and on the
catalyst diffraction profiles (lower part) on the shape of resulting
profile from the metal (upper part).

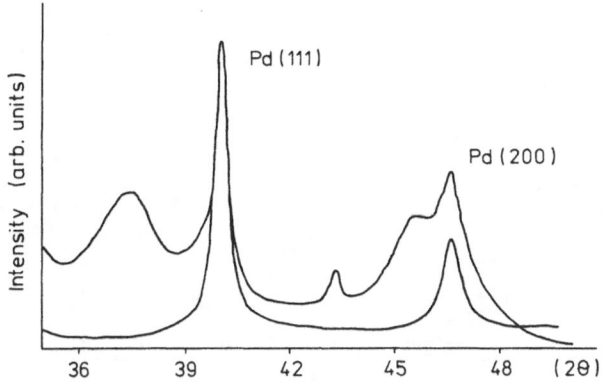

Fig. 2. Part of diffraction pattern from the 3 wt.% Pd/γ-Al$_2$O$_3$ and profile from palladium after subtraction of the support.

the figure). Then both profiles are normalized at an arbitrary selected angle $2\Theta_1$ and subtracted (upper part of the figure). It is implicitly assumed that the $2\Theta_1$ position on both spectra corresponds to the background level. To verify this the procedure is repeated for some other values of 2Θ ($2\Theta_2$, $2\Theta_3$, etc.) and the resulting profiles compared. The 2Θ value at which the resulting profile does not change any more is chosen as a true background position. This procedure can be applied either to low or high angle side of a selected diffraction peak. In the case of no overlap with other peaks the background levels on both sides are the same. Contrary, the non-matching of the background levels is indicative of the overlap. It can happen that the counting range is chosen too narrow and the procedure should be repeated, especially when peaks are strongly asymmetrical. When the diffraction peaks from the metal are well resolved and reasonably strong this procedure is equivalent to those of the normalization to the maximum of a background peak (ref. 8,10).

RESULTS

The catalysts have undergone different treatments. Figure 2 shows part of the diffraction pattern of the 3 wt.% Pd/Al$_2$O$_3$ catalyst and the one after subtraction of the profile of 0.1 wt.% Pd/γ-Al$_2$O$_3$ catalyst. It was found that the pure γ-Al$_2$O$_3$ can not be used as a catalyst support standard because the effect of a controlled atmosphere and/or catalytic reaction was different on the pure γ-Al$_2$O$_3$ and the same component in a catalyst. This is why the catalyst with 0.1 wt.% of metal loading was chosen as a support standard. It was subjected to the same treatment as the 3 wt.% catalyst. The catalyst was one of a series studied in the process of hydride formation and decomposition[9] and the Fourier analysis has shown that the transformation of the metal into hydride and its decomposition lowers the average particle size from 140 Å to about 80 Å and increases the microstrains $\langle \varepsilon_{50}^2 \rangle^{1/2}$ from 0.0067 to 0.011.

Part of a diffraction pattern from a 1 wt.% Pd/γ-Al$_2$O$_3$ catalyst together with corresponding pattern of the support (0.1 wt.% Pd catalyst) and the resulting (111) profile are shown in Fig. 3a. When studying this catalyst in the reaction of acetylene hydrogenation it was possible to correlate quantitatively the change in its selectivity versus the amount of hydride phase formed[11]. Typical pattern of Pd partially transformed into hydride is shown in Fig. 3b, together with the catalyst and support diffraction profiles. Usually it is assumed that silica gel support used for catalyst manufacturing

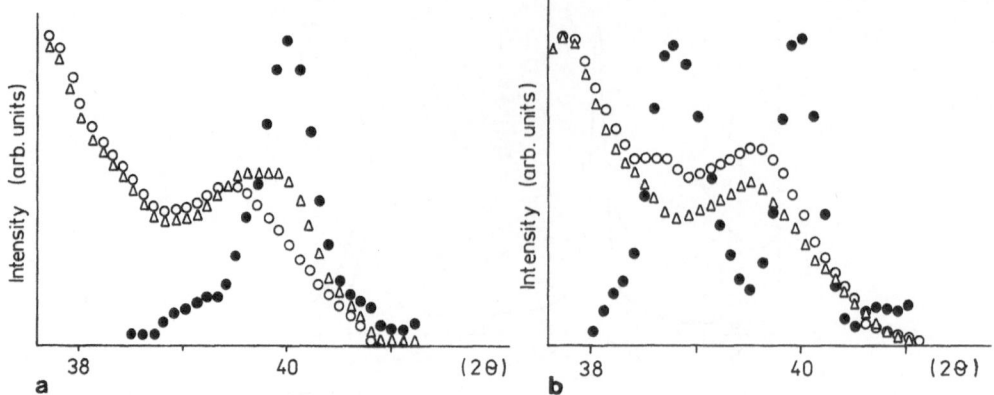

Fig. 3a.   Part of diffraction pattern from 1 wt.% Pd/Al$_2$O$_3$ and 0.1 wt.%
           Pd/γ-Al$_2$O$_3$ together with resulting profile of the Pd(111) peak;

   3b.   as 3a for the catalyst partially transformed into hydride.

are amorphous. Close inspection of its diffraction spectra reveals however
that possible irregularities reflecting in fact the quasi crystalline state
are not to be neglected when the intensity diffracted by the deposited metal
is very low. In most cases the peak to background ratio of the strongest
(111) reflection is about 0.1 and the next (200) reflection is not distinguishable. This was also the case of the 0.69 wt.% Pd/SiO$_2$. Nevertheless,
when using this catalyst in the reaction of acetylene hydrogenation it was
possible to follow the formation of the hydride phase. In Fig. 4 are presented resulting profiles of the metal phase of the starting catalyst, and
of partially and fully transformed into hydride. Particle size estimated
from the FWHM is about 90 Å.

    Of special interest in catalyst studies is a problem of possible metal-
support interaction. For some supports this is a well established fact
(see e.g. [12] for the case of TiO$_2$). The interaction of palladium with silica
after heating a 10 wt.% PD/SiO$_2$ catalyst in hydrogen at 600 °C was also
reported [13].

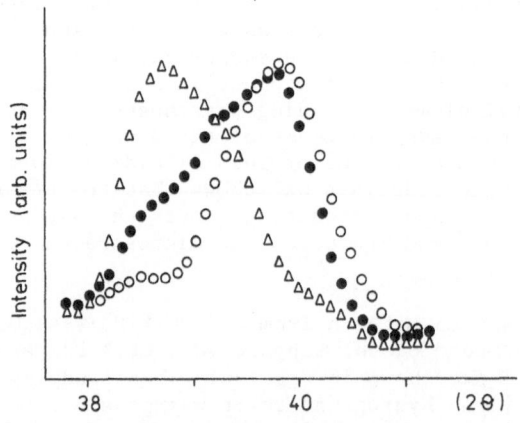

Fig. 4.   0.69 wt.% Pd/SiO$_2$: profiles of palladium after support background
          subtraction. o – pure palladium, ● – palladium partially trans-
          formed into hydride, Δ – palladium hydride. Particle size (FWHM)
          about 90 Å.

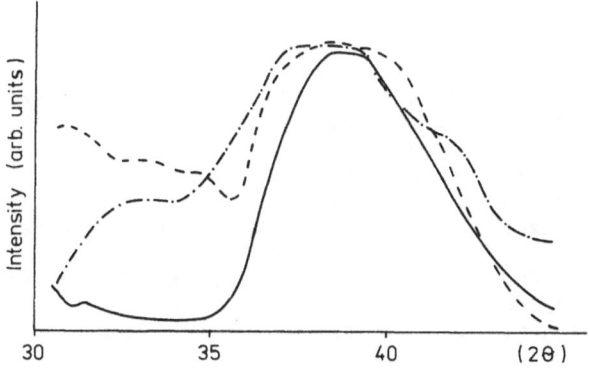

Fig. 5. 0.76 wt.% Pd/SiO$_2$: profiles of palladium after support background subtraction, (——) palladium after hydride decomposition, (---) after reduction at 450 °C, (-.-) after reduction at 300 °C. Particle size (FWHM) about 30 Å.

The 0.76 wt.% Pd/SiO$_2$ catalyst was subjected to high temperature reduction at 450 and 300 °C and to hydride transformation at room temperature Figure 5 shows the difference between the shape of the resulting (111) spectra. After high temperature reduction some shoulders appear at the low angle side of (111) reflection. They are suppressed by the transformation of palladium into hydride. The possible explanation of the shoulders is the formation of palladium silicides which up to now was confirmed only at higher temperature reduction and for catalyst with much higher metal loading. It should be however pointed out that the possible effect of the reaction conditions on the SiO$_2$ (support) structure although not found yet cannot be completely ruled out. The particle size estimated from the FWHM of the profile after hydride decomposition is about 30 Å.

All the presented spectra were subjected to smoothing procedure. Due to high statistical error of the resulting subtracted profiles smoothing by the method of least squares of a polynomial of selected order over chosen number of points or spline functions did not give good results. Also smoothing by the Fourier transformation method recommended for data with high statistical error was giving profiles with significant oscillations at the tails. The best results were obtained for smoothing with the moving average over the preselected number (here 11) of points. This method should however be used with some care since although its influence on the position of the centroid and on the FWHM of the profile is negligible the possible flattening of the maximum of the profile can change the value of the integral half width of the profile.

CONCLUSIONS

It was shown that for metals supported on quasi-amorphous supports it is possible to extract diffraction profiles originating from the metal even for samples with low metal loading. The proper background subtraction needs both appropriate normalization procedure and choice of background standard sample. For 3 wt.% of Pd on γ-Al$_2$O$_3$ and particle size of the order of 140 Å the resulting pattern was good enough to perform Fourier analysis of the profiles. For catalysts with less then 1 wt.% of metal loading the resulting profiles, although are subjected to high statistical error and need significant smoothing procedure, can be used for calculation of particle size and qualitative intensity measurements as well as for observation of phase transformation caused by chemical and thermal treatment.

REFERENCES

1. B. E. Warren, "X-Ray Diffraction", Addison-Wesley Publ. Co. (1969).
2. J. B. Cohen, P. Georgopoulos, J. B. Butt, and R. L. Burtwell, Jr., Diffraction from Supported Metal Catalysts, ASC Symp. Ser. (Catal. Character Sci.) 288:385 (1985).
3. J. Pielaszek, to be published in Catalysis Letters.
4. W. Juszczyk, Z. Karpiński, J. Pielaszek, I. Ratajszykowa, and Z. Stanasiuk, Characterization of Supported Palladium Catalysts, submitted for publication.
5. A. Borodziński, R. Duś, R. Frąckiewicz, A. Janko, W. Palczewska, in: "Proceedings 6th International Congress on Catalysis", G.C. Bond, P. B. Wells, F. C. Tompkins, Eds, 1:150, The Chem. Soc. London (1977).
6. J. Zieliński, A. Borodziński, A Novel-Design Camera for Simultaneous X-Ray and Kinetic Studies, Appl. Catal. 13:305 (1985).
7. M. A. Blokhin, Methods of X-Ray and Spectral Investigations, Moscow (1959).
8. A. G. Dhere, R. J. DeAngelis, A Method for Separation of X-ray Diffraction Patterns from Co-ZSM-5 Cat5alysts, J. Catal. 81:464 (1983).
9. A. D. Dhere, R. J. DeAngelis, In Situ Diffraction Investigation of the Structural Characteristics of two Co/YSM-5 Catalysts, J. Catal. 192:145 (1985).
10. J. Pielaszek, J. B. Cohen, A Low-High Temperature Camera for In-situ X-ray Diffraction Studies of Catalysts, Adv. X-Ray Anal. 27:285 (1984).
11. J. Pielaszek, J. Sobczak, W. Palczewska, Supported Palladium Phase Transformation in a $H_2+C_2H_2$ mixture, Communication presented at XI Polish-Czechoslovak Coll. on Catalysis, Karniowce, Poland, Oct. (1985).
12. S. J. Tauster, S. C. Fung, and R. L. Garten, Strong Metal-Support Interactions: Groups 8 Noble Metals Supported on TiO_2, J. Amer. Chem. Soc. 100:170 (1978).
13. R. L. Moss, D. Pope, B. J. Davis, D. M. Edwards, The Structure and Activity of Supported Metal Catalysts, J. Catal. 58:206 (1979).

EFFECTIVE DEPTH OF X-RAY PENETRATION

AND ITS ORIENTATION DEPENDENCE

I. Tomov

Institute of Physical Chemistry
Bulgarian Academy of Sciences
Sofia 1040, Bulgaria

A new diffraction method for calculating the effective depth of X-ray penetration is proposed. The effect of pole density on the practical effective depth of X-ray penetration t* is analysed. It is shown that t* may be considered as a superposition of the contribution of two components. The first component is due to a random crystallite distribution and the second component is due to the pole density. The effective depths of textured nickel specimens as well as of simulated random crystallite distribution in the same specimens are determined.

INTRODUCTION

In X-ray studies of polycrystalline surface layers and bulk metal materials, it is necessary to know the thickness of the layer from which the diffraction pattern has been obtained. Hence, it is important to know the effective depth of X-ray penetration, which concerns reflection techniques.[1] Cullity's analysis reveals a possibility for an a priori determination of the effective depth and takes into consideration a trend of increase in its value with the increase in the geometrical factor, i.e. the Bragg angle Θ.

The information obtained by[1] is useful, but not precise, since it ignores the integrated intensity I in a selected <hkl> direction. It is known, however, that I is dependent on the real structure of the polycrystalline material (for example the crystallite orientation distribution and perfection) and on instrumental conditions. This means that by ignoring I, the effect of the aforesaid factors on the effective depth of X-ray penetration is also ignored in practice. The progress of the problem, with taking into account the effect of experimental conditions, has been presented in a previous paper[2], which revealed a possibility to establish an orientation dependence of the effective depth of X-ray penetration[3]. The aim of the present paper is to summarize these studies of ours, emphasizing at the same time the physical essence of the established dependence.

EXPRESSING THE EFFECTIVE DEPTH OF X-RAY PENETRATION WITH THE HELP OF INTEGRATED INTENSITY

Definition of the Problem

The present study concerns both random and textured specimens, the orientation distribution of which is uniform (homogeneous texture). This means that the orientation distribution in each thin layer is identical to the distribution in its neighboring layer.

Proceeding from Cullity's work, it is easily established that the integrated intensity diffracted by an infinitely thin layer dt located at a depth t under the specimen surface may be expressed by

$$dI = I \frac{2\mu}{\sin \Theta} \exp(-\frac{2 \mu t}{\sin \Theta}) \, dt, \tag{1}$$

where I is the total integrated intensity, diffracted by an infinitely thick specimen, μ is the linear absorption coefficient of the material. When this expression is integrated for each chosen depth t of the material, the integrated intensity $I_{(t)}$ diffracted by a layer with a thickness of t is obtained

$$I_{(t)} = \int_{t=0}^{t=t} dI = I[\, 1 - \exp(-\frac{2\mu t}{\sin \Theta})]. \tag{2}$$

The following equation may be written for each

$$I = I_{(t)} + \Delta I_{(t)}, \tag{3}$$

where $\Delta I_{(t)}$ $(0 \leq I_{(t)} \leq I)$ is the quantity of integrated intensity which is "ignored" beyond the effective depth t. After substituting eq. (3) in eq. (2) the following expression can be derived for the effective depth

$$t = \frac{\sin \Theta}{2\mu} \ln \frac{I}{\Delta I_{(t)}}. \tag{4}$$

This expression is valid for determining the effective depth t in the interval $[0, \infty]$. It is evident from eq. (2) that each effective depth t is responsible for the integrated intensity $I_{(t)}$.

Expressing the Practical Effective Depth of X-Ray Penetration

To be able to define the effective depth t*, which is responsible for almost the entire integrated intensity, the intensity $\Delta I_{(t*)}$ must be negligibly small. Let us denote t* as the practical effective depth of X-ray penetration. Its determination requires that a criterion be formulated for a negligibly small measurable intensity.

It is known that the sensitivity for detecting weak peaks is determined by the statistical fluctuations of the background. It is impossible to detect a diffraction peak, if the background fluctuations are comparable to the peak height. We will express the negligibly small intensity $\Delta I_{(t*)}$ by means of the background fluctuation in connection with the above indicated physical fact. Moreover, we assume the value of the fluctuation to be $\sqrt{B}$, i.e.

$$\Delta I_{(t*)} = \sqrt{B} , \tag{5}$$

where B is the registered value for the background. This definition (5)
takes into consideration the fact that measurement sensitivity is dependent
not only on the precision of the X-ray instrument, but also on the structure
of the specimen, i.e. on its texture, density and type of lattice defects.
By substituting eq. (5) in eq. (4) an expression for the practical effective
depth of X-ray penetration is obtained

$$t* = \frac{\sin \Theta}{2\mu} \ell n \frac{I}{\sqrt{B}} . \tag{6}$$

The equation is valid for all $I \geq \sqrt{B}$. The background B is approximately
equal for all diffraction lines of the specimen, measured under identical
experimental conditions.[4] The t* thus expressed is dependent on all factors
the integrated intensity and background are dependent on. In general the
integrated intensity is influenced by the crystallographic and real structure of the polycrystalline material and the instrumental conditions. (This
influence affects all parameters the integrated intensity is a function of,
namely: integrated reflection from unit crystal volume, linear absorption
coefficient, intensity of incident X-ray beam and its cross-section (total
energy falling per unit time on the specimen)[5] and the pole density.) From
a posteriori considerations these same factors, not counting the crystallographic structure, affect the value of the background, and hence the measurement sensitivity. The effective depth t* is directly connected to the whole
set of experimental conditions, which are taken into account by means of a
number defined by the $(I/\sqrt{B})$ ratio. What is more, t* is responsible for
almost the total measured integrated intensity, ignoring only the contribution which is equal to the measurement sensitivity (a measure of which is the
chosen background fluctuation $\sqrt{B}$). Consequently, t* defines the lower level
of the layer from which nearly the entire diffraction information on the respective <hkl> crystalline direction under the given experimental conditions
is obtained. Since t* is connected to concrete measurement results, it can
be determined only after the experiment, i.e. we are concerned with an a
posteriori method for its determination.

RELATION BETWEEN EFFECTIVE DEPTH OF X-RAY PENETRATION AND POLE DENSITY

It is known from texture analysis that the measured intensity $I_{i(\varphi)}$ in
φ direction of the i-pole figure is proportional to the pole density $P_{i(\varphi)}$
$P_{i(\varphi)}$ in the same direction (ref.[6]).

$$I_{i(\varphi)} = N_i P_{i(\varphi)} , \tag{7}$$

where φ is the angle between the normals of the reflecting planes and the
normal of the surface of the flat specimen, N_i, is the normalization factor,
which reduces the measured intensity values to multiples of the random density[6,7]. In practice, the measured effect from the random distribution of
a given crystalline direction is simulated by means of the respective normalization factor. At $\varphi = 0$, which indicates a diffraction vector parallel to
the normal of the specimen surface,

$$I_i = N_i P_i . \tag{8}$$

Substituting eq. (8) in eq. (6), we obtain

$$t_i^* = \frac{\sin \Theta}{2 \mu} \ell n \frac{N_i P_i}{\sqrt{B}} . \tag{9}$$

217

The lower limit of the pole density $P_{i,min}$ may be defined from the condition $t_i^* = 0$, i.e.

$$\sqrt{B} = N_i P_{i,min}.$$ (10)

The physical essence of $P_{i,min}$ is that each $P_i \leq P_{i,min}$ causes a diffraction contribution, which is commensurable to the background fluctuations under the respective experimental conditions.

Formally, t_i^* may be considered as a superposition of the contributions of two components. The first component t_i^r, due to a random crystallite distribution is defined by

$$t_i^r = \frac{\sin\Theta}{2\mu} \ln \frac{N_i}{\sqrt{B}}.$$ (11)

It is influenced by the experimental conditons through the parameters N_i and B.

The second component t_i^p, due to the pole density, is defined by

$$t_i^p = \frac{\sin\Theta}{2\mu} \ln P_i.$$ (12)

It is influenced only by the crystallite orientation distribution (texture) of the specimen. For homogeneous textures P_i is a constant for the relevant crystal direction, independent of the depth. Eq. (12) is valid for all $P_i > 0$, but t_i^p is positive for $P_i > 1$. At a random distribution $P_i = 1$, and then $t_i^* = t_i^r$. In the cases of $P_i < 1$ the contributions are negative. Consequently, there actually exists an effective depth t_i^*, based on the superposition of two indivisibly bound components t_i^r and t_i^p.

EXPERIMENTAL AND DISCUSSION

Electrolytically deposited nickel layers were employed as models for the present study. They represent fiber textures with varying texture components: < 100> + <221> (specimens No. 1, 2), <110> + <411> (specimen No. 3), <110> + <411> + <100> (specimen No. 4) and <110> + <100> + <221> (specimen No. 5).

The X-ray measurements of the pole figures were performed with a texture goniometer (CuK_α radiation and nickel filter) and an X-ray diffractometer (CuK_α radiation and LiF focussing monochromator). The lines measured by an X-ray diffractometer are equivalent to the pole figures at $\varphi = 0$. In the present study both terms are employed.

We have illustrated that the effective depth is a function of the integrated intensity. If experimentally determined, however, its values depend on the scanning time (respectively, on the parameters of the scan mode) of the diffraction lines. The measure for the diffracted X-ray intensity during diffractometric measurements is the counting rate. The value of the counting rate depends on the real structure of the specimen and on the whole set of instrumental conditions, not counting the scan mode. This is why in the case of effective depth determination the integrated intensity has to be expressed through the counting rate.

Table 1. Effective Depth of X-ray Penetration $t^*_{hkl} \times 10^6$ m and Pole Density P_{hkl} of 111, 200 and 220 Diffraction Lines of Electrolytically Deposited Nickel Layers (Specimens No. 1 to 5); CuK_α Radiation, LiF Focussing Monochromator and X-ray Tube Rating 800 Watts.

hkl	$t^*_{hkl} \times 10^6$ m, P_{hkl}									
	1		2		3		4		5	
111	18.2	0.07	23.9	0.25	10.6	0.10	19.9	0.11	20.3	0.13
200	47.9	22.90	46.6	17.90	19.8	0.12	30.6	0.92	41.7	7.90
220	-	0.003	11.0	0.02	68.0	38.50	64.0	22.70	59.8	13.90

If we are to illustrate the dependence of the effective depth of X-ray penetration from texture, we will have to determine the difference in t^*_i and t^r_i for equivalent directions of the same specimen. To this end the values of I_i and N_i have to be found from diffractometric measurements by the method described in[3]. The effective depths are calculated in <111>, <100> and <110> directions, by measuring 111, 200 and 220 diffraction lines, respectively. The values are highest in the directions of the main texture components. This fact becomes evident from the t^*_i and P_i values listed in Table 1. They are obtained by measuring five specimens with a diffractometer at an X-ray tube rating 800 Watts. The comparison of the t^*_i values of equivalent directions of the specimens indicates that the higher the P_i value, the greater the effective depth. The low background, due to the use of a focussing monochromator, also leads to high effective depth values t^*_i. We should note here that $P_{110} < P_{110,min}$ for <110> direction (diffraction line 220) of specimen No. 1, which was the reason why it was not possible to calculate the value of t^*_{110}.

In order to compare the magnitudes t^r_i for the same specimens (No. 1 to 5) Table 2 lists their values relevant for X-ray tube rating 800 Watts. The differences in the t^r_i values for equivalent directions are insignificant. The absolute error Δt_i of the results listed in Tables 1 and 2, which is due to the counting statistics is approximately 0.1×10^{-6} m.

Table 2. Effective Depth of X-ray Penetration $t^r_{hkl} \times 10^6$ m of 111, 200 and 220 Diffraction Lines of Electrolytically Deposited Nickel Layers (Specimens No. 1 to 5); Simulated for CuK_α Radiation, LiF Focussing Monochromator and X-ray Tube Rating 800 Watts.

hkl	$t^r_{hkl} \times 10^6$ m				
	1	2	3	4	5
111	30.5	30.3	30.2	30.1	29.8
200	31.2	31.1	31.1	31.0	30.6
220	40.0	40.5	40.4	40.4	39.9

CONCLUSION

The essence of the treated phenomenon becomes clear if we take into consideration the fact that the pole density is proportional to the volume fraction of crystallites which corresponds to it. This means that for each particular value of effective depth of X-ray penetration the volume fraction of crystallites is responsible, which has correct orientation for reflection. At higher pole density, the measurable diffraction contributions are due to crystallites located in relatively deep layers below the specimen surface. At $P_i = P_{i,min}$ the corresponding volume fraction of crystallites supplies diffracted intensity commensurable to the selected background fluctuation $\sqrt{B}$. This is the physical essence of the treated phenomenon.

The a priori determination of the effective depth according to Cullity

$$t = \frac{\sin \Theta}{2\mu} \ell n \frac{1}{1-G_{(t)}} = \frac{\sin \Theta}{2\mu} \ell n \, G_{(t)} \qquad (13)$$

where $G_{(t)}$ is the postulated part of the integrated intensity which is due to the penetration of the X-rays to a depth t, leads to physically incorrect corollaries. For example, if $G_{(t)} > (I/\sqrt{B})$, then $t > t^*$, because a contribution is added to the integrated intensity which is less than the measurement sensitivity and the registering of such a contribution is physically imposs- ible. What is more, if for a single phase diffraction pattern of textured specimen we accept $G_{(t)}$ = const, then we will be determining an effective depth even in the directions in which no intensity has been measured (there are no diffraction contributions). Moreover, in the a priori determination of the effective depth we cannot take into consideration the orientation de- pendence since the portion $G_{(t)}$ is postulated apart from the real structure of the specimen (in particular its texture).

The orientation dependence of t^* enables us to determine in some cases the quantities pole density and normalization factor – well known from tex- ture analysis – in selected directions by the use of an X-ray diffractometer.

The performed study also offers a possibility in assessment of the ef- fective irradiated volume to be made in the case of a diffractometer. If S is the cross-section area of the incident X-ray beam, the effective irra- diated volume V_i^r of a specimen with a random distribution of crystallites may be expressed by

$$V_i^r = \frac{S}{\sin \Theta} \, t_i^r \qquad (14)$$

At substitution of t_i^r from Eq. (11) it follows

$$V_i^r = \frac{S}{2\mu} \ell n \frac{N_i}{\sqrt{B}} \qquad (15)$$

so that the effective irradiated volume is not dependent on the Bragg angle, V_i^r, however, is a logarithmic function of the ratio $N_i/\sqrt{B}$, which decreases with the decrease in N_i (respectively with the increase in $P_{i,min}$ (see Eq. (10)), since the background is approximately constant for all diffraction lines of the specimen, measured under particular experimental conditions[4]. Consequently, the presence of a background determines the change of the V_i^r values with different hkl. This result does not correspond to the result of the idealized treatment of Cullity, which states that the effective irra- diated volume is constant and does not depend on Θ. (The treatment does not take into account the presence of a background during experimental measurements.) The constant value of the absorption factor $1/2\ \mu$ is only a prerequisite for a constant value of the effective irradiated volume, but the experimental conditions impede this possibility. However, if the measurement is carried out under ideal experimental conditions, which ensure a background $B \rightarrow 0$, then the effective irradiated volume will be a constant, the value of which, according to Eq. (15), will be $V_i^r \rightarrow \infty$. In this sense, the result of our treatment should be understood as a supplement to Cullity's treatment. This supplement concerns the case when the experimental conditions are taken into account.

REFERENCES

1. B. D. Cullity, Element of X-Ray Diffraction, Addison and Wesley Publ. Co., London (1967).
2. I. Tomov, A new diffraction method for the determination of the effective depth of X-ray penetration, Phys.Stat. Sol. (a) 95:397 (1986).
3. I. Tomov, Orientation dependence of the effective depth of X-ray penetration, Phys. Stat. Sol. (a) 98:43 (1986).
4. W. Parrish and T. R. Kohler, Use of counter tubes in X-ray analysis, Review Sci. Instrum. 27:795 (1956).
5. R. W. James, The Optical Principles of the Diffraction of X-Rays, G. Bell and Sons, Ltd., London (1965).
6. I. T. Tomov and H.-J. Bunge, An analytical method for the quantitative determination of the volume fraction in fibre textures, Texture Cryst. Sol. 3:73 (1979).
7. H.-J. Bunge, Texture Analysis in Material Science - Mathematical Methods, Buttersworth, London (1982).

STRUCTURE OF LASER MODIFIED SURFACE LAYERS OF AlZn ALLOYS

V. Synecek, J. Lasek, P. Bartuska, and M. Simerská

Institute of Physics, Czechoslovak Academy of Sciences

Prague 6, Czechoslovakia

INTRODUCTION

Specific features of laser treatment of metals arise from the possibility of supplying large amount of energy on a small area of metal surface [1-4]. The heat input is also substantially faster than its output by heat conduction of the material. Only very thin surface layers are therefore heated up during a locally short interval of laser exposure. Both the fraction of incident beam energy available for rapid heating and the rate of subsequent cooling determine the influence of laser treatment on the structure of material. The final structure is thus affected by both the optical (absorptivity) and the thermal (thermal conductivity and thermal diffusivity) properties of the material. The absorptivity of clean smooth solid metal surfaces is rather low (for polished aluminum at room temperature it is about two per cent only) and depends on both their temperature and the wavelength of laser radiation.

Metal surfaces are often mechanically roughened or provided with anti-reflecting coatings (oxides, phosphates, carbon, etc.) to increase the efficiency of laser beam heating of materials below their melting temperature. The absorptivity rapidly increases at the melting temperature which allows an easy laser welding (below the boiling temperature) and cutting (above the boiling temperature) to be performed.

Laser treatment of aluminum and of its alloys is somewhat more difficult than of other metallic materials due to their low absorptivity for laser radiation. Many papers have been published dealing with problems of laser treatment of aluminum alloys both in the basic research, e.g.[5-8] and in the technical applications including welding, cutting[8-11], and surface treatment (refs 12-16).

Laser surface melting of aluminum alloys enables us to study processes proceeding during rapid solidification of the melted surface layers and during fast quenching of alloy materials laser annealed below the solidus temperature . The improvement of mechanical properties of age-hardenable aluminum alloys is based on the formation of metastable structure states during the proper heat treatment. The high rates of heating and cooling during laser treatment may thus lead to surface layer structures with considerably different properties from those of bulk material.

Laser treatment of AlZn based alloys presents additional difficulties due to the substantial increase of vapor tension of zinc on heating (pure zinc has the boiling point at 907 $^{\circ}$C, aluminum at ca 2500 $^{\circ}$C). It may lead to an explosive vaporization of zinc followed by immediate oxidation of its vapor already at small laser induced superheating of the irradiated area above the boiling temperature of the alloy. The product of power density and of interaction time should be therefore kept within much more narrow limits than for other aluminum alloys. The oxide deposits along the laser path initiate a substantial and uncontrolled increase of energy absorption leading to severe damages of surface layer.

The aim of the present paper is to gain the basic knowledge of structure changes introduced by laser surface treatment of pure AlZn alloys both below and above the melting temperature of the alloy.

The model system Al-Zn used often for the basic investigations of the precipitation phenomena in alloys has a great advantage in a very broad concentration region of the solid solution of zinc in aluminum at elevated temperatures. The mutual miscibility of zinc and aluminum in the solid state extends to more than 63 at.% Zn and in about 100 $^{\circ}$C broad temperature interval for alloys containing more than 30 at.% Zn. This temperature interval is expanding to lower zinc contents being larger than 270 $^{\circ}$C for the Al 15 at.% Zn alloy employed in the present study. This broad region of solid solution and large miscibility gap in the Al-Zn system present extremely suitable conditions for the investigations of the mechanisms and kinetics of various phase transformations in wide temperature and concentration ranges[17-21]. Such processes comprise e.g. the decomposition of supersaturated solid solution leading to the formation of solute enriched coherent Guinier-Preston (G.P.) zones responsible for the precipitation hardening of alloys, transformations of coherent precipitates into metastable partially coherent rhombohedral α_R' and semicoherent fcc α' phases, and the direct discontinuous nucleation and growth of equilibrium phases from the supersaturated solid solution. The latter process is related to eutectoid decomposition of the alloy into the lamellar structure of α matrix and of ß phase which can proceed in two-phase region in alloys bearing more than 10 at.% Zn.

The phase transformations in AlZn alloys allow us to get a very detailed information on the processes and structure changes proceeding during laser processing from X-ray diffraction analysis and SEM examination. The results of the present investigation clearly indicate that the distribution of ß precipitates at various positions of the solidification front reveals the existence and changes of the extensive segregation of zinc at this front and provides us with details of the mechanism of epitaxial regrowth of substrate grains. The precipitation of ß gives us also valuable information on the mode of protrusion of grain boundaries from the substrate into the melted pool during the growth of neighboring grains as well as on the growth of grains nucleated ahead of the solidification front. In the heat affected zone the grain boundaries and high-angle boundaries of lamellar regions as well as deformation bands can be also easily recognized by enhanced segregation and subsequent precipitation of ß. On the contrary, the heat affected zone alone can be resolved due to dissolution or spheroidization of ß lamellae formed during the prior heat treatment of the alloy.

We have not found any published results concerning the laser treatment of binary AlZn alloys. The paper[22] deals with the diffusion of zinc (implanted into polycrystalline aluminum) during the pulsed ruby laser irradiation.

The above mentioned brief survey of phenomena dealt with in the present paper indicates the advantage of the Al-15 at.% Zn alloy for revealing the details of structure forming processes in laser treated surface layers of the AlZn based alloys.

EXPERIMENTAL

The study was carried out on the Al-15 at.% Zn alloy prepared from 5N purity materials. The castings (11 x 20 x 100 mm^3) were subjected to various thermomechanical treatments and surface modifications to get specimens suitable for laser beam irradiation. Some castings were machined to the final thickness of 2 mm by milling but most of them were hot-rolled from 11 to 6 mm with subsequent cold-rolling to 2 mm. Several samples were cold-rolled to 0.2 mm thick foils. The surface of planar specimens was mostly ground with metallographic paper 600 to enhance the absorption of laser radiation. The homogenization involving recrystallization of cold-worked samples (490 $^\circ$C/1h) followed by their water quenching led to the decomposition of solid solution resulting in the formation of zinc enriched G.P. zones coherent with the zinc impoverished α matrix. Some of homogenized and quenched samples were subsequently annealed (220 $^\circ$C/3d) to form spheroidal fcc α' precipitates or converted (90 $^\circ$C/4d) into coarse lamellar structure (alternating lamellae of fcc α solid solution rich in aluminum and of hcp ß phase composed of almost pure zinc). Much finer lamellar structure was prepared by storing the quenched specimens for several months at room temperature. The fine lamellar structure was also obtained already after storing the specimens at room temperature for several days if the rate of lamellar decomposition was enhanced by previous surface grinding of quenched samples. The use of different initial structures of the specimens led to a better understanding of structure changes within the heat affected zone formed during the laser treatment.

The specimens with dimensions of 10 mm x 20 mm x 2 or 0.2 mm used for laser treatment were mounted on a cross table permitting the fine feed of the specimen in two perpendicular horizontal directions using micrometric screws. The cross table was placed on a computer controlled manipulation support for scanning the specimen perpendicularly to the laser beam with velocities ranging from 0.2 to 10 m/min. The cw CO_2 laser (Coherent General EFA 51) emitting light radiation with wave length 10.6 μm has the output beam diameter of 16 mm with the power from 200 to 1400 W. The beam is focussed by ZnSe lenses (f=5" or 2.5"). The focus is placed 1 mm below jet orifice permitting to blow gas on the specimen during laser treatment. Individual or overlapping laser traces with spacings from 0.1 to 0.5 mm were produced by moving the support with the specimen perpendicularly to the laser beam.

The overlapping traces provide a large enough continuous area suitable for X-ray diffraction from surface layers. Ni-filtered CuK$_\alpha$ radiation and a cylindrical camera were used. The photographs were taken with beam inclination of 25° to sample surface to get 80-90 % of scattered intensity from 100 μm thick surface layer for reflections with lower diffraction angles. The information on structure changes due to laser treatment was accordingly drawn from reflections with low diffraction angles ranging from 36 to 50°. This region comprises the reflections 111 and 200 from the matrix of the alloy and 002, 100 and 101 from the stable ß phase. Some X-ray photographs were taken with additional oscillations ± 7.5°. The diffraction pattern of laser treated specimen was always compared with the pattern of the untreated specimen.

Scanning electron microscopic observations of the laser irradiated alloys were performed on JEOL 733, 15 kV, 10^{-10} A. The study was carried out either on electropolished (HNO$_3$ conc. + methanol, 1:3, -30 $^\circ$C, 100 mA/cm^2, 3 min) surfaces of samples or on transverse cross-sections of samples cut perpendicularly to the laser beam scanning direction. The cutting surfaces were first wet ground on metallographic paper (F17-600 Albis, GDR) and then mechanically (AP Alumina Paste F, Struers, Denmark) and electrolytically polished.

RESULTS OF LASER TREATING THE Al-15 at.% Zn ALLOY

Influence of Specimen Preparation

The thermomechanical treatment of samples and the preliminary modification of their surface may significantly affect both the absorption of laser radiation and the reproducibility of its structure modifying effect.

As cast alloys are not suitable for the research of laser surface treatment due to their content of numerous inhomogeneities (contractions, cracks and cavities). The associated local changes of energy absorption lead to sudden uncontrolled local overheating of specimen surface. Such overheating initiates the eruptions of zinc vapor due to high vapor tension of zinc followed by the formation of oxide surface coatings. The latter increase immediately the energy absorption which is enhanced also during laser scanning along adjacent paths.

Deep-worked samples yield better results irrespective of their heat treatment preceding laser irradiation. The energy absorptivity amounting only several per cents for a glossy surface can be appreciably enhanced e.g. by grinding with metallographic papers, by etching in 30 % KOH, by anodizing, or by graphitizing. The previous dry grinding of the surface in the direction perpendicular to laser scanning proved to be most effective for both the increase and the surface uniformity of energy absorption.

Structure Examination by X-ray Diffraction

Samples for X-ray diffraction study were hot- and cold-rolled to the thickness of 2 mm. Three preliminary thermomechanical treatments leading to different initial structures were applied: i) Homogenization (490 $^{\circ}$C/1 h) followed by surface grinding led to fine lamellae of α and β phases formed by storing the specimen at room temperature. ii) Homogenization, annealing 4 days at 90 $^{\circ}$C and surface grinding yielded coarse lamellae formed at 90 $^{\circ}$C. iii) Grinding followed by homogenization and quenching resulted in α matrix with coherent G.P. zones enriched in zinc.

The initial structure with fine lamellae reveals almost continuous diffraction lines corresponding to lamellae of α and β phases formed at room temperature due to the grinding of sample surface. Laser annealing of these samples leads to disappearance of β lines due to the dissolution of β lamellae during the laser heating to temperatures within the one-phase region of α below the solidus line. Small spots of α reflections are due to a rapid recrystallization of mechanically distorted α matrix in surface region. Large spots of α reflections are from large grains recovered from strains introduced by initial lamellar structure and situated below a very thin recrystallized surface layer.

Laser melted layer reveals distinct diffraction spots of α from large grains grown epitaxially from substrate grains during solidification. Small diffraction spots of α correspond to small new grains nucleated ahead of liquid/solid interface. Continuous diffraction lines of β are due to fine β precipitates nucleated and grown in zinc rich regions. The latter have been formed during the solidification process alongside of the zinc impoverished columnar cells aligned within each grain in the direction of the greatest velocity of grain growth.

The diffraction pattern of the sample annealed to form coarse lamellae displays reflections corresponding to α and β lamellae. The diffraction pattern from specimens with deeply melted surface layer (more than 100 μm thick) is analogous to that from samples with laser melted surface layer comprising initially fine lamellae. Continuous diffraction lines of β are again from β

precipitates formed on cooling in regions highly enriched on zinc, i.e. within the cell boundary regions. The comparison with diffraction pattern of initial structure indicates the presence of a smaller amount of ß phase. Somewhat higher but still small amount of ß reveals the diffraction pattern of slightly melted specimens (less than 50 μm thick). The reflections from only partially dissolved or spheroidized thick ß lamellae below the melted layer may in this case contribute to the intensity of ß reflections.

The initial structure of ground and afterwards homogenized and quenched samples is apparent from its diffraction pattern. Distinct spots from large grains of α formed during homogenization are surrounded by anisotropic dif- fuse scattering due to ellipsoidal G.P. zones nucleated homogeneously during quenching to room temperature. Diffraction pattern of laser melted surface layer is again analogous to that of the laser melted sample with initial structure consisting of fine α and ß lamellae. This proves that the struc- ture of solidified melt does not depend on initial structure of investigated alloy.

It is convenient to blow gas stream onto the specimens of AlZn alloys during their laser treatment. Otherwise overheating of the melted surface arises, zinc evaporates, its vapor oxidizes and oxides deposit on and near the treated surface. Continuous lines on diffraction pattern from such sur- face layer namely reveal that there are fine crystallites of ZnO along with coarse reflections of the substrate; ß reflections are not present due to the evaporation of zinc from surface regions. Air blowing increases the cooling rate of the treated region and also removes the zinc vapor from the neighborhood of the laser trace. The absence of ZnO deposits reserves then the surface clean and thus also the uniformity of its laser treatment.

Structure Examination by SEM

Only heat affected zone appears within the surface layer after laser irradiation provided the temperature does not reach the solidus temperature on the surface of the material (Fig. 1, A). The surface layer starts to melt on exceeding this temperature by absorption of higher amount of energy.

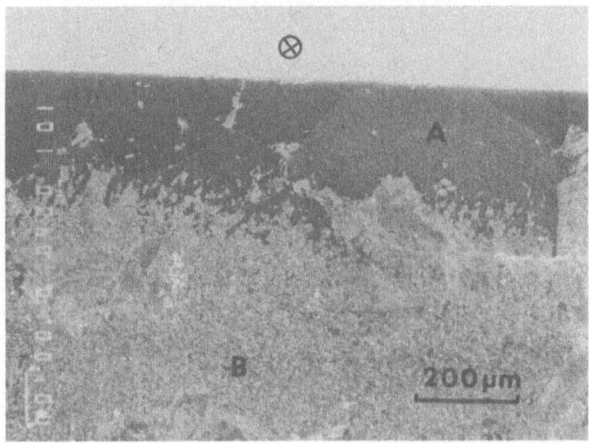

Fig. 1.   Dissolution of fine lamellae by laser annealing - cross section;
· P = 1300 W, v = 1 m/min, air jet 0.2 MPa. A - heat affected zone in which fine lamellae were dissolved, B - original fine lamellae, Θ - downward direction of the laser beam passage forming the traces; the same symbol is used also in following figures.

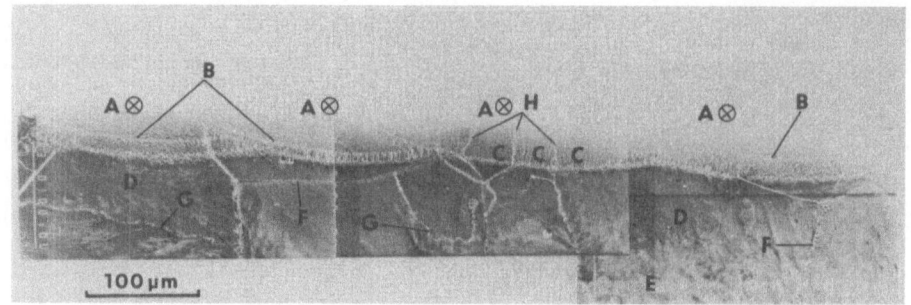

Fig. 2. Continuous surface layer of laser melted zone in coarse lamellar
structure - cross section; P = 1000 W, v = 3 m/min, air jet 0.2 MPa.
A - overlapped individual traces, scanning distance = 0.2 mm, B -
secondary solidification interfaces, C - epitaxial growth of grains
into the melted zone from individual grains of substrate, D - heat
affected zone with partially dissolved lamellae, E - original lamellar structure, F - original grain boundary decorated by ß precipitates, G - high-angle boundary of lamellar region decorated by ß
precipitates, H - high-angle boundary between neighboring grains
during their epitaxial growth.

The heat affected zone is then situated between the melted region and the
unheated material with initial structure (Fig. 2, D). In the heat affected
zone the straight grain boundaries and irregular high-angle boundaries of
lamellar regions are decorated by small globular ß precipitates (Fig. 2, F,
G). The parts of individual unmelted grains of the substrate regrow epitaxially into the melted zone with high-angle boundaries protruding into the
resolidified region (Fig. 2, C, H). Cracks along original grain boundaries
and thick oxide layer on the surface often arise during deep laser melting
(Fig. 3, C, A).

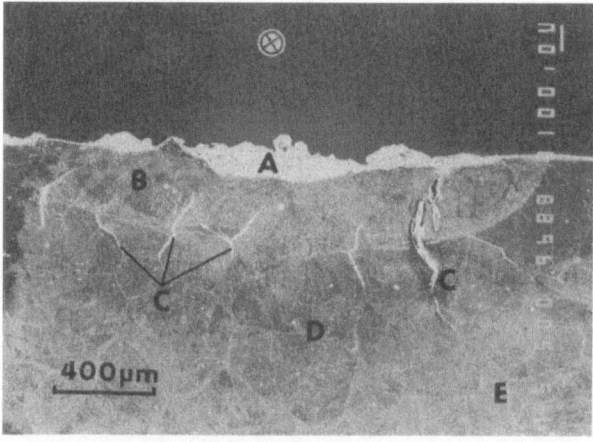

Fig. 3. Oxide layer and cracks formed during laser melting - cross section;
P = 1000 W, v = 3 m/min, without gas jet. A - ZnO layer, B - overlapping laser traces, C - crack along grain boundaries, D - heat
affected zone with partially dissolved coarse lamellae, E - original
lamellar structure.

Only the sizes and spatial distribution of G.P. zones unresolved by
SEM may be changed in the heat affected region in samples quenched after
homogenization. Also temporary dissolution of G.P. zones may proceed during
laser treatment. Clearly apparent changes of the structure proceed within
the heat affected zone if the initial structure comprise incoherent particles.
Fine lamellae (formed during storing the homogenized and quenched samples for
several months at room temperature or during short storing of samples in
which the rate of lamellar reaction was enhanced by surface grinding) are
dissolved almost completely in the heat affected zone. Distinct difference
between SEM images of the heat affected zone and of primary structure comprising fine lamellae is well apparent from Fig. 1. Coarse lamellae grown
during annealing (90 $^{\circ}$C/4d) either dissolve completely near the deepest
liquid/solid interface or spheroidize during their partial dissolution. The
samples annealed to form spherical fcc α' precipitates (220 $^{\circ}$C/3d) exhibit
partial melting and coarsening of these particles in the heat affected zone
during laser treatment. Deformation bands formed within heat affected zones
during solidification of melted pools are decorated by small globular ß precipitates.

The appearance of the deeply melted region is demonstrated in Fig. 4.
The melted region is in contact with the heat affected zone at the deepest
liquid/solid interface the features of which depend on the initial structure
of the sample. In cold-rolled samples or samples annealed to form either
thick lamellae of α and ß or α' precipitates, irregular clusters of spheroidal particles (of ß or α') appear close to the deepest liquid/solid interface. These clusters penetrate into the heat affected zone irregularly
though preferentially in the neighborhood of high-angle boundaries.

The deepest liquid/solid interface is very sharp in homogenized and
quenched samples. One, two, or three distinct layers of small globular ß
precipitates can be seen on the transverse cross section perpendicular to
the laser trace (Fig. 4, D; Fig. 5, A). The remaining features of the melted
region are practically independent on the previous heat treatment. Their
changes along the direction of the solidification start with a band of irregular cells (about 4 μm x 6 μm) adjoined to layers of precipitates, demon-

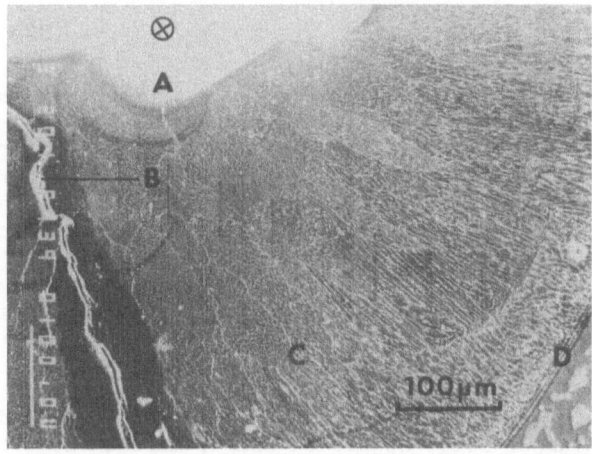

Fig. 4. Laser deep-melted layer - cross section; P = 1300 W, v = 1 m/min,
without gas jet. A - oxide layer, B - crack in the direction of
solidification, C - grain nucleated in the melt ahead of solidification front, D - deepest liquid/solid interface with three layers of
globular ß precipitates.

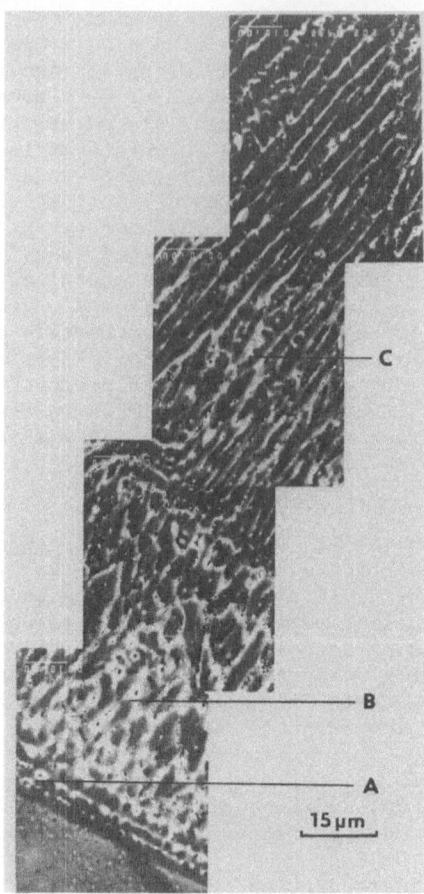

Fig. 5. Structure of solidified melt in homogenized and quenched specimen with G.P. zones - cross section; P = 1300 W, v = 1 m/min, without gas jet. A - deepest liquid/solid interface with layers of globular ß precipitates, B - region of irregular cells, C - region of columnar cells elongated approximately in the direction of solidification.

strating small local increase of cooling rate and thus an inexpressive directionality of heat transfer (Fig. 5, B). This band is followed by a region of long coarse columnar cells (about 4 μm x 20 um; Fig. 5, C) reducing their width with increasing distance from the deepest interface. Parallel arrays of gradually finer columnar cells are then established. The shape and growth direction of the cells are marked by subtle globular ß particles precipitated out of the intercellular zinc enriched melt. The region of elongated cells proceeds up to the irradiated surface in slightly melted samples. The arrangement of columnar cells in a plane approximately perpendicular to cell axes is apparent from the top view of a laser trace in Fig. 6 revealing the regular honeycomb arrangement of neighboring cells.

New non-epitaxial grains elongated in the solidification direction and composed of long columnar cells appear in deeply melted laser traces. The direction of cell axes is the same within any particular grain (Fig. 4, C).

The laser melted surface is covered by oxide layer (Fig. 3, A; Fig. 4, A) formed from oxidation products of evaporated material unless gas stream is blown onto the specimen during its laser treatment.

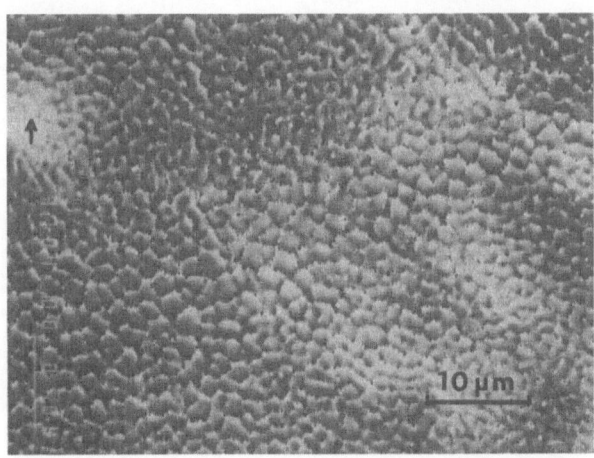

Fig. 6.   Arrangement of columnar cells in a plane approximately perpendicu-
          lar to cell axes - specimen surface; P = 1400 W, v = 1 m/min, with-
          out gas jet. The cell boundaries are decorated by ß precipitates.
          The arrow indicates the direction of the laser beam passage.

INTERPRETATION OF RESULTS AND DISCUSSION

     The characteristic features of the resolidification process of the
rapidly laser melted surface region are associated with the fact that the
solidification starts and proceeds as epitaxial and directional regrowth of
partially melted grains of the bulk substrate without any nucleation barrier.
The grains of the substrate grow in the direction of the temperature gradient
which is perpendicular to the liquid/solid interface. Rapid advance of the
liquid/solid interface proceeds by simultaneous cooperative growth of neighboring cells within each grain in the direction perpendicular to planes of
relatively low reticular densities. One of the directions of the most rapid
growth making the smallest angle with the local direction of the temperature
gradient may thus assure the rapid advance of the liquid/solid interface.
Two crystallographically equivalent directions of the greatest axial velocity
of growth may be alternately operative in the grain growth if they make locally almost equal angles with the normal to the liquid/solid interface as apparent from Fig. 7. The competitive epitaxial regrowth of unmelted parts of
neighboring grains into the melted zone leads to the protrusion of their
common boundary from the substrate into the resolidified region (Fig. 2, H).
Rapid alternating and irregular jerky advances of liquid/solid interface in
neighboring grains give rise to corrugations of their common grain boundary
facets. The simultaneous growth of all grains proceeding with equal instan-
taneous temperature and velocity along the whole liquid/solid interface is
the reason for the tendency of grain boundaries to develop in directions
parallel to the local temperature gradient. Both mentioned features of grain
boundaries within the solidified melt are clearly apparent in Fig. 2, H.

     The mechanism of resolidification in homogenized and quenched specimens
will be now discussed in greater detail. The melting of laser treated alloy
is stopped with zero rate of temperature change at the deepest liquid/solid
interface below which the solidus temperature has not been reached by laser
heating. Above this interface a thin layer of partially melted alloy is
situated where the temperature is raising from the value of solidus to that
of liquidus. The resolidification is associated with redistribution of alloy
constituents between the growing zinc impoverished solid solution and the
adjacent melt with gradually increasing content of zinc. The solid interface
is advancing in the direction of positive gradient of temperature pointing

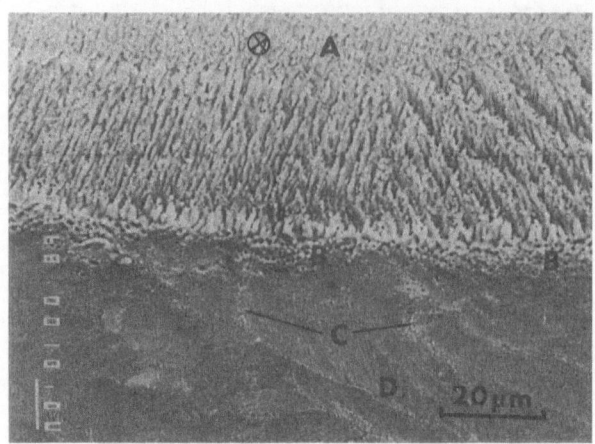

Fig. 7. Effect of growth anisotropy on epitaxial growth of grains in laser
melted region - cross section; P = 900 W, v = 3 m/min, air jet
0.2 MPa. A - laser melted zone with columnar structure oriented in
the direction of temperature gradient modified by two alternating
preferred growth directions, B - regions of ß precipitates penetrat-
ing irregularly into the heat affected zone, C - ß precipitates at
grain boundaries, D - heat affected zone with partially dissolved
lamellae.

to the laser heated surface. During the initial slow regrowth with low
cooling rate the solid will enclose the zinc enriched droplets at a distance
close to the deepest interface. These droplets are thus situated along an
internal surface parallel to this interface. The above mentioned redistribution of alloy constituents allows the process to be repeated several times
during the period of the onset of solidification. The slow drop of temperature near the advancing interface keeps it relatively smooth due to the steep
temperature gradient. The repeated simultaneous formation of zinc rich droplets ahead of the deepest interface is apparent from Fig. 5, A where two
arrays of ß precipitates nucleated during subsequent cooling are apparent.
The rapid drop of energy input leads to an increased rate of solidification
at the more distant locations of advancing interface from its initial position. This is associated with increasing rate of the release of the latent
heat of crystallization at advancing interface decreasing there locally the
temperature gradient. This enhances the constitutional supercooling within
the solute rich liquid ahead of fast advancing interface. The separation
of this liquid into droplets becomes more difficult and local projections
of faster growing grains impoverished in zinc start to advance into the melt
from the interface. The supersaturated liquid starts to form interconnected
channels at subsequent positions of the interface. This leads to the establishment of the cellular growth. The projections givve rise initially to
irregular cells surrounded by grooves filled with zinc rich liquid. The
grooves are formed by deepening of channels during the growth of neighboring
cells. The irregular cells are only slightly elongated in the direction of
solidification. The shape of the cells is well apparent from Fig. 5, B due
to decoration of cell contours by ß precipitates formed later within the
grooves. Further increase of solidification rate is facilitated by the cooperative growth of parallel arrays of columnar cells. The cells are distinctly elongated in the direction of the greatest velocity of grain growth
and mutually separated by deep grooves filled up with supercooled liquid from
which ß particles are later precipitating. The columnar shape of cells is
apparent by comparing Fig. 5 with Figs 6 and 8. Diffusion paths perpendicular to the axial direction of cells are shortened at higher cooling rates.

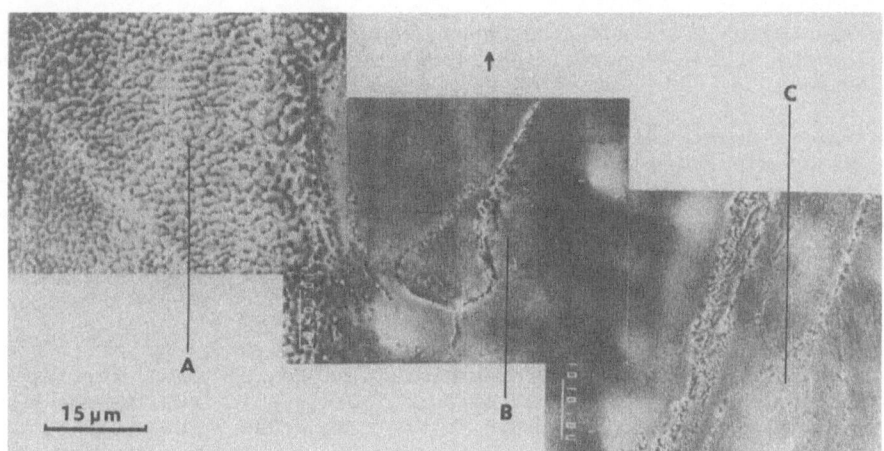

Fig. 8. Effect of laser treatment on coarse lamellar structure - specimen
surface; P = 1400 W, v = 1 m/min, without gas jet. A - laser melted
zone with cellular structure, B - heat affected zone with partially
dissolved or spheroidized lamellae, C - original lamellar structure.
The arrow indicates the direction of the laser beam passage.

This leads to the decrease of cell diameters with the increase of cooling
rate usually at larger distances from the deepest interface in the direction
of solidification. The epitaxial growth of grains may proceed up to the
specimen surface if the melted pools are shallow. The epitaxial growth is
abandoned in the deeper laser melted pools by nucleation and growth of new
randomly oriented crystals ahead of solidification front. This nucleating
is due to large constitutional supercooling at the advancing interface.
The growth of new grains elongated in the direction of the temperature gradient is rapid immediately after their nucleation. This is apparent from the
arrays of parallel thin columnar cells developed within each new grain
(Fig. 4, C).

The above described processes of solidification are independent on the
initial structure of the investigated alloy except of their initial stage
at the deepest liquid/solid interface. The latter is far from being sharp
if the preliminary structure is highly inhomogeneous in solute distribution
e.g. when treating specimens containing coarse ß lamellae within the α
matrix. This can be seen in Figs 7 and 8. The structure differences at the
deepest interface may be associated with the retardation of temperature
changes near this interface during the time interval in which melting ceases
and solidification commences. The temperature changes are there further
slowed down by both the absorption of heat of melting at the end of the melting period and the subsequent initial release of heat of solidification.
The retarded drop of temperature within the heat affected zone close below
the deepest interface may facilitate the heterogeneous precipitation of ß
especially if the distribution of solute is there highly inhomogeneous after
the previous dissolution of large zinc rich precipitates (e.g. of coarse ß
lamellae or of large α' precipitates). The heterogeneous precipitation of
ß will then proceed preferentially in regions highly strained by rapid cooling of that part of heat affected zone which is close to adjacent layer of
solidified melt. The ß precipitates will also nucleate along structure dis-
continuities such as high-angle boundaries of grains and of lamellar regions.

Inhomogeneity of the distribution of solute leading to local inhomogeneities of strains will also affect the structure of regrown grains at the
onset of solidification. It results in the absence of parallel layers of
small globular precipitates which are forming in previously homogenized and

quenched specimens. This and other above mentioned features of the structure in the vicinity of the deepest liquid/solid interface are clearly revealed in Figs 6 and 8.

Incomplete dissolution of large ß particles and their enhanced rate of growth during cooling may be also operative within the regions of the heat affected zone more remote from the deepest interface. This is demonstrated by larger ß particles along high-angle boundary of lamellar region apparent in Fig. 8, B. The outer rim of the heat affected zone contains larger ß particles arisen from partially dissolved ß lamellae by their rapid growth and spheroidization.

Similar results were found also for samples containing larger α' particles grown during previous heat treatment. The nucleation of ß particles in strained regions close below the deepest interface is facilitated in this case by local compositional inhomogeneities introduced by the rapid dissolution of α' particles during the laser treatment. The onset of only partial dissolution of α' particles is again immediately followed by their rapid growth. The region comprising ß particles, situated close below the deepest interface, is in the heat affected zone followed by a band of large α' precipitates. Theis sizes decrease with the increasing distance from the deepest interface due to gradually lower temperatures reached locally within the heat affected zone.

CONCLUSIONS

The main experiences and results gained from the present investigation of the surface laser treatment of an Al-15 at.% Zn alloy can be summarized as follows:

1. The formation of ZnO coatings on the specimen surface due to oxidation of zinc vapors can be prevented by gas blowing which keeps the surface clean and ensures the uniformity of surface laser heating.

2. The influence of numerous macroscopic inhomogeneities is as cast alloys must be avoided by hot and cold rolling to prevent local overheating associated with uncontrolled zinc vapor eruptions and with severe local surface damages.

3. The energy absorptivity can be substantially enhanced by grinding of specimen surface with metallographic paper 600 perpendicularly to the scanning direction.

4. Only heat affected zone is formed in surface layers during laser annealing below the solidus temperature, i.e. by lower rate of energy supply. Dissolution of small precipitates is completed within this zone. A distinctly sharp resolution of the heat affected zone from the bulk can be thus achieved e.g. by laser induced dissolution of the fine ß lamellae.

5. The higher rate of energy supply leads to the melting of surface layer surrounded by the heat affected zone. The latter is separated from the resolidified melt by the deepest liquid/solid interface which is very sharp for previously homogenized and quenched samples and considerably diffuse if the initial structure contains large zinc rich precipitates.

6. The heterogeneous precipitation of ß within the band of heat affected zone adjacent to resolidified melt is facilitated by quenched-in strains and by local fluctuations of composition formed after rapid dissolution of large precipitates. High-angle boundaries and deformation bands are then clearly displayed by ß precipitates. A complete or partial dis-

solution of precipitates is found in more remote distances from the deepest interface within the heat affected zone. The temperature in its outer part is approaching that of the solvus of the precipitated phase. Larger precipitates of α´ or coarse ß lamellae may be there dissolved only slightly. Their increased sizes are due to their subsequent rapid growth at elevated temperatures.

7. The directional solidification of the melt proceeds without nucleation barrier by epitaxial regrowth of unmelted parts of substrate grains. Solute (zinc) impoverishment of growing grains is accompanied by solute enrichment of the liquid adjacent to the advancing smooth liquid/solid interface. The low cooling rate at the onset of solidification leads in previously homogenized and quenched samples to a planar solidification front. The supersaturated liquid is accumulating ahead of this front into droplets forming several layers parallel with the advancing interface. Later solidification of droplets is accompanied by precipitation of ß particles. This layer structure is disturbed or even missing when larger zinc rich precipitates (e.g. ß or α´) are present within the initial structure.

8. The projections of growing grains protrude into the melt at increased rate of solidification. This leads first to the establishment of irregular cells followed by the formation of arrays of parallel columnar cells. The latter are elongated in the direction of rapid growth which is only slightly inclined to the local direction of solidification. The cells are mutually separated by grooves filled up by supersaturated liquid. The diameter of cells decreases with the gradual increase of cooling rate at rapidly advancing interface. Precipitation of ß from later solidified zinc rich liquid is distinctly displaying the structure forming process of solidification.

9. The large constitutional supercooling in deep melted pools may lead to nucleation and subsequent cellular growth of new grains ahead of advancing interface. The new grains are thus terminating the epitaxial regrowth of substrate grains.

10. The competitive epitaxial regrowth of neighboring substrate grains leads to the protrusion of grain boundaries into the solidifying melt in the form of corrugated facets approximately parallel with the local direction of solidification.

REFERENCES

1. W. W. Duley, "CO_2 Lasers", Academic Press, New York-San Francisco-London (1976).
2. W. W. Duley, "Laser Processing and Analysis of Materials", Plenum Press, New York and London (1983).
3. "Laser Materials Processing" ed. by M. Bass, North-Holland, Amsterdam - New York - Oxford (1983).
4. J. T. Luxon and D. E. Parker, "Industrial Lasers and their Applications", Prentice-Hall, Englewood Cliffs, New Jersey (1985).
5. J. W. Zindel, J. T. Stanley, R. D. Field, and H. L. Fraser, Microstructures of Rapidly Solidified Aluminum Alloys, in: "Rapidly Solidified Metastable Materials", ed. by B. H. Kear and B. C. Giessen, North-Holland, New York-Amsterdam-Oxford (1984).
6. R. Mehrabian, S. Kou, S. C. Hsu, and A. Munitz, Laser Surface Melting and Subsequent Solidification, in: "Laser-Solid Interactions and Laser Processing 1978", ed. by S. D. Ferris, H. J. Leamy, and J. M. Poate, American Institute of Physics, New York (1979).

7. K. N. Rao and J. A. Sekhar, Solidification of the quasi crystalline phase in the Al-Cu-Li system, Scripta Metall. 21:805 (1987).
8. M. K. El-Aldavi, Laser melting of solids, J. Appl. Phys. 60:2256, 2260 (1986).
9. A. V. La Rocca, Laser applications in manufacturing, Scientific American 246:80 (1982).
10. Y. Arata and I. Miyamoto, Laser welding, Technocrat 11:33 No 5 (1978).
11. J. Mazumder, Laser welding: State of the art review, J. Metals 34:16 July (1982).
12. P. L. Antona, S. Appiano, and R. Moschini, Laser Surface Remelting and Alloying of Aluminium Alloys, in: "Laser Treatment of Materials", ed. by B. L. Mordike, Deutsche Gesellschaft f. Metallkunde, Oberursel (1987).
13. U. Luft, H. Bergmann, and B. L. Mordike, Laser Surface Melting of Aluminium Alloys, in: "Laser Treatment of Materials", ed. by B. L. Mordike, Deutsche Gesellschaft f. Metallkunde, Oberursel (1987).
14. H. Volmer and E. Hornbogen, Microstructure of Laser Treated Al-Si-Alloys in: "Laser Treatment of Materials", ed. by B. L. Mordike, Deutsche Gesellschaft f. Metallkunde, Oberursel (1987).
15. G. Coquerelle and J. L. Fachinetti, Friction and Wear of Laser Treated Aluminium-Silicon Alloys, in: "Laser Treatment of Materials", ed. by B. L. Mordike, Deutsche Gesellschaft f. Metallkunde, Oberursel (1987).
16. K. V. Rao and J. A. Sekhar, Surface solidification with moving heat source: A study of solidification parameters, Acta Metall. 35:81 (1987).
17. J. Lašek, On the influence of the average composition on the position of the coherent miscibility gap in the Al-Zn system, Czech. J. Phys. B15:848 (1965), in German.
18. M. Simerská and V. Syneček, The mechanism of structure transformations in super-saturated Al-Zn alloys, Acta Metall. 15:223 (1967).
19. M. Simerská and P. Bartuška, The X-ray diffraction and electron microscopic investigation of stable and metastable equilibria in Al-rich Al-Zn alloys, Czech. J. Phys. B24:553 (1974).
20. M. Simerská, P. Bartuška, and V. Syneček, Structure transformations in supersaturated Al-Zn alloys, Acta Cryst. A34:S304 (1978).
21. H. Löffler, V. Syneček, M. Simerská, G. Wendrock, P. Bartuška, and R. Kroggel, On the mode of decomposition of Al-Zn alloys, Phys. Stat. Solidi (a) 65:197 (1981).
22. W. R. Wampler, D. M. Follstaedt, and P. S. Peercy, Pulsed Laser Annealing of Aluminum, in: "Laser and Electron-Beam Solid Interactions and Materials Processing", ed. by J. F. Gibbons, L. D. Hess, and T. W. Sigmon, North-Holland, New York-Oxford (1981).

STRUCTURE CHARACTERIZATION OF LASER TREATED WC-Co

BY MEANS OF POSITION SENSITIVE DETECTORS (PSD)

Martin Ermrich

Central Institute of Solid State Physics and Materials
Research Dresden
Helmholtzstr. 20, Dresden 8027, GDR

INTRODUCTION

Main application fields of position sensitive detectors (PSD) are time
resolved investigations, high speed diffractometry and stress measurements[1-3]
in order to reduce measuring time and to improve statistical reliability.
It is the aim of this work to use the advantages of PSD for local X-ray investigations with a view to a complete characterization of these small
regions without changing the diffraction equipment during various experiments. Laser treated WC-Co hardmetal was used as an example of application
for qualitative phase analysis and characterization of states of stresses
and texture of different irradiated areas.

EXPERIMENTAL

The experimental equipment consists of an X-ray diffractometer using
Co-K_α-radiation (U= 40 kV, I = 30 mA), a curved quartz-monochromator (10$\bar{1}$1),
a position sensitive proportional detector[4,3], its electronic device and
a multichannel analyzer with an 8k analog-digital converter, time control-
ler and a 4096 multichannel counter. The system includes a microprocessor
MPS 4944 with 64K of memory. The terminal is an alphanumerical keyboard
with colour display. The peripherals are cassette as external storage
medium and a digital plotter. The used PSD is a closed detector with an
energy and linear resolution of about 20 % and $FHWM_x$ of 150 µm, respectively. The principle of work is the risetime difference measurement of the
pulses at the ends of the resistive anode wire. In the present case the
angular range recorded simultaneously was about $\Theta \approx 3.1°$ (sample-to-detector
distance R = 250 mm). The detector was filled with Ar/CH_4 = 9:1, p = 1.04MPa.
Therefore, the quantum efficiency is about 50 %.

The investigated WC-Co (10 vol.% Co) ground hard metal was produced
along the usual metallurgical line with grain sizes of 1 to 5 µm. The sample
was irradiated by a neodymium-glass laser (λ = 1.06 µm) in free generation
(power about 5J, impulse length about 1.2ms). Mean power density averaged
over the laser spots was up to 10^5 W/cm^2. Three laser spots (A,B,C) were
generated. The ratio of power densities of the spots was 1:0.6:1.4, the
diameters were about 4, 5, 3 mm, respectively.

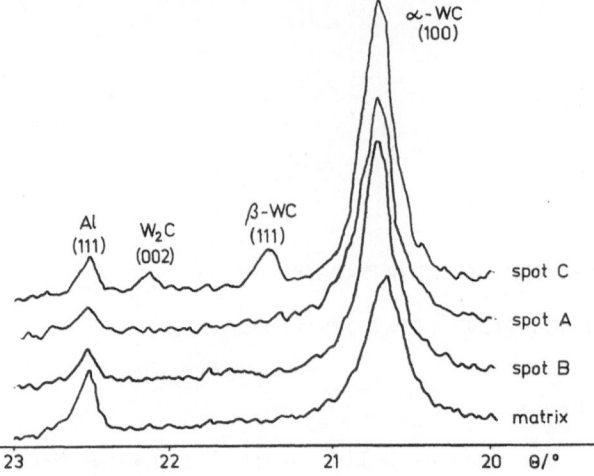

Fig. 1. Results of qualitative phase analysis of the laser spots
(t_{spot} = 600 s).

RESULTS

The original WC-Co hardmetal consists of the hexagonal α-WC and cubic
Co. In Fig. 1 the simultaneously detected reflexes of the α-WC(100), the
cubic ß-WC(111) and the hexagonal W_2C(002) are plotted. Further the (111)
peak of aluminium is shown, which was used for making the actual observed
sample region. Only in the laser spot C, ß-WC and W_2C were observed.
Fig. 2a shows a rough drawing of the laser spot C. For the investigations

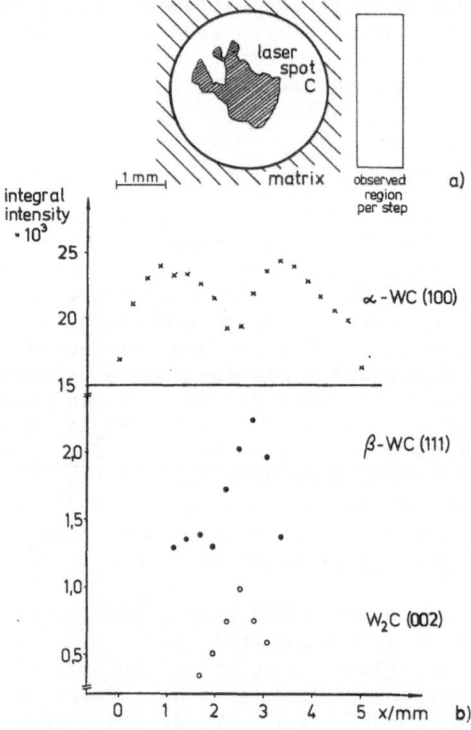

Fig. 2. Rough drawing of laser spot C (2a) and phase distribution (2b)
of α-WC, ß-WC and W_2C (t_{step} = 1800 s).

of the distribution of ß-WC, W_2C and α-WC a region of 0.8 x 3.0 mm^2 was
irradiated by the X-ray beam. The sample was moved in steps of Δx = 0.25 mm.
The integral intensities of the reflections were recorded simultaneously.
They are shown in Fig. 2b against the position in x-direction. In the centre
of the laser spot C (dark region) the maximum content of ß-WC and W_2C appears, corresponding to the fact, that the α-WC(100) diffraction line has
a minimum. Texture measurements in these regions point to grain-coarseness
or a weak (100) texture of the ß-WC.

The qualitative results of stress investigations of the α-WC (reflection
112), Θ about 61.8°, only $K_{\alpha 1}$-radiation) in the three spots are given in
Fig. 3. A small effect of overlapping of untreated regions is taken into
consideration, especially for spot C. The laser treatment reduces the strong
compressive stresses of the matrix even to a small tensile stresses (spot B).

Our results show, that it is useful to take advantage of the PSD not
only with respect to time but also for special diffraction applications,
which are very difficult using standard diffractometer devices.

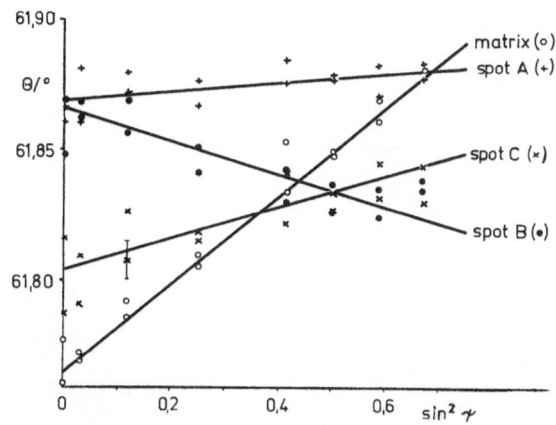

Fig. 3. Stress measurements of α-WC (reflex (112), $K_{\alpha 2}$ - radiation was
separated by monochromator, t = 1200 s).

REFERENCES

1. H. E. Goebel, Adv. X-ray Anal. 24:123 (1981).
2. Y. Yoshioka, K. Hasegawa, K. Mochiki, Adv. X-ray Anal. 22:233 (1978).
3. U. W. Arndt, J. Appl. Cryst. 19:145 (1986).
4. K. Richter, Thesis A, Techn. Univ. Dresden, GDR (1985).
5. K.-H. Kleinstück, K. Richter, Preprint 05-11-84, Techn. Univ. Dresden,
   GDR (1984).

DIFFRACTION STUDIES OF SnO_x THIN FILMS PREPARED BY VACUUM METHODS

G. Beensh-Marchwicka and L. Król-Stępniewska

Technical University

50-372 Wrocław, Janiszewskiego 11-17, Poland

INTRODUCTION

Tin oxide thin films have been studied for many years due to their potential industrial applications. The tin oxide films may be produced by a number of methods including chemical or vacuum methods. Oxidation of deposited thin films of tin has also been used to prepare SnO_x films. The effects of annealing are complex, and various phenomena may be observed.

In this paper the effect of heat treatment of tin films on their phase composition is described. The films were prepared by low energy sputtering and by evaporation on various substrates. The heat treatment in air or oxygen in the temperature range of 360-893 ^{o}C was performed after deposition. The films were examined by X-ray and electron diffraction methods.

EXPERIMENTAL

The tin films were sputtered in argon atmosphere on fused quartz and Corning 7059 glass substrates at the temperature of 150 ^{o}C. The details of the preparation conditions have been described in earlier works[1,2]. Thickness of all films measured with profilometer after annealing was about 350 nm. Tin was evaporated in vacuum of about 10^{-5} Torr on unheated quartz substrates. Both sputtered and evaporated films were annealed in air at two temperatures 360 ^{o}C and 510 ^{o}C. However, the sputtered films on quartz substrates were heated in oxygen atmosphere at four temperatures 360 ^{o}C, 510 ^{o}C, 675 ^{o}C and 893 ^{o}C. The samples were inserted in the constant temperature zone of the furnace with oxygen flow rate of about 32 $l.h^{-1}$. The annealing was carried out for about 4 hours at each temperature and after each annealing the films were brought to room temperature for measurements. The film structure and phase composition were investigated by X-ray and electron diffraction analysis. Scanning electron microscopy provided information about film morphology. For the X-ray measurements the flat 20 x 30 x 0.7 mm substrates were used while for the electron microscope study the small disks with diameter of about 3 mm and thickness about 0.2 mm were prepared. Glass and quartz disks were mechanically thinned up to perforation[3]. In such way we can observe as-deposited and annealed films on substrates.

Table 1. ß-Sn Phase in As-deposited Films Identified by X-ray Diffraction

| | Interplanar distances   d[Å] | | |
| ß-Sn standard | Evaporated | Sputtered | films |
	Quartz	Quartz	Corning 7059
2.91	2.9029	2.9218	2.9214
2.79	2.7798	2.7802	2.7967
2.05		2.0582	2.0578
2.01	2.0144	2.0063	2.0144
1.65		1.6580	1.6584
1.48	1.4796		1.4838
1.45	1.4547		1.4547
			(1.3052)
1.289			1.2899
1.20	1.20		1.2043
1.092			
1.04			
1.022			
0.98			

The film structure was investigated using X-ray diffraction with CuK$_\alpha$ radiation, λ = 1.54178 Å. The X-ray diffraction patterns of the films were compared with ASTM data. All as-deposited films consisted of tetragonal ß-Sn phase as presented in Table 1. The "d" values observed match those accepted for tetragonal ß-Sn, although the intensity ratios are different and some reflections are absent especially in evaporated films. Scanning electron microscope investigations revealed considerable differences in film morphology (Fig. 1). The evaporated films showed fine granular structure in comparison with the sputtered films what was connected with lower substrate temperature during deposition.

Annealing in 360 °C caused significant changes in the phase composition of films. The X-ray diffraction patterns after annealing are collected in Table 2. Besides ß-Sn phase reflections in oxygen-annealed films the reflections from tetragonal α-SnO phase were visible.

However, in air-annealed films the reflections from other oxides, namely Sn_3O_4 and SnO_2, were observed independently of deposition conditions. Annealing in oxygen in elevated temperature yielded the SnO_2 phase (Table 3). While in air-annealed films all earlier discovered phases are present. The intensity and number of reflections from SnO_2 phase increased, from ß-Sn and Sn_3O_4 phases decreased. The results of electron diffraction investigations confirmed the presence of those phases as shown in Table 4. Compared to electron diffraction pattern of the films air-annealed at 360 °C, the films annealed at 510 °C have had stronger and spotty rings. Some additional reflections were visible. They can be ascribed to tin oxide hydrates. The changes in film structure were accompanied with significant changes in film morphology. The scanning electron micrographs obtained after two steps of annealing are presented in Figure 2. It can be seen that in evaporated

Table 2. X-ray Diffraction Measurements for Sn Films after Annealing in the Temperature of 360 $^{\circ}$C during 4 h

Interplanar distances d[Å]

Evaporated	Sputtered films		Identified
Quartz	Quartz	Corning 7059	phases
Air	Oxygen	Air	
8.17			Sn_3O_4
		4.85	α-SnO
		3.34	SnO_2
	2.998	2.99	α-SnO
2.90	2.992	2.90	ß-Sn
2.788	2.814	2.78	ß-Sn, Sn_3O_4
2.71	2.714	2.71	α-SnO, Sn_3O_4
2.64		2.64	SnO_2
	2.429	2.42	α-SnO, Sn_3O_4
2.06	2.067	2.07	ß-Sn
2.01	2.023	2.02	ß-Sn
	1.797	1.797	α-SnO, Sn_3O_4
	1.664	1.65	ß-Sn, SnO_2
	1.610		α-SnO
1.48	1.488	1.48	ß-Sn, α-SnO, SnO_2
1.45			ß-Sn
1.44		1.44	SnO_2
1.32			SnO_2
1.2		1.2	ß-Sn, α-SnO, SnO_2
0.97			SnO_2

films the coalescence phenomenon took place during annealing and then the grains of metallic tin were present up to 510 $^{\circ}$C. The SnO_2 phase which have been created during annealling in oxygen at the temperature of 510 $^{\circ}$C recrystallized in higher temperatures (Table 5). During annealing in the temperature range of 510 to 893 $^{\circ}$C the intensity of the reflections increased from 0.5 to 4 times. In these films the reflection from planes (101) was strongest, what indicated the preferential growth of crystallites. After annealing the mean grain size achieved the value of 50 nm.

Table 3. X-ray Diffraction Measurements for Sn-Films after
Annealing in the Temperature of 360 °C and 510 °C

Interplanar distances d [Å]			
Evaporated	Sputtered films		Identified
Quartz	Quartz	Corning 7059	phases
Air	Oxygen	Air	
	3.3522	3.3515	SnO_2
		2.992	$\alpha-SnO$
2.915		2.9179	$\beta-Sn$
2.7932		2.7929	$\beta-Sn$, Sn_3O_4
2.7183			$\alpha-SnO$, Sn_3O_4
2.6433	2.6366	2.6445	SnO_2
		2.4168	$\alpha-SnO$, Sn_3O_4
	2.3678	2.3678	SnO_2
2.0578		2.0623	$\beta-Sn$
2.0144		2.0229	$\beta-Sn$
1.7585	1.7648	1.7648	SnO_2
	1.669		SnO_2
	1.590		SnO_2
	1.494		SnO_2
1.4796			$\beta-Sn$, $\alpha-SnO$, SnO_2
1.4527			$\beta-Sn$
1.4347			SnO_2
1.3178			SnO_2

CONCLUSION

A comparative oxidation study of tin thin films deposited by evaporation
and sputtering methods on amorphous substrates (quartz, Corning 7059 glass)
was undertaken. These types of substrates are commonly used in microelec-
tronic devices but they are unsuitable for electron diffraction method.

It was necessary to do all investigations on the same substrate to
avoid the influence of the surface carrier on the film structure and the
oxidation process[5,6,7]. In this paper the results of examination of film
structure by electron diffraction method on especially prepared glasses are
described and compared with those which obtained by X-ray diffraction method.

Table 4. Electron Diffraction Measurement for Sn-Films after Annealing in Air

Interplanar distances d [Å]				
360 °C		510 °C		Identified
Evaporated	Sputtered	Evaporated	Sputtered	phases
Quartz	Corning 7059	Quartz	Corning 7059	
3.66				Sn_3O_4
			3.44	Sn_3O_4
	3.38	3.34	3.34	SnO_2
2.99		2.99		α-SnO
2.91				ß-Sn
2.77				ß-Sn, Sn_3O_4
2.70	2.71			α-SnO, Sn_3O_4
		2.64	2.58	SnO_2
		2.32	2.32	SnO_2
	2.23			
	2.00		2.03	ß-Sn
1.96				
1.92				α-SnO
	1.80			α-SnO, Sn_3O_4
	1.76	1.77		SnO_2
	1.73			Sn_3O_4
			1.69	
	1.64			Sn_3O_4
1.55				Sn_3O_4
	1.43	1.42		SnO_2, Sn_3O_4
			1.40	SnO_2, Sn_3O_4
	1.15			SnO_2
	1.05			SnO_2
	0.96			SnO_2

Table 5. X-ray Diffraction Data for Sputtered Films on Quartz Substrates after Annealing in Oxygen

Temperature of annealing	Intensity from SnO_2 planes [arbitrary units]					
	110	101	200	111	211	220
510 °C	100	61	47		44	
675 °C	57	100	22		26	
893 °C	58	100	19	2	25	5
ASTM data	100	80	25	6	65	18

All investigated as-deposited films were polycrystalline with some amorphous background and were composed of ß-Sn phase. The films deposited by sputtering showed improved crystallinity in comparison with the evaporated films. The temperature and the gaseous atmosphere of annealing have a marked effect on the oxidation process independently of the deposition parameters. The oxidation in air up to 510 °C yielded the coexistence of four phases: ß-Sn, SnO, Sn_3O_4 and SnO_2. The possibility of presence of oxides with a stoichiometry between SnO and SnO_2 was reported by authors[8,9]. Metallic tin can be also observed during oxidation process due to the decomposition of SnO above 400 °C. Our ED study showed that air-annealed films can contain tin oxide hydrates. The presence of such compounds were detected on polycrystalline tin oxide by SIMS depth profiling method[10].

The completed oxidation to SnO_2 was observed only during annealing in oxygen atmosphere. Just after annealing in 510 °C the films consisted of SnO_2 cassiterite type phase. During annealing in higher temperatures the amorphous phase disappeared and crystallinity improved, the grain size increased similarly as mentioned in paper[11]. As shown by the X-ray results the oxidation of two-phase films (α-SnO and ß-Sn) in 510 °C yielded the untextured SnO_2 films. Higher temperatures of heat treatment caused substantial reordering in the films and appearing of the preferred growth in the direction [101]. Existence of various textures in thermal oxidized thin films were described by other investigators (ref. 9,10).

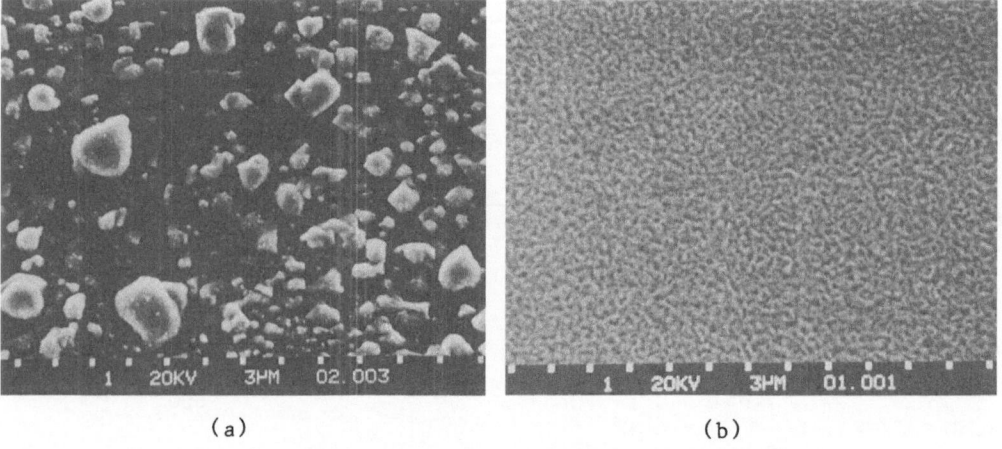

(a)   (b)

Fig. 1. Scanning electron micrographs of as-deposited films
(a) sputtered Sn; (b) evaporated Sn.

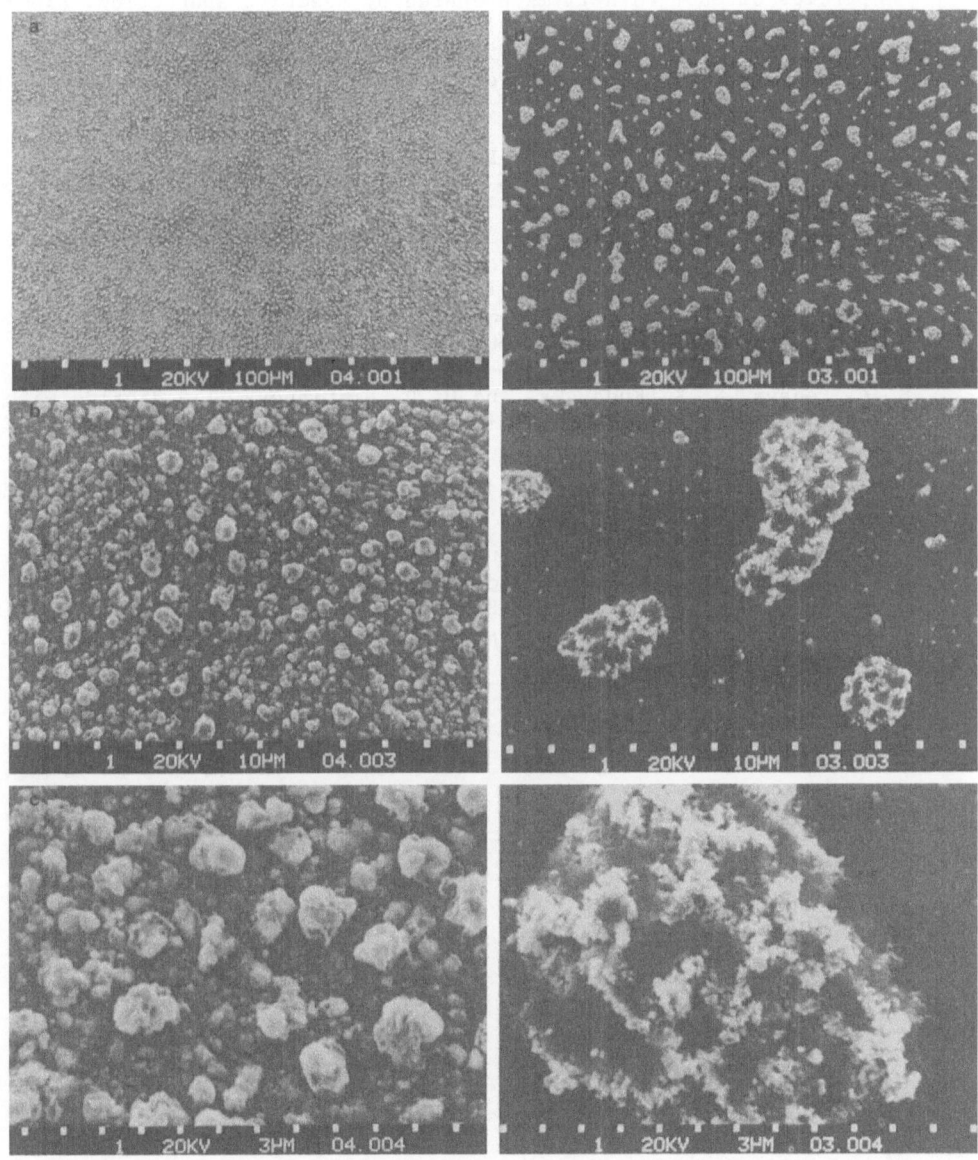

Fig. 2. Scanning electron micrographs of air-annealed films
(a)-(c) sputtered Sn; (d)-(f) evaporated Sn.

REFERENCES

1. G. Beensh-Marchwicka, L. Król-Stępniewska, and A. Misiuk, Influence of annealing on the phase composition, transmission and resistivity of SnO_x thin films, Thin Solid Films 113:215 (1984).

2. G. Beensh-Marchwicka and L. Król-Stępniewska, Reproducibility of properties of SnO_x thin films prepared by reactive sputtering, Electrocomp. Sci. and Tech. 11:271 (1985).

3. G. Beensh-Marchwicka, L. Król-Stępniewska, and M. Łukaszewicz, Obserwacja warstw cienkich łącznie z podłożem w transmisyjnym mikroskopie elektronowym, in: "III Konferencja Naukowa Technologia Elektronowa", Z.G. VAM, Poznań (1987).

4. Powder Diffraction File, ASTM, Philadelphia, PA, Cards 21-1250, 20-1293, 25-1259, 24-1342, 14-140, 25-1303, 18-1386, 11-694, 6-0395, 7-195, 13-111, 16-737.

5. M. Fantini, I. L. Torriani, and C. Constantino, Influence of the substrate on the crystalline properties of sprayed tin dioxide thin films, J. Cryst. Growth 74-439 (1986).

6. V. Kraševec, A. Prodan, M. Hudomalj, and S. Sulčič, A new form of SnO_2 found during oxidation of α-SnO thin films, Phys. Stat. Sol. (a) 87-127 (1985).

7. S. K. Peneva, R. K. Rudarska, and D. D. Nihtianova, Tin dioxide with the CaF_2 structure in thin tin oxide films, Thin Solid Films 112:247 (1984).

8. N. S. Murty and S. R. Jawalekar, Structural studies of chemically vapour-deposited tin oxide films, Thin Solid Films 100:219 (1983).

9. J. Geurts, S. Rau, W. Richter, and F. J. Schmitte, SnO films and their oxidation to SnO_2, Raman scattering, IR reflectivity and X-ray diffraction studies, Thin Solid Films 121:217 (1984).

10. D. F. Fox and G. B. Hoflund, A SIMS depth profiling study of the hydration layer formed at polycrystalline tin oxide surfaces by atmospheric exposure, Appl. Surf. Sci. 26:239 (1986).

11. K. B. Sundaram and G. K. Bhagavat, High-temperature annealing effects on tin oxide films, J. Phys. D: Appl. Phys. 16:69 (1983).

X-RAY DIFFRACTIONAL INVESTIGATION

OF THE COPPER-ALUMINIUM INTERFACE REACTION

Ehrenfried Zschech and Andreas Kaiser

Forschungsinstitut für Nichteisenmetalle

Freiberg, G.D.R.

INTRODUCTION

For the application of the copper wire ball/wedge bonding it is necessary to analyze the wire/substrate transition regions. The investigation
of the interface reaction between copper ball and aluminium bonding pads
serves for the assessment of the contact reliability at the copper ball
bonds from the points of view of solid state physics.

In this paper, thermally induced interdiffusion and the formation of
intermetallic Cu-Al phases (cf. Fig. 1) are studied for the case of "bulk
Cu with a thin Al or AlSi1 film" using X-ray diffraction.

TRANSPORT PROCESSES AND PHASE FORMATION

The copper-aluminium interface reaction has been studied very exten-
sively over the past five years, in particular by the use of the X-ray diffraction technique[1-4]. But there is a controversal situation concerning
the experimental data and their interpretation, caused especially by different geometrical conditions of the Cu-Al couples.

Vandenberg et al.[1-4] have shown that not all intermetallic phases occurring in the equilibrium diagram are generated by diffusion in thin film systems. In addition, they investigated the formation of stable and metastable
Cu-Al intermetallic phases for several thickness ratios of the primary components, i.e., for several material supplies.

At a temperature of 130 $^{\circ}$C, there exists only the $CuAl_2$ intermetallic
phase independent of the material supply. At temperatures above 200 $^{\circ}$C,
there are formed the final-state phase Cu_9Al_4 and pure α-Cu for a high copper supply and the final-state phases $CuAl$ and $CuAl_2$ for a high aluminium
supply[5].

By in-situ X-ray diffractional investigations Vandenberg et al.[1,2]
could detect two phases not occurring in the equilibrium diagram: $CuAl_x$
and Cu_3Al.

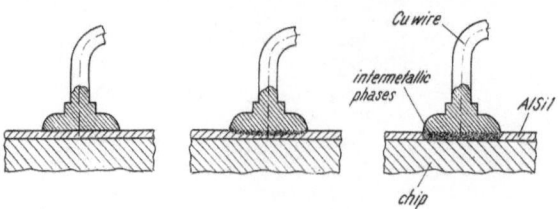

Fig. 1. Formation of intermetallic Cu-Al phases at copper ball bonds

SAMPLE GEOMETRY AND PREPARATION

X-ray diffractional investigations require relatively large specimen amounts. Therefore, it is necessary to simulate the geometrical conditions at a copper wire ball bonding contact. The samples were prepared by evaporation or sputtering of 1 μm Al or AlSi1, respectively, on thick copper sheets, cf. Fig. 2. The bulk copper material (Cu4N, CuNi1 or CuZn1) corresponds to the copper ball and the aluminium film corresponds to the bonding pads.

The samples thus prepared were submitted to a temperature-step-stress regime (150 ^{o}C $\leq$ T $\leq$ 450 ^{o}C, ΔT = 50 K, in each case 24 h, inert gas) to study the diffusion and phase formation processes.

EXPERIMENTAL PROCEDURE

An X-ray diffractometer HZG-4 produced by the VEB Freiberger Präzisionsmechanik (G.D.R.) was used for the ex-situ X-ray analysis. The measurements were carried out with copper radiation. X-ray diffraction patterns were recorded in the primary state of the samples and after each step of the temperature treatment to follow up phase formation and phase growth in a large 2 angular range and to study diffusion processes in an angular range around the Al(111) and Cu(111) diffraction lines.

RESULTS

In the primary state, there exist only Cu and Al diffraction lines marked distinctly for all samples.

From an X-ray profile analysis - after background correction and Rachinger separation - we can obtain a broadening of the Al(111) diffraction line with increasing temperature treatment. This effect is caused by copper atoms

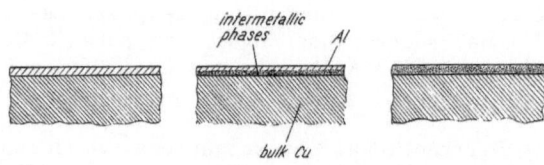

Fig. 2. Formation of intermetallic Cu-Al phases in simulated samples for X-ray studies.

solved in the thin Al film after the thermally induced interdiffusion and
to a smaller degree by the reduction of the Al film thickness in consequence
of the formation of intermetallic phases. The amount of Cu atoms solved in
the Al matrix and consequently the change of the lattice parameters depend
on the distance from the Cu-Al interface.

After the annealing at 200 $^\circ$C, the intermetallic phases $CuAl_2$ and Cu_9Al_4
can be seen (cf. Fig. 3). At temperatures above 250 $^\circ$C the $CuAl_2$ intermetallic phase is transformed into other ones, and at 300 $^\circ$C only the Cu_9Al_4
intermetallic phase grows with additional temperature treatment and consumes
fractions of the thin Al or AlSil films, respectively.

In the case of the CuNil/Al and CuNil/AlSil couples the consumption
of the thin film is smaller than for the other samples. This fact can be
interpreted with a reduction in diffusion and phase formation rate caused
by the Ni atoms. An influence of l wt.% Si in the thin Al film could not
be detected.

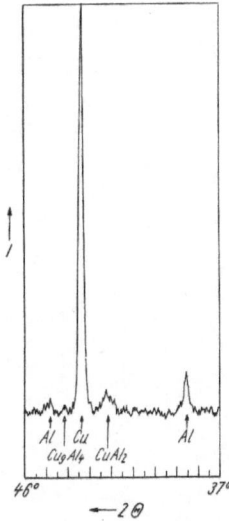

Fig. 3. X-ray diffraction pattern of a Cu/Al couple after a temperature
treatment (150 $^\circ$C/24 h + 200 $^\circ$C/24 h).

The recorded Cu-Al intermetallic phases ($CuAl_2$, Cu_9Al_4) and the disappearance of all Cu-Al intermetallic phases except of the Cu_9Al_4 phase in
the temperature range from 200 $^\circ$C to 300 $^\circ$C are in good agreement with the
studies by Vandenberg et al.[1,2] for samples with a great Cu supply. However, the CuAl intermetallic phase could not be detected.

The results published in this paper will be completed by in-situ X-ray
diffractional experiments on Cu-Al couples[6].

REFERENCES

1. J. M. Vandenberg, F. J. A. Den Broeder, and R. A. Hamm, Formation of intermetallics and grain boundary diffusion in Cu-Al and Au-Al thin film couples, Thin Solid Films 93:285 (1982).

2. J. M. Vandenberg and R. A. Hamm, An in-situ X-ray study of phase formation in Cu-Al thin film couples, Thin Solid Films 97:313 (1982).

3. F. J. A. Den Broeder, M. Klerk, J. M. Vandenberg, and R. A. Hamm, A comparative study of diffusion induced grain boundary migration, recrystallization and volume diffusion during the low temperature diffusion of Al into Cu and Au, Acta Metall. 33:97 (1985).

4. R. A. Hamm and J. M. Vandenberg, A study of the initial growth kinetic of the copper-aluminium thin-film interface reaction by in-situ X-ray diffraction and Rutherford backscattering analysis. J. Appl. Phys. 56:293 (1984).

5. H. T. G. Hentzell, R. D. Thompson, and K. N. Tu, Interdiffusion in copper-aluminium thin film bilayers, J. Appl. Phys. 54:6923 (1983).

6. E. Zschech, J. Uhlig, and I. Geleji-Neubauer, A study of phase formation in copper-aluminium couples by in-situ X-ray diffraction, in preparation.

X-RAY EXAMINATION OF $(PbTe)_{1-x}(GeTe)_x$ CRYSTALS

M. Leszczynski

High Pressure Research Center "Unipress"
Polish Academy of Sciences
Sokolowska 29/41, 01-142 Warsaw, Poland

Perfection of narrow gap semiconductor crystals of PbGeTe grown by vapour deposition technique was estimated by careful shape examining of rocking curves measured with X-ray diffractometer.

The experimental data revealed:
(a) broadening of rocking curves versus GeTe content till low angle boundaries occurrence in crystals of x higher than 65 %,
(b) fully reversible phenomenon of rocking curves narrowing when rising temperature.

The experimental results are related to the values of high temperatures phase transition from cubic (NaCl) to rhombohedral (As-like structure) that took place when cooling after growth process and at the experiment. The qualitative microscopic model of real crystal structure below and above phase transition is proposed.

INTRODUCTION

The narrow gap IV-VI semiconductors attract much attention recently because of their interesting physical properties, e.g. a tendency to polymorphic pressure and temperature phase transitions, high values of dielectric constant, high value of the effective charge, L-point band inversion[1], and also because of their applications as infrared lasers and detectors tunable by temperature[2] or pressure[3].

The PbGeTe alloys are particularly intensively studied as some of their properties are different than of the other IV-VI compounds, e.g. non-linear dependence of

(i) phase transition temperature from rhombohedral to cubic structure[4] and (ii) phase transition pressure from cubic to orthorhombic structure[5] versus the alloy composition.

The IV-VI compounds monocrystals are grown by two methods: Bridgeman[6] and vapour deposition technique[7]. Regretfully, there are very few works on crystallographic perfection of IV-VI crystals (ref. 6,8,9), but from our measurements we are convinced that the second growth method is so far the best one. The aim of the presented work was to estimate the perfection of

PbGeTe single crystals grown by vapour deposition technique and to state if any of their physical features can be explained by the crystal defect presence.

EXPERIMENTAL

    $Pb_{1-x}Ge_xTe$ crystals of x=0;0.25;0.37;0.48;0.65;0.78;1, as was estimated by lattice constant measurements, were examined. The crystals were of p-type with carrier concentration about $10^{18}/cm^3$. The faces of crystals were of (100), (110) and (111) directions. Polishing and etching in HBr+Br 2 % solution did not change the X-ray diffraction, so as grown faces of size 1-2 mm were examined.

    X-ray examinations were done with the diffractometer of high angular accuracy (1 sec of arc) and fine collimated beam of divergence about 2'. Lattice constant was measured by Bond method[10]. The examinations of rocking curves were carried out with a set of narow slits put before the counter and with the double crystal set up. Aa an analyzing crystal Si wafer adjusted to (333) reflection was used.

    High temperature up to 500 ^{o}C was obtained in resistance furnace mounted on the diffractometer. The stability of temperature was better than 0.1 ^{o}C. The samples of x=0.37-0.78 could be heated only up to 200 ^{o}C because of their decomposition. Additional measurements of the temperature phase transition were done with HZG4 powder diffractometer.

RESULTS

1. Rocking curve width, that can be used as a qualitative measure of crys-. tal perfection was an order smaller for crystals of low GeTe content, than in the case of the other IV-VI compounds crystals obtained from various sources (1-2° as compared with 4' in (400) reflection of PbTe crystal). It agrees with the reports on studies of Bridgeman grown crystals where low angle boundaries were present (refs 6,8,9).

2. The rocking curve widths increased versus GeTe content till low angle boundaries occurrence in crystals of x>65% (Fig. 1). These cystals consisted of small crystallites of size 0.01-0.05 mm inclined to each other at angles 1-3°. The direction of the neighbor crystallites were (111) and (11$\bar{1}$) for (111) type faces and (220) and ($\bar{2}$20) for (220) type, as could be stated from lattice constant measurements. The rhombohedral angle and the lattice constant measured in these imperfect monocrystals were identical with the data obtained from powder diffractometry.

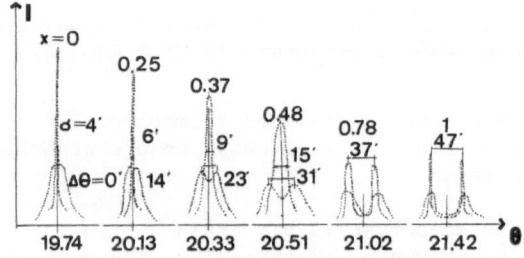

Fig. 1. Rocking curves of $Pb_{1-x}Ge_xTe$ for various x. The upper lines were taken from single crystals, lower than from powder diffractometry ((220) and ($\bar{2}$20) reflections).

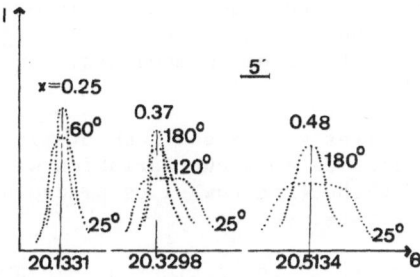

Fig. 2. Rocking curves of Pb$_{1-x}$Ge$_x$Te single crystal (220) reflections at elevated temperatures.

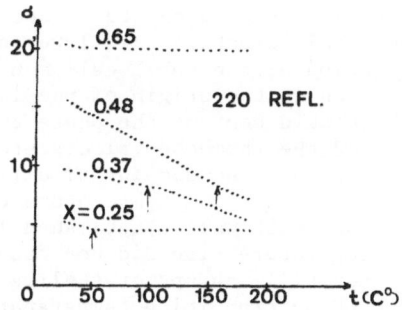

Fig. 3. Rocking curve widths vs. temperature for Pb$_{1-x}$Ge$_x$Te single crystals. Arrows show phase transition temperature.

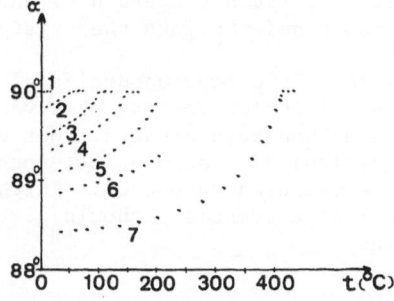

Fig. 4. Temperature dependence of rhombohedral angle of Pb$_{1-x}$Ge$_x$Te for x=0 (sample no 1); 0.25; 0.37; 0.48; 0.65; 0.78; 1 measured with powder diffractometry.

3. In the case of crystals of x=0.25; 0.37; 0.48 no evidence of rhombohedral
    angle in monocrystals could be found. Analysis of (200), (220) and (222)
    reflections taken from the same and various spots gave similar value of
    lattice constant, but no trace of rhombohedral angle of the value that had
    been measured on powder.

    The width of rocking curves was caused both by misorientation of crystal-
    lographic planes and lattice constant variation what could be stated from
    measurements with and without narrow slits placed before the counter and
    also by using double crystal set up.

4. Taking (200), (220) and (222) reflections at elevated temperatures the
    narrowing of the peaks was observed (Fig. 2). The phenomenon was fully
    reversible and could be repeated for several temperature cycles. In the
    crystals of x=0; 0.80 and 1 a small increase of rocking curve widths
    versus temperature was observed. At 300 $^{\circ}$C GeTe crystals experienced
    sudden irreversible peaks widening. The parts of 65 % crystal containing
    no distinguishable grain boundaries gave small narrowing of rocking
    curves.

DISCUSSION

    As it was mentioned earlier PbGeTe crystals experience the phase transi-
tion from cubic to rhombohedral structure when lowering the temperature after
crystal growth. The elongation of the cubic cell can be in any of eight < 11 ⊳
directions and this elongation is the origin of the internal stresses observed. In crystals of high GeTe content the phase transition takes place at
high temperature (Fig. 4) and the rhombohedral distortion is high enough to
produce small angle boundaries between domains of different < 11 ⊳ elongation.
Thus stresses can be relaxed and at room temperature the crystallites are of
comparatively good perfection, with well established lattice constant and
rhombohedral angle. The temperature rise did not cause any significant
changes in the boundaries net till abrupt crystallographic quality worsening
when again approaching the phase transition temperature.

    In the crystals of x lower than 65 % the peaks were narrow as compared
with the splitting of peaks characteristic for rhombohedral structure that
could be observed in powder diffractography. It meant that the average rhombohedral angle was smaller in monocrystals. This could be explained by the
assumption that the internal stresses between domains were not high enough
to produce dislocations forming low angle boundaries, but could suppress some
of the crystal regions. The rhombohedral distortion in such regions could be
smaller. It would be in agreement with the experimental data that the temperature of phase transition is lowered (what should be tantamount to dimin-
ution of rhombohedral angle at certain temperature) when applying hydrostatic
pressure[11] or introducing point defects into the crystal[12].

    When rising the temperature the rhombohedral angle got lower and a re-
versible improvement of crystal perfection was observed, what meant that no
stable defects between domains had been created. Not explainable yet seems
to be the narrowing of peaks from 37 % sample that took place also above
phase transition temperature measured on powder. Further examinations that
are carried out and theoretical estimations should give more quantitative
model in the nearest future.

ACKNOWLEDGEMENTS

    I am indebted to Dr A. Szczerbakow for supplying splendid crystals and
to Dr T. Suski for stimulating discussions.

REFERENCES

1. T. Suski, Phase transitions and resistivity anomalies in (Pb, Sn, Ge)Te compounds, Materials Sci. XI:3 (1985).
2. W. Ralston, I. Malngailis, A. R. Calawa, W. T. Lindley, Stripe-geometry PbSnTe diode lasers, IEEE J. Quantum Electron. 9:350 (1973).
3. J. M. Besson, W. Paul, A. R. Calawa, Variable wavelength laser spectrometry, Phys. Rev. 173:699 (1968).
4. W. Jantsch, "Dynamical Properties of IV-VI Compounds", Springer Tracts in Modern Physics 99, Springer Verlag, Berlin-New York.
5. T. Suski, M. Leszczynski, K. Murase, S. Katazama, S. Takaoka, High Pressure Investigations of PbGeTe Semiconductor Crystals in: "Proc. Jap. High Press. Conf. Kyoto" 20 (1982).
6. J. Morawiec, J. Auleytner, Z. Furmanik, A. Jedrzejczak, TEM and X-ray investigations of defect structure of PbSnTe crystal, Cryst. Res. Techn. 20,7:931 (1985).
7. T. C. Harman, J. P. McVittie, Horizontal unseeded vapour growth of IV-VI compounds and alloys, J. Electron. Mater. 3:843 (1974).
8. J. Morawiec, E. Mizera, X-ray and TEM investigations of PbTe crystal, Phys. Stat. Sol. (a) 63:K1 (1981).
9. Z. Golacki, M. Gorska, T. Wazrminski, A. Szczerbakow, Vapour phase growth and properties of PbSnTe single crystal, J. Cryst. Growth 74:129 (1986).
10. W. L. Bond, Precision lattice constant determination, Acta Cryst. 74:129 (1960).
11. T. Suski, S. Takaoka, K. Ishi, K. Murase, High pressure investigations of ferroelectric phase transition in PbGeTe, J. Phys. C, 17:2181 (1984).
12. T. Suski, M. Konczykowski, M. Leszczynski, D. Lesueur, J. Dural, Ferroelectric phase transition in electron irradiated PbSnTe crystal, J. Phys. C, 16:1358 (1982).

V. SINGLE CRYSTAL THIN LAYERS

STRUCTURAL STUDIES OF GARNET FILMS

S. Lagomarsino

Istituto Elettronica Stato Solido, CNR

V. Cineto Romano 42, 00156 Roma, Italy

INTRODUCTION

Garnets are known since ancient time as natural crystals, often considered as gem stones. The first artificial ferrimagnetic garnets have been
grown in 1956,[1,2] and since then their impact in physics and technology
became more and more important. Among the most important applications are
the magnetic bubble memories, the magneto-optical devices, the microwave
devices. Non-magnetic applications include garnet crystals as matrices
for lasers and for cathodoluminescent screens. Very good reviews on these[3,4]
applications and on the basic properties of garnets have been published.
In the applications, thin films are of fundamental importance, due to their
specific properties. A special issue dedicated to garnet films has been
published by "Thin Solid Films".[5]

In the following some overviews on the basic structural properties
will be given in next chapter. Characteristics peculiar to thin films and
structural modifications due to ion implantation and thermal treatments,
together with techniques to study it, will be presented in the two following chapters. Finally in the last section a very powerful technique for
the study of surfaces and interfaces, the X-ray standing wave technique
(XSW), will be applied to garnet crystals.

STRUCTURE

The general formula of the garnets is $A_3B_2C_3O_{12}$ where A, B and C represent cations and O the oxygen anions. The crystals are essentially ionic
in nature. The unit cell contains 8 unit formula and therefore 160 atoms.
Their spatial group is the (Ia3d) (1,2). The cations are distributed into
three different crystallographic sites, i.e. tetrahedral, octahedral and
dodecahedral, distinguished by their coordination with oxygen. Fig. 1 shows
the cation arrangement in four octants of the unit cell. Considering the
$\{Y_3\}[Fe_2](Fe_3)O_{12}$, the prototype of the ferrimagnetic garnets, the 24 Y
cations occupy the dodecahedral (c or $\{\}$) sites (8-fold coordinated with
oxygen), 16 Fe occupy the octahedral (a or []) sites (6-fold coordinated)
and 24 Fe the tetrahedral (d or ()) ones (4-fold coordinated). Each oxygen
ion is at the corner of four polyhedra, a tetrahedron, an octahedron and
two dodecahedron, as shown in Fig. 2 (the second dodecahedron is not dis-

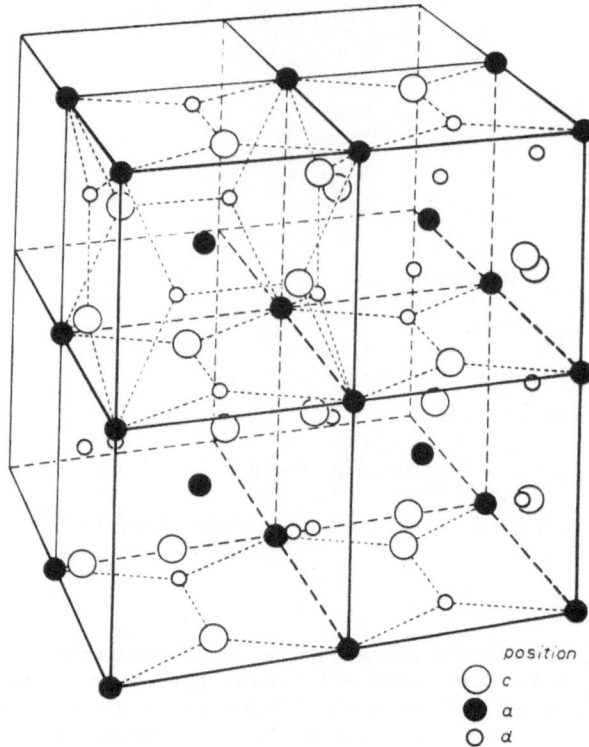

Fig. 1. Geometry of cations in four octants of the garnet unit cell.
c = dodecahedral sites; a = octahedral sites; d = tetrahedral sites
(After S. Geller[3]).

played). Note the difference in distance between the oxygen ions and the
three cation sites, and the difference in volume of the three sites. The
polyhedra in fact are not regular, but are distorted along a local symmetry
axis. A very important peculiarity of garnets is their capability of selectively incorporate in their different sites a great number of cations. For
magnetic garnets the most important basic materials are the Yttrium Iron
Garnet (YIG) and the Gadolinium Iron Garnet (GdIG). Very important is also
the Gadolinium Gallium Garnet (GGG) which is not magnetic but is utilized
as substrate for epitaxial film. YAG (Yttrium Aluminum Garnet) is utilized
as a matrix for laser active ions (generally Nd). On these basic materials,
several hundreds of compositions have been reported which tailor the wanted
properties[6].

EPITAXIAL FILMS

The most common growth technique for garnet films is the Liquid Phase
Epitaxy on substrates grown by the Czochralski technique[8]. If, as very often
happens, the lattice parameter a_f of the film is different from the substrate
lattice parameter a_s, a stress raises at the interface. The stress can give
raise to an elastic strain or can be partially or totally relieved by dislocations or cracks (S. Lagomarsino et al. in [5]). In the first case the
growth is really epitaxial and the film is coherent with the substrate. The
elastic strain causes a curvature of the system substrate-film and a deformation of the film crystal lattice. For example, in the case of a (111) substrate orientation the cubic lattice becomes rhombohedral[9]. This implies
that both the interplanar spacing and the orientation of corresponding lat-

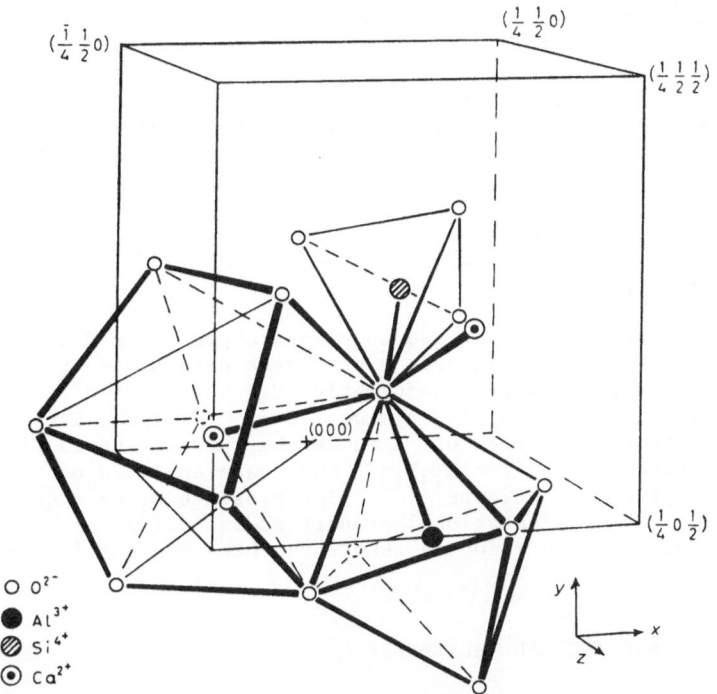

$(\frac{\bar{1}}{4}\frac{1}{2}0)$ $(\frac{1}{4}\frac{1}{2}0)$

$(\frac{1}{4}\frac{1}{2}\frac{1}{2})$

(000)

$(\frac{1}{4}0\frac{1}{2})$

○ O^{2-}
● Al^{3+}
◍ Si^{4+}
◉ Ca^{2+}

Fig. 2. Tetrahedral, octahedral and dodecahedral coordination around an oxygen atom for a $\{Ca\}_3 [Al]_2 (Si)_3 O_{12}$ garnet. A second dodeca- hedral site is not shown for clearness purpose. (After S. Geller[3]).

tice planes are different for film and substrate. X-ray double-crystal dif- fraction (XDC) is the primary technique to measure lattice mismatch and de- formation. In the case of coherence between film and substrate the lattice mismatch can be measured directly in the direction perpendicular to the sur- face. The strain-free lattice parameter of the film can be obtained, just by consideration of the elastic properties, as:

$$(\frac{\Delta a}{a})_r = \frac{1-\nu}{1+\nu} (\frac{\Delta a}{a})_\perp , \qquad (1)$$

where ν is the Poisson's ratio. In this case the lattice mismatch in the direction parallel to the surface $(\Delta a/a)_\parallel$ is zero. However, if partial or total stress relief is present, $(\Delta a/a)_\parallel$ is no more zero. To obtain the strain-free lattice parameter of the film is necessary to measure indepen- dently at least two lattice mismatches for planes making different angles with the surface. The relation between the relative mismatch for a plane making an angle Φ with the surface $(\Delta d/d)_\Phi$ and the parallel and perpendicu- lar mismatches is the following:

$$(\Delta d/d)_\Phi = (\Delta a/a)_\perp \cos^2\Phi + (\Delta a/a)_\parallel \sin^2\Phi \qquad (2)$$

With two reflections at different Φ values it is possible to obtain $(\Delta a/a)_\perp$ and $(\Delta a/a)_\parallel$. The lattice parameter pertaining to the strain-free film is given in this case by:

$$\left(\Delta a / a \right)_r = \frac{1-\nu}{1+\nu} \left(\Delta a / a \right)_\perp + \frac{2\nu}{1+\nu} \left(\Delta a / a \right)_{\shortparallel} \tag{3}$$

Experimentally, one can obtain the values of $(\Delta d/d)_\Phi$ from the angular separation $\Delta\omega$ of the film and substrate peaks in the double-crystal rocking curves:

$$\left(\Delta d / d \right)_\Phi = \frac{\Delta \omega}{\mathrm{tg}\, \Phi - \mathrm{tg}\, \vartheta_B}, \tag{4}$$

where ϑ_B is the Bragg angle.

Limiting values in mismatch and thickness above which dislocations at the interface and/or cracks are formed have been determined by Matthews et al.[10]. In practice pure elastic deformation can be expected for films few microns thick with a mismatch of the order 10^{-3}. Precise measurements of the lattice parameter have been utilized in the study of doped-YIG. In particular, Ca-doped YIG epitaxial films have been analyzed both by X-ray and by magneto-optical techniques[11,12]. The presence of Fe^{4+} centers with at least two different orientational properties and different behaviors under thermal treatment in reducing atmosphere has been put in evidence[12].

ION IMPLANTATION AND THERMAL TREATMENTS

The main reason for the importance of garnets in science and technology is due to their very peculiar magnetic and anisotropic properties. Garnets are ferrimagnetic and due to their very great flexibility hey can incorporate a great number of different ions which can vary their magnetic properties[3] (see Chapter 2). On the other hand, even if the structure is cubic, anisotropic properties are present. In particular, uniaxial anisotropy is of fundamental importance for a large number of applications[3-5]. The uniaxial anisotropy can be growth-induced or stress-induced. The bubble domains are generated in garnet films containing rare-earth ions with a large uniaxial growth-induced anisotropy. In these films the magnetization is perpendicular to the surface. However, better performances can be obtained, both by eliminating "hard" bubbles (bubbles with a reduced mobility), and by creating propagation patterns with improved characteristics if a thin surface layer with the magnetization flipped in an in-plane direction is formed on top of the film. Since the fundamental work of North and Wolfe[13] an important number of papers have been dedicated to the effect of ion implantation on the structural and magnetic properties of garnet films. A recent review can be found in[5] by P. Gerard. The basic mechanism for the flipping of the magnetization is the following: the ion beam which penetrates in the garnet film causes a strain and a damage to a surface layer few thousands of Ångstrom thick. The strain tends to increase the lattice parameter, but the expansion can take place only in the perpendicular direction due to the presence of the substrate. Therefore, the surface layer is in a state of lateral compression. If the magnetostriction constants of the material are negative and sufficiently large, this compressive stress induces an in-plane anisotropy which brings the magnetization in-plane. X-ray Double Crystal Diffraction has given an important contribution to the analysis of the structural properties of implanted films. In fact, this non-destructive technique can give the strain and damage profile caused by the implantation with very good accuracies (of the order of few percent). The starting point is the analysis of the Pendellösung fringes which arise on the tails of the rocking curves when diffraction from a thin layer takes place[14]. Due to the distortion and the damage present in the implanted layers the dynamical theory for imperfect crystals should be used[15,16]. However, in the case of layers few thousand of Ångstrom thick, the kinematical theory can be used as well[17], following the well verified assumption that for very thin crystals the dynamical and the kinematical theories give essentially the same results. A confirmation comes from a recent paper which compare the results of these two theories in non uniform InGaAs and InGaAsP layers[18].

In accordance with the kinematical theory, the basic formula for the fitting of the experimental diffraction curves are given in the following.

If d_0 is the undistorted lattice parameter and δ_n the strain at the n-th plance, the (distorted) lattice parameter d_n at the n-th plane is:

$$d_n = d_0(1+\delta_n) \tag{5}$$

The total reflected amplitude at an angle ϑ is the sum of the contributions from each coherently scattering plane:

$$A(\vartheta) = C \sum_{n=1}^{N_{Tot}} \exp(i\varphi_n) \tag{6}$$

$$\varphi_n = \frac{4\pi}{\lambda} \sum_{k=1}^{n-1} \frac{d_n}{(h^2+k^2+l^2)^{1/2}} \sin\vartheta \tag{7}$$

where h,k,l are the Miller indexes of the reflecting planes and C is an amplitude constant factor. Instead of calculating the phase φ_n for each plane, we divided the implanted layer in M laminae of N atomic planes each, and consider the strain δ_n constant in each lamina. $A(\vartheta)$ becomes:

$$A(\vartheta) = C \sum_{m=1}^{M} A_m \exp(i\psi_m) \tag{8}$$

$$A_m = \frac{\sin(N\varphi_m/2)}{\sin(\varphi_m/2)} \quad ; \quad \varphi_m = \frac{4\pi}{\lambda} \frac{d_0}{(h^2+k^2+l^2)^{\frac{1}{2}}} (1+\delta_m) \sin\vartheta \tag{9}$$

$$\psi_m = N \sum_{k=1}^{m-1} \varphi_k + (N-1) \frac{\varphi_m}{2} \tag{10}$$

In implanted crystals the strain is determined essentially by the energy lost by the ions during nuclear collisions[19]. The distribution of the energy loss being an asymmetric gaussian[20], this is also the shape of the strain profile. In Fig. 3 the stopped ion distribution and the strain distribution are shown for an implantation of neon ions of energy 200 KeV at a dose of 2×10^{14} ions/cm^2. Therefore we can assume δ_n as:

$$\delta_n = \delta_M \exp[-(D_n - \rho)^2/\sigma^2], \tag{11}$$

where D_n is the perpendicular distance of the n-th plane from the surface, ρ is the depth at which the maximum distortion δ_M occurs, and $\sigma = \sigma_1$ for $D_n \leq \rho$, $\sigma = \sigma_2$ for $D_n > \rho$. The implanted layer extends up to a depth $\rho + 2.65\ \sigma_2$ from the surface. The remaining of the sample can be considered as a constant background. The fitting procedure is a least-square one, based on iterative and independent variations of the four parameters δ_M, ρ, σ_1, σ_2 and a normalization parameter for the amplitude A. In Fig. 4 a dif-

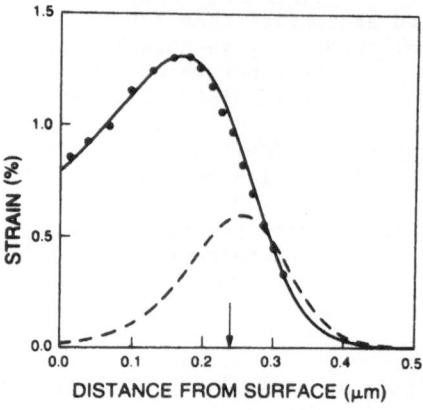

Fig. 3. Stopped-ion distribution (dashed line) and strain distribution calculated (full line) and experimentally determined (closed points) (After MacNeal and Speriosu[19]).

fraction pattern for a pure Yttrium Iron Garnet sample implanted with Ne ions at a dose of 2×10^{14} ions/cm^2 is shown (dashed line). The diffraction peaks of the bulk of the film and of the substrate are not shown. The full line indicates the best fit following the above procedures. The agreement is very good.

Besides the strain, ion implantation induces a damage. The damage level can also be determined by X-rays, in the sense that it can be treated in a similar way as the Debye-Waller factor. The damage profile can be determined with enough accuracy by a step-wise etching method[20]. A general feature, determined by X-rays is that the damage distribution has a maximum at a depth closer to the surface with respect to the strain distribution[20]. Another general feature is that both the strain and the damage are proportional to the implanted dose. An important point is that the reconstruction mechanism can be induced on the implanted samples by thermal treatment. Both structural and magnetic measurements by FMR[21,22] and Conversion Electron Moessbauer Spectroscopy (CEMS)[23] have shown that at least two different mechanisms are involved: one on a short range involving oxygen diffusion and another on a long range involving both geometrical rearrangement of

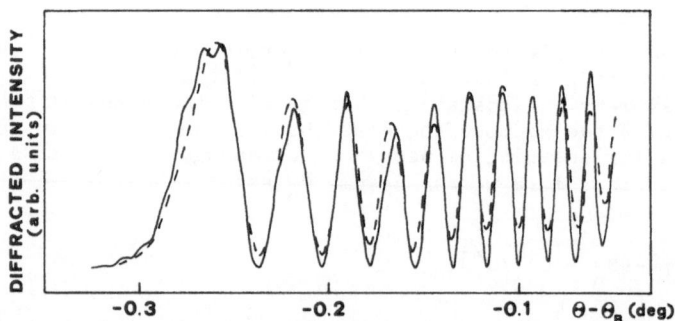

Fig. 4. X-ray double-crystal interference fringes of an implanted YIG film. Dashed line: experimental curve. Full line: best fit following a kinematical approximation. The angular scale represents the deviation from the Bragg angle of the bulk of the film (not shown in the figure). Reflection: (444) CuK$_\alpha$. Implantation; Ne$^+$ ions 200 KeV 2×10^{14} ions/cm^2. (After G. Balestrino et al.[23])

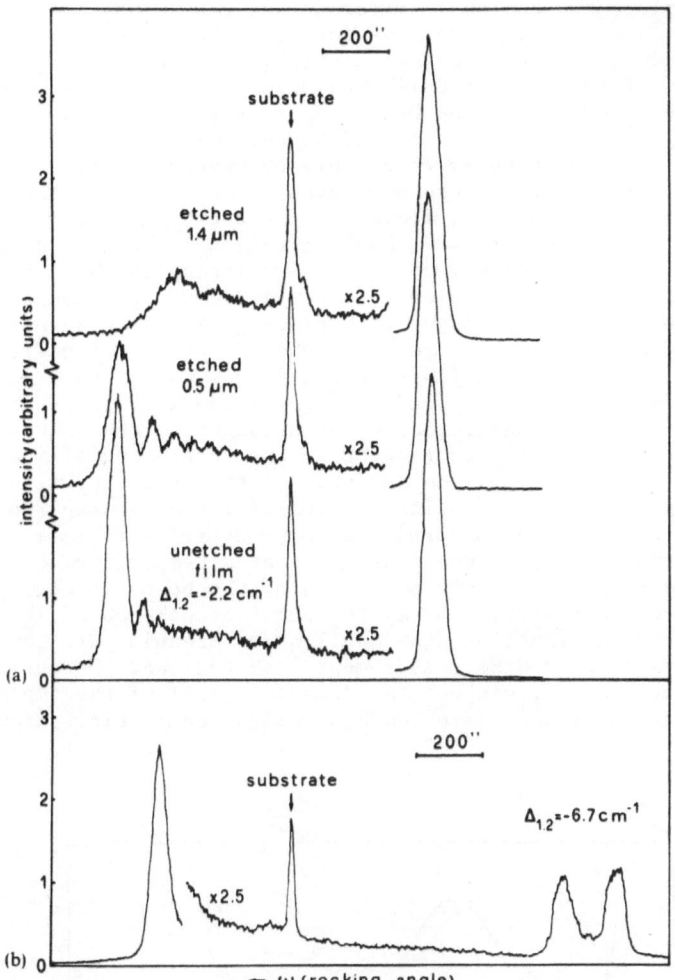

Fig. 5.   X-ray double crystal diffraction curve of a Ca-doped YIG film after
annealing in H_2 atmosphere at (a) 470 °C for 9 hours - (b) at 540 °C
for 4 hours. The peak at the right-hand side of the substrate peak
belongs to the bulk of the film, while that at the left-hand side
to a surface layer with a contracted lattice parameter (see text).
The three diffraction spectra in (a) refer to different steps of
an etching experiment. Reflection: asymmetric (12 6 0) - CoK_α
(After B. Antonini et al.[24]).

oxygen polyhedra and/or cation diffusion. The two mechanisms are activated
at different temperatures, the lower one being that relative to the oxygen
diffusion.

Structural variations in the surface layers are also important in diffusion processes in garnet films. The first occurrence of the formation of
a surface layer with structural properties different with respect to the
underlying bulk film caused by thermal diffusion was detected in a Ca-doped
YIG film[24]. The crystal was annealed for several hours in a hydrogen atmosphere at temperatures higher than 450 °C. In Fig. 5 are shown the double-
crystal rocking curves for the sample annealed at two different temperatures
and after two successive chemical etchings.

From comparison between the different spectra it has been possible to conclude that a thin surface layer was formed on top of the bulk of the film with a contraction of the lattice parameter by about 0.2 %. The contraction of the lattice started at an annealing temperature of about 470 °C. The thickness of the layer was at the beginning nearly 1000 Å but became larger and larger for longer annealing time and higher temperatures. The formation of the surface layer was totally reversible by oxygen annealing. Magnetic measurements by CEMS showed on the same sample that the orientation of the easy axis of magnetization was different for the two layers: it was in plane for the film bulk (in the $[011]$ or $[01\bar{1}]$ directions) and lied along the $\langle 111 \rangle$ directions in the thin surface layer[25]. The flipping can be explained by pure magnetostriction considerations. In pure YIG we obtained similar re-sults: formation of a surface layer took place at the same annealing tempera-ture as for Ca-doped YIG in hydrogen atmosphere[26,27]. Some differences have to be noted: while in Ca-doped YIG an expansion of the lattice parameter took place probably due to the formation of oxygen vacancies, in stoichiometric YIG no relevant expansion was detected. Consequently, the value of the strain of the surface layer with respect to the underlying film was much smaller than for Ca-doped YIG. An analysis of the strain profile was carried out[26], utilizing the same kinematical approximation above reported. Fig. 6 shows in detail the diffraction profiles (only the relevant parts of the tails are shown) for two different annealing times at a temperature of 510 °C. Fig. 7 shows the strain profiles as derived by the best fit of the curves: a layer with a constant lattice parameter and a transition region were formed. The thickness was different, but the qualitative behavior was the same for the two annealing times. CEMS measurement[27] showed that the surface layer was a good quality garnet, without the damage typical of the implanted crys-tals. The formation of this layer with a contracted lattice parameter can be

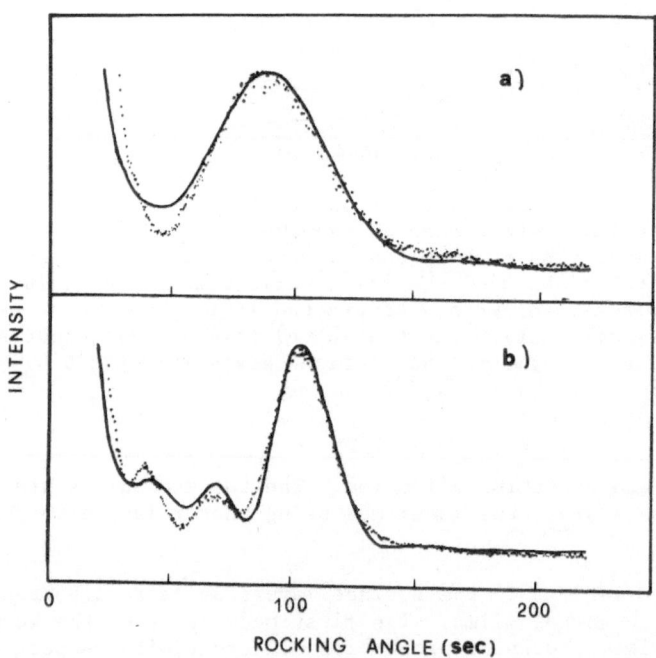

Fig. 6. Experimental data (points) and theoretical fit (full line) of the high-angle tail of the rocking curve of a pure YIG crystal annealed at 510 °C for (a) 4 hours – (b) 18 hours. Reflection: symmetric (888) – CuK$_\alpha$ (After G. Balestrino et al.[26]).

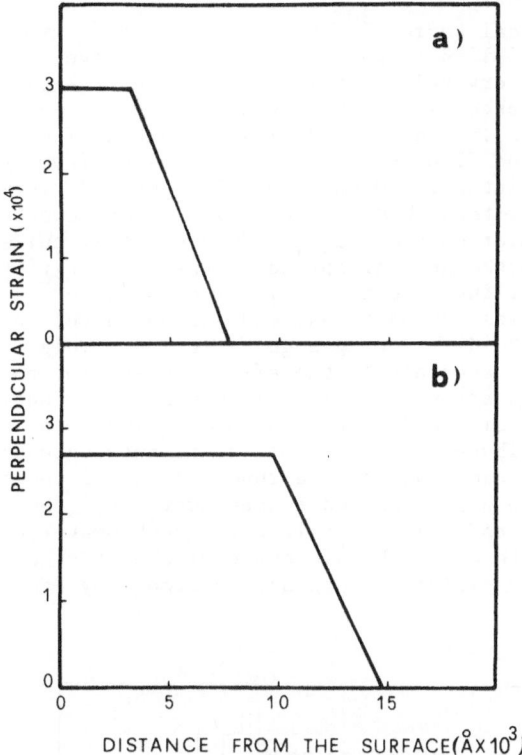

Fig. 7. Strain distributions obtained by the best fits shown in Fig. 6(a) and (b). The fittings are calculated following a kinematical theory approximation. (After G. Balestrino et al.[26]).

of great practical interest. In fact, it could substitute the damaging procedure of ion implantation to obtain a layer with the magnetization in plane.

STANDING WAVES

The X-ray standing wave technique (XSW) is receiving, from its first applications[28], a growing interest due to its potentiality in the structural studies of surfaces and interfaces. The technique is based on the interference between the incident and the diffracted waves in the case of Bragg diffraction of an X-ray plane wave by a perfect crystal. The electromagnetic field into the crystal has the form of a standing wave with the same periodicity of the diffracting planes. The standing wave pattern changes its phase with respect to the diffracting planes by π when the crystal is scanned through the diffracting region. In other words, the standing wave nodes coincide with the diffracting planes in the low-angle side of the rocking curve, while the antinodes coincide with the diffracting planes in the high-angle side of the rocking curve. Therefore, there is a modulation of the electromagnetic intensity at the crystallographic planes when the crystal is rocked through the diffracting angles. The modulation of the intensity obviously implies a corresponding modulation of the secondary effects such as the production of fluorescent radiation or the emission of photoelectrons. If these secondary effects, both from constituent atoms and/or from foreign atoms, are monitored together with the diffracted intensity, which gives the angular scale, an accurate determination of the lattice position can be given. The foreign atoms can be both distributed in a surface layer[29] or

269

adsorbed at the physical surface[30] or even epitaxially grown[31]. This technique has also been applied to garnets[32]. Garnets have, as discussed in second section, three crystallographic sites. When the (444) planes are considered, the dodecahedral and the tetrahedral sites coincide with the diffracting planes, while the octahedral sites are half-way between them. Thus by monitoring the fluorescent radiation coming from atoms occupying different sites, the intensity modulation will have different behavior for different occupation sites. However, the first measurements carried out on garnets[32], in particular on a $Gd_3Ga_5O_{12}$ (GGG) crystal, suffered both for low resolution due to the angular spread of the reflectivity curve of the first crystal, and for the effect of primary extinction[14]. The effect of primary extinction gives an additional modulation in the intensity of fluorescent radiation in the diffracting angular region. This is due to the fact that when diffraction takes place, the effective absorption coefficient of the incident beam is a strong function of the incidence angle and has its maximum at the center of the diffraction curve. Thus also the thickness of the layer which contributes to the fluorescence, and consequently the intensity of the fluorescence, change as a function of the rocking angle. In our last experiment both problems have been overcome. The first one has been resolved by utilizing an asymmetrically cut monochromator which strongly reduces the angular width of the Darwin curve of the first crystal[29,14]. The problem of primary extinction has been overcome by detecting, by means

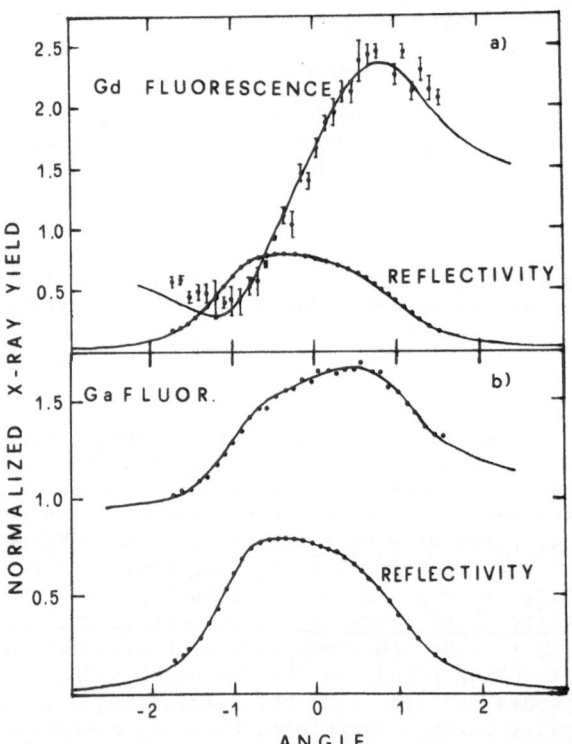

Fig. 8. Reflectivity and fluorescence angular yield modulation of (a) Gd L_α fluorescent radiation and (b) Ga K_α fluorescent radiation. The acceptance angle was limited by a slit. The low and high limits were respectively 3.5 and 13.5 mrad. The Gd L_α fluorescence angular yield was corrected by taking into account the secondary interelement fluorescence effect (see text). (After S. Lagomarsino et al.[33])

of a slit, only the Ga Kα and Gd Lα fluorescent radiation which comes out at a very glancing angle. By this way the layer which contributes to the detected fluorescence is less than a thousand Ångstrom thick, thus eliminating quite completely the effectiveness of the primary extinction. Fig. 8 shows the reflectivity curve and the fluorescence modulation for the Ga Kα and Gd Lα radiation. The Ga occupies both the tetrahedral and octahedral sites, and therefore the fluorescence is the weighted contribution from the two sites. Furthermore, a correction to the Gd fluorescence has been applied which takes into account the secondary interelement fluorescence effect that is the fact the Gd Lα fluorescence can be excited by the Ga Kα fluorescence. After the correction, the agreement between the theoretical (full line) and experimental results (closed points) is very good. These measurements open very good perspectives concerning the possibility of studying by this technique structural properties of thin layers such as site occupancy, damage profile by ion implantation and interfaces.

REFERENCES

1. F. Bertaut and F. Forrat, C.R.Acad.Sci. 243:382 (1956).
2. S. Geller and M. A. Gilleo, Acta Cryst. 10:239 (1957).
3. Physics of Magnetic Garnets, in: Proceedings of LXX Course of Varenna School, Ed. A. Paoletti - Publ. North Holland (1978).
4. G. Winkler, "Magnetic Garnets", Vieweg, Braunschweig (1981).
5. Thin Solid Films, Special Issue on Magnetic Garnet Films 114 (1984).
6. S. Geller, Z. Kristallogr. 125:1 (1967).
7. A. H. Bobeck, E. Della Torre, "Magnetic Bubbles", North Holland Publ. Co., Amsterdam (1975).
8. R. C. Linares, J.Cryst.Growth 3/4:443 (1968).
9. S. Isomae, S. Kishino, and M. Takahashi, J.Cryst.Growth 23:253 (1974).
10. J. W. Matthews and E. Klockholm, Mater.Res.Bull. 7:213 (1972).
11. B. Antonini, S. L. Blank, S. Lagomarsino, A. Paoletti, P. Paroli, and A. Tucciarone, IEEE Transactions on Magnetics, MAG-17:3220 (1981).
12. B. Antonini, S. L. Blank, S. Lagomarsino, A. Paoletti, P. Paroli, and A. Tucciarone, J.Appl.Phys. 53:2495 (1982).
13. J. C. North and R. Wolfe, in:"Ion Implantation in Semiconductors and Other Materials", ed. B.L. Crowder, Plenum, N.Y. (1973).
14. R. W. James, "The Optical Principles of the Diffraction of X-rays", Bell, London (1954).
15. S. Takagi, Acta Cryst. 15:1311 (1962).
16. D. Taupin, Bull.Soc.Franc.Minéral.Cryst. 87:469 (1964).
17. V. S. Speriosu, H. L. Glass, and T. Kobayashi, Appl.Phys.Lett. 34:539 (1979).
18. A. T. Macrander, E. R. Minami, and D. W. Berreman, J.Appl.Phys. 60:1364 (1986).
19. B. E. MacNeal and V. S. Speriosu, J.Appl.Phys. 52:3935 (1981).
20. W. H. de Roode and J. W. Smits, J.Appl.Phys. 52:3969 (1981).
21. V. S. Speriosu and C. H. Wilts, J.Appl.Phys. 54:3324 (1983).
22 W. H. de Roode and H. A. Algra, J.Appl.Phys. 53:2507 (1982).
23. G. Balestrino, S. Lagomarsino, E. Milani, P. Gerard, and A. Tucciarone, submitted for publication to J.Appl.Phys.
24. B. Antonini, C. D. Brandle, S. Lagomarsino, A. Paoletti, P. Paroli and A. Tucciarone, J.Appl.Phys. 55:2179 (1984).
25. G. Balestrino, S. Lagomarsino, and A. Tucciarone, Thin Solid Films 125:263 (1985).
26. G. Balestrino, S. Lagomarsino, and A. Tucciarone, J.Appl.Phys. 59:424 (1986).
27. G. Balestrino, S.Lagomarsino, A. Tucciarone, and P. Gerard, J.Magn.Mat. 62:103 (1986).
28. A. Golovchenko, B. W. Batterman, and W. L. Brown, Phys.Rev. B10:4239 (1974).

29. S. K. Andersen, J. A. Golovchenko, and G. Mair, Phys.Rev.Lett. 37:1141 (1976).

30. J. R. Patel, J. A. Golovchenko, J. C. Bean, and R. C. Morris, Phys.Rev. B31:6884 (1985).

31. E. Vlieg, A. E. M. J. Fisher, J. F. Van der Veen, B. N. Dev, and G. Materlik, Surf.Sci. 178:36 (1986).

32. S. Lagomarsino, F. Scarinci, and A. Tucciarone, Phys.Rev. B29:4859 (1984).

33. S. Lagomarsino, J. R. Patel, and J. A. Golovchenko, submitted to Phys. Rev.

LOCATION OF IMPURITY ATOMS IN THE VOLUME AND SURFACE LAYERS

OF SILICON CRYSTALS, BY X-RAY STANDING WAVE IN THE LAUE GEOMETRY

A. Yu. Kazimirov, M. V. Kovalchuk

Institute of Crystallography
Academy of Sciences of the USSR
Moscow 117333, USSR

V. G. Kohn

I. V. Kurchatov Institute of Atomic Energy
Kurchatov Square 46
Moscow 123182, USSR

The dynamical theory of X-ray diffraction predicts the formation of standing wave fields inside crystal. The nodal and antinodal planes of these standing waves are parallel to the diffracting planes and have the same period. The angular dependence of the fluorescent yield is defined by the position of the atom under study with respect to nodes and antinodes of the standing wave. Thus, by comparing the experimental curve of the fluorescent yield with the theoretical one, it is possible to obtain information about the position of impurity atoms in the crystal lattice.

By now, the problem of localization of the impurity atoms in a thin surface layer or adsorbed atoms on the surface of a crystal has been solved with the use of the Bragg geometry when the formation of standing waves takes place in a relatively thin surface layers [1-3]. Meanwhile, as it is shown in the present work, the Laue diffraction geometry, where the fluorescent yield of impurity atoms from the exit surface of a crystal is measured, has several advantages. First, it is possible to study the position of impurity atoms distributed through the bulk of the crystal. In this case the sensitivity of this method in the Bragg geometry decreases because of the well known extinction effect. Moreover, in the Laue geometry it is easy to use different reflections (including asymmetrical ones) to study the positions of the impurity atoms in different crystallographic directions.

Let us consider the asymmetrical case in the Laue diffraction with a plane wave incident at the angle $\Theta = \Theta_B + \Delta\Theta$ to the diffracting planes and at the angle ψ to the surface of the crystal. It can be shown from the general theory that for the case of $L_a \gg L_{yi} \gg L_{ex}$ ($L_{ex} = \lambda\gamma_o/\pi|\chi_{rh}\beta|^{1/2}$ is the extinction length, L_{yi} is the depth of the fluorescent yield, $L_a = \gamma_o/\mu_o$ is the depth of X-ray absorption) the angular dependence of the fluorescent yield is described by the following equation (ref.[4]):

$$\varkappa(\Delta\Theta) = \sum_{i=1,2} I_i(\Delta\Theta) \exp(-\Delta\mu_i(\Delta\Theta).t) \tag{1}$$

here

$$I_i = (1 + \xi_i^a \beta + \xi_i B)(1 + \xi_i^2)^{-2}$$

$$\Delta\mu_i = \xi_i \left[2C \beta^{1/2} \frac{|\chi_{ih}|}{\chi_{io}} - \xi_i(1-\beta) \right]/L_A(1+\xi_i^2) \ ,$$

$$\xi_{1,2} = C^{-1}[y \mp (y^2 + c^2)^{1/2}],$$

$$\gamma_o = \sin \psi, \quad \beta = \gamma_o/\sin(\psi + 2\Theta_B) \ , \tag{2}$$

$$y = -\beta^{1/2} |\chi_{rh}|^{-1} \sin 2\Theta_B [\Delta\Theta - \chi_{ro}(1-\beta)(\beta \sin 2\Theta_B)^{-1}],$$

$$B = 2C \beta^{1/2} |\chi_{ih}| |\chi_{io}|^{-1} \cos(2\pi \Delta d/d),$$

where t is the crystal thickness, λ is the wavelength, μ_o is the normal absorption coefficient, C is the polarization factor, $\chi_{ro}, \chi_{io}, \chi_{rh}, \chi_{ih}$ are the Fourier components of the real χ_r and imaginary χ_i parts of the crystal polarizability, $\Delta d/d$ is the displacement of the impurity atom from the diffracting plane. As follows from (1), the angular dependence of the fluorescent yield is formed by both standing wave fields: one with nodes on the atomic planes (a weakly absorbed field, i=1) and another with antinodes on the atomic planes (a strongly absorbed field, i=2). It can be seen that the angular dependence does not depend on the values of L_{yi} and L_{ex}.

The scheme of the experiment is shown in Fig. 1. C_1 is an asymmetrically cut crystal with the asymmetry factor b=25, C_2 is a sample crystal. We have studied Si (100) single crystals with different thicknesses (2.2 mm and 0.48 mm) uniformly doped by Ge to $7.5 \cdot 10^{19}$ at/sm^3 concentration and a Si (111) crystal with Si epitaxial layer of 1.6 thickness doped by Ge to 10^{22} at/sm^3 concentration. The asymmetrical (III) reflection of MoK$_\alpha$ radiation was used.

The angular dependence of the GeK$_\alpha$ fluorescent yield from a Si crystal of t=2.2 mm thickness uniformly doped by Ge is shown in Fig. 2. Since the crystal is rather thick (t=4.3 L$_a$), the angular dependence is formed only by a weakly absorbed filed (i=1 in equation (1)). It is a specific feature of the Laue geometry in this case that the intensity of the standing wave field increases or decreases with changing $\Delta\Theta$, but there is no movement of the nodal and antinodal planes with respect to diffracting planes, as it is in the Bragg geometry.

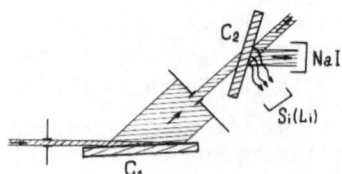

Fig. 1. Scheme of the experiment: C_1 is a monochromator crystal, C_2 is a sample crystal.

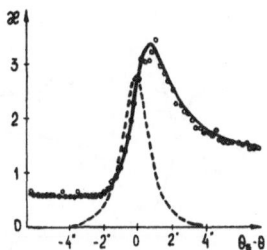

Fig. 2. Angular dependence of the GeK$_\alpha$ fluorescent yield from a Si crystal of t=2.2 mm uniformly doped with Ge: the circles are experimental, solid line is calculated for substitutional impurity atoms $\Delta d = 0$.

For the substitutional impurity atoms lying on the atomic planes ($\Delta d = 0$) the increase of the fluorescent yield is explained by the anomalous transmission of X-rays. The fluorescent yield increases (an impurity atom occurs in the region of an increased standing wave field) with increasing displacement of an impurity atom from the diffracting planes. The solid line in Fig. 2 is the theoretical curve for $\Delta d = 0$ calculated for the real experimental conditions where the angular divergence of the incident beam is taken into account. There is a good agreement between the experimental curve and the theoretical one. For the impurity atom between the diffracting planes ($\Delta d/d = 0.5$), the maximum fluorescent yield would be equal to 13.9. It should be noted that the sensitivity to the position of an impurity atom increases with the increase of the thickness t of a crystal.

The formation of the fluorescent yield curve in the case of a thin crystal is more complicated because of the existence of both standing wave fields at the exit surface of a crystal. The experimental curve of the GeK$_\alpha$ fluorescence yield from the thin Si crystal uniformly doped with Ge is shown in Fig. 3. In the same figure, the theoretical curve (1) is for the substitutional impurity atoms ($\Delta d = 0$), the curve (2) is for the random distribution of impurity atoms. The minimum of curve (1) is explained by the anomalous absorption of the strongly absorbed wave field. For curve (2) this effect is compensated by the weakly absorbed field. Again, as it is in the case of a thick crystal, there is a good agreement between the experimental curve and a theoretical one for substitutional impurity atoms. This result can be explained by the fact that germanium is the isolated impurity in silicon.

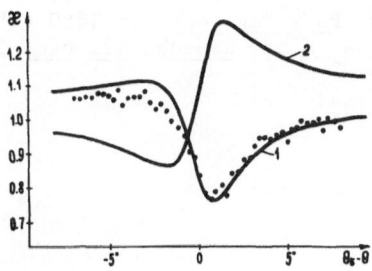

Fig. 3. Angular dependence of the GeK$_\alpha$ fluorescent yield from a Si crystal of t=0.48 mm uniformly doped with Ge:
the circles are experimental; (1) the calculation for substitutional impurity atoms ($\Delta d=0$); (2) calculation for the random distribution of impurity atoms.

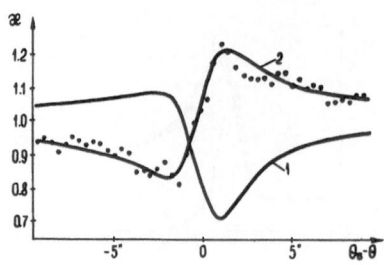

Fig. 4. Angular dependence of the GeK$_\alpha$ fluorescent yield from a Si crystal
with an epitaxial layer:
the circles are experimental; (1) the calculation for substitutional
impurity atoms ($\Delta d=0$); (2) the calculation for the random distribution of impurity atoms.

     In Fig. 4 one can see the experimental result and theoretical curves
for the Si crystal with an epitaxial layer. In this case the theoretical
calculation was made taking into account that the depth of the fluorescent
yield L_{yi} is limited by the thickness of an epitaxial layer. The experimental curve is in good agreement with the calculation for the random distribution of impurities. The real situation is more complicated. The analysis
of the diffraction rocking curve showed that in addition to the main maximum
(diffraction by the matrix) there is a weak maximum of diffraction by the
thin epitaxial layer. Thus, the epilayer has a different lattice constant
($\Delta d/d = 4.10^{-3}$). This is why the standing waves formed in the bulk of a
crystal "see" the impurity atoms in the layer as random distributed. Our
additional experiments with the X-ray standing wave formed by the diffraction in the layer showed that impurity atoms have correct positions in the
lattice of the epitaxial layer.

REFERENCES

1. J. A. Golovchenko, B. W. Batterman, and W. L. Brown, Observation of
     internal X-ray wave fields during Bragg diffraction with an applica-
     tion to impurity lattice location, Phys.Rev.Ser.B 10:4239 (1974).
2. A. Materlik and J. Zegenhagen, Phys.Lett.Ser.A 104:47 (1984).
3. P. L. Cowan, J. A. Golovchenko, and M. F. Robbins, X-ray standing waves
     at crystal surfaces, Phys.Rev.Lett. 44:1680 (1980).
4. M. V. Kovalchuk and V. A. Kohn, Uspechi fiz.nauk 29(5):426 (1986).

X-RAY STANDING WAVES IN THE STUDY OF CRYSTALS

AND SURFACE LAYERS

M. V. Kovalchuk

Institute of Crystallography
Academy of Sciences of the USSR
Moscow 117333, USSR

V. G. Kohn

I. V. Kurchatov Institute of Atomic Energy
Kurchatov Square 46
Moscow 123182, USSR

The physical nature, potentialities and prospects of application of the new method of studying the structure of crystals and thin surface layers (standing wave technique) is discussed. Applications of the external photoeffect excited by the X-ray standing wave to structure studies of crystal subsurface layers are discussed. In experiments conducted with an epitaxial Si film doped by B and Ge a change of phase of the scattering amplitude on the photoemission curve has been found. This change is caused by the total surface displacement due to a change in the interplanar spacing in the disturbed layer. The potentialities of the depth-selective analysis of X-ray standing waves are analyzed. Theoretical foundations of the secondary-radiation yield under conditions of multiple diffraction have been developed.

A new field in the physics of diffraction of hard electromagnetic radiation has emerged and taken shape in the past 20 years. It is based on studying and using X-ray standing waves that appear in a perfect crystal under conditions of dynamic diffraction. Apart from a general physical interest involving the anomalously sharp change in the character of the interaction of X-rays with an atom in the crystal, this field is highly promising for analyzing the structure of crystals and surface layers.

The X-ray standing wave is a coherent superposition of incident and reflected plane waves with the wave vectors $\vec{k}_o$ and $\vec{k}_h = \vec{k}_o + \vec{h}$, that differ in a reciprocal lattice vector $\vec{h}$, and its magnitude is $2\pi \, {}^{o}n/d$, where d is the interplanar spacing, n is the reflecting order. Accordingly, the period of the standing wave d/n copies the periodicity of the crystal lattice in the crystal volume being studied and near its surface. In a symmetric case the intensity of the X-ray standing wave field is

$$I(z) = |E_o|^2 \left[1 + \frac{|E_h|^2}{|E_o|^2} + 2 \frac{|E_h|}{|E_o|} \cos \left(\frac{2\pi \, n}{d} \cdot z + \alpha \right) \right] , \qquad (2)$$

where E_o and E_h are the amplitudes of incident and reflected waves, respectively, $\alpha = 2\pi \, n\Delta \, z/d$ is the phase of the ratio E_h/E_o, z is the coordinate along the h vector.

In the Bragg diffraction geometry and in the angular range of the total X-ray diffraction reflection the ratio of E_h/E_o is close to unit and α-phase changes from $-\pi$ to 0. This results in a deviation of the nodes and antinodes of the wave in the magnitude of $d/2n$. Thus by varying the angle of incident rays on a crystal in a very small range of order of one to ten angle seconds one can sharply change the degree of interaction of the radiation field with atoms of the perfect crystal lattice. To record such angular dependence, it is necessary to measure any secondary radiation yield with escape depth of $L_{yi} \ll L_{ex}$, where L_{ex} is the extinction length equal to the minimal X-ray penetration depth in the crystal under conditons of total reflection.

Measurement of the fluorescent radiation from the layer of impurity or adsorbed atoms, disposed at a distance z_o from the crystal surface, is most suitable for this purpose. In this case the fluorescent intensity would be directly proportional to equation (1), where one should change z for z_o. In Fig. 1 the angular dependent curves of the fluorescence for the first reflection order are shown, $z_o = d$ (Fig. 1a), $z_o = d/2$ (Fig. 1b), and for a case where z_o does not have a precisely defined position (Fig. 1c). One can see how sharply the character of the curve varies. The real pattern is an intermediate one, i.e., atoms with a definite probability F_o occupy some mean position z_o. Here, the angular dependence of the secondary radiation yield is described by an equation:

$$\varkappa = 1 + \frac{|E_h|^2}{|E_o|^2} + 2\,\frac{|E_h|}{|E_o|}\,F_o\,\cos(P_o + \alpha) \ . \tag{2}$$

The parameters $P_o = 2\pi\,nz_o/d$ and $F_o = e^{-W}$ introduced here are usually called a coherent position and a coherent fraction.

Equation (2) is applicable to a description not only of the angular dependence of the fluorescent yield from the impurity atom layer, but to all the cases where $L_{yi} \ll L_{ex}$. For example, when the impurity is distributed in a surface layer of a crystal, and also in the measurement of the external X-ray photoeffect from the thin disturbed crystal layer under a condition that the thickness of the layer $L \ll L_{ex}$. In this case the parameters P_o and F_o are the characteristics, averaged over the thickness of the layer and, respectively, over the escape depth of the electrons.

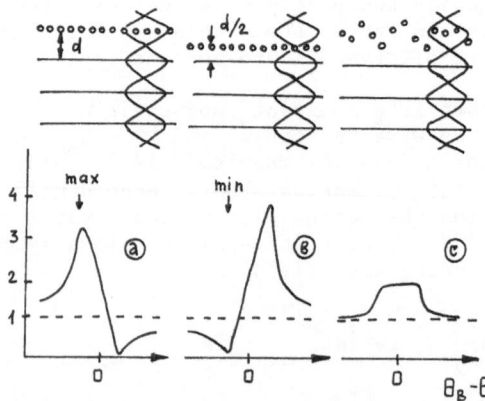

Fig. 1. Scheme illustrating the possible positions of the impurity layer with respect to the crystal surface and the angular dependent fluorescent yield curves corresponding to them.

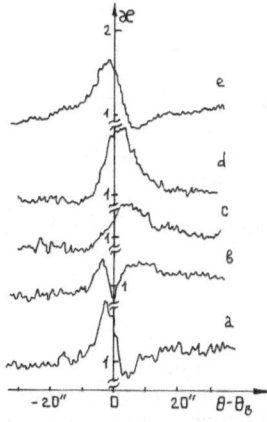

Fig. 2. Angular dependence of the photoelectron emission yield under conditions of the 444-reflection of the CuK_α X-rays for Si with the epitaxial films of thickness of 1.5 μm, which differ in Ge-N concentration of atoms. Curve a) N = 0; b) N = 3.7; c) N = 7.0; d) N = 9.0; e) N = 15.0 x 10^{19} at/m^3.

The experiment demonstrating advantages of XRSW was carried out together with Dr Lobanovich. The external photoeffect from silicon single crystals doped by B, where Ge was also added at various concentrations, was measured. The 444-diffraction of CuK_α radiation was used. Here, L_{yi} = 0.45 μm, L = 1.5 μm, L_{ex} = 10.5 μm. i.e. the necessary condition $L_{yi} < L < L_{ex}$ is fulfilled. The curves obtained are shown in Fig. 2. The fitting of curves according to equation (2) by the least squares method enabled us to find the values of P_o and F_o parameters in connection with the Ge concentration, and then to calculate the mean deformation in the epitaxial film. In addition to the fact that this experiment was carried out using a real object of semiconductor microelectronics, the optimum Ge concentration necessary for the compensation of undesirable tensions in epitaxial structures was also determined.

The instrumental basis of the standing wave method is an X-ray spectrometer which consists of a block of monochromators, a radiation source and a precise goniometer for the adjustment and rotation of the crystal under study. Modification of this method, based on the measuring the photoelectron emission, is based on different systems of the electron detection. In the first stage of the study the secondary electron multiplier (SEM) was used for this purpose. It was placed in the vacuum volume together with a specimen. The recording had an integral character, that is, all the electrons emerging from the crystal were detected, regardless of their energy. The experiment described above was carried out according to such scheme.

The next step in the use of photoemission was the invention of a new vacuum free method of measurement based on a gas proportional flow-through counter of a speical design. In this method the crystal under study is placed directly inside a gas proportional counter (Fig. 3). The incident and reflected X-ray beams pass the walls of the counter through special windows covered with a thin organic film. The counter is blown with the gas mixture of 90% He + 10% CH_4. The electrons leaving the crystal are accelerated by an electric field created by high voltage applied between the crystal and a thin filament, ionize the gas and are recorded as an electric pulse.

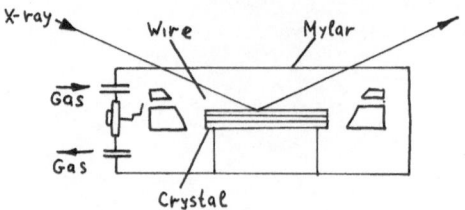

Fig. 3. Scheme of the gas proportional low-through counter for registering the photoelectrons.

A great advantage of this method, except the high effectiveness of electron couonting, is the fact that the counter used as an low-resolution (20 %) electron spectrometer served as the basis for a new approach – a depth-selective standing wave method, which allows one to carry out a layer non-destructive structure analysis of the surface layers.

A further development of this method is connected with the use of not only the Bragg geometry, but also the Laue geometry, and also with measurements under the conditions of multiwave X-ray diffraction, when collimation of the incident beam in two directions, $\Delta\Theta$ and $\Delta\varphi$, is necessary.

The development of the theory of secondary radiation yield (SRY) under the conditions of multiwave diffraction is very important for an interpretation of experimental results in the latter case. Following the analysis, we came to a conclusion that the main equations of this theory may be those given below for the intensity SRY:

$$\varkappa_\nu^{(s)}(\Delta\Theta,\Delta\varphi) = \int\limits_o^t dz\, P_\nu(z)\, \frac{d\, N_\nu^{(s)}(z,\Delta\Theta,\Delta\varphi)}{dz}\ , \tag{3}$$

where t is the crystal thickness, s, ν are the polarization indices and the types of the SRY, respectively, $P_\nu(z)$ is the probability function of the SRY yield in a crystal from the depth z towards its surface. The function

$$\frac{d\, N_\nu^{(s)}(z)}{dz} = \frac{c}{8\,\pi}\,\frac{\mu_{o\nu}}{h\omega} \sum_{\substack{m,m'\\s's''}} \frac{\chi_{imm'}^{s's''}(\nu)}{\chi_{io(\nu)}}\, f_{m\,m'}(z)\, E_{m\,s'}^{(s)}(z)\, E_{m's''}^{(s)}(z) \tag{4}$$

describes the number of X-ray quanta absorbed at a depth z with the following generation of the SRY of the ν-type. Amplitudes of the direct (m = 0) and diffracted (0 < m < N_h – 1) waves in a crystal with a distorted surface layer satisfy the generalized Takagi-Taupin system of equations,

$$\frac{d\, E_{ms}}{d\, z} = \frac{i\pi}{\lambda\gamma_m} \sum_{m's'} [\chi_{m\,m'}^{s\,s'}\, f_{m\,m'}(z) - \alpha_m\, \delta_{m\,m'}^{s\,s'}] E_{m's''} \tag{5}$$

where $\chi_{m\,m'}^{s\,s'} = \chi_{m-m'}(\vec{e}_{ms}\, \vec{e}_{m's'})$, $\vec{e}_{ms}$ are polarization vectors, $\chi_{m-m'}$ is the Fourier component of the polarization of a crystal ($\chi = \chi_r + i\chi_i$) on the reciprocal lattice vector $\vec{h}_{mm'} = \vec{K}_m - \vec{K}_{m'}$. The function

$$f_{mm'}(z) = \exp(-i\vec{h}_{mm'}\cdot\vec{u}(z) - \tfrac{1}{2}h_{mm'}^2 <\Delta u^2(z)>) \tag{6}$$

280

describes distortions of the crystal lattice. The parameters

$$\alpha_m = (K_m^2 - K_o^2)/K_o^2 = C_{mi}\, \Delta\Theta + C_{mz}\, \Delta\varphi \tag{7}$$

specify deviations from Bragg's conditions, and the parameter $\gamma_m = \pm\sin\varphi_m$, where φ_m is the angle between the direction of the m-beam and the crystal surface. For the Laue geometry $\gamma_m > 0$, and for the Bragg geometry, $\gamma_m < 0$.

The rest of the notations are standard. In particular, $\delta_{mm'}^{ss'}$ is the symbol of Kroneker and is equal to unity at $m = m'$ and $s = s'$. The amplitudes $E_{ms'}^{(s)}(z)$ satisfy the following boundary conditions: $E_{os'}^{(s)}(0) = A_s\delta_{ss'}$, $E_{ms'}^{(s)}(0) = 0$ for the Laue beams $(\gamma_m > 0)$ and $E_{ms'}^{(s)}(t) = 0$ for the Bragg beams $(\gamma_m < 0)$.

The obtained equations allow one to solve a wide range of problems involved in the study of the angular dependence yield of different secondary radiations in ideal crystals as well as in real crystals. Let us analyze only one particular case, namely, the problem of a direct observation of the Borrmann effect on the X-ray photoelectron yield. The photoelectron emission from the exit surface of a thick perfect absorbing crystal should be recorded in this case. Here, one can show, using the formalism developed above, that for $\varkappa^{(s)}$ the following approximate equation is valid:

$$\varkappa^{(s)} = \frac{1}{\mu_o} \sum_j \mu_j\, e^{-\mu_j t}\, |E_{os}^{(j)}|^2 \,, \tag{8}$$

where μ_j and $E_{os}^{(j)}$ are the absorption coefficients and the amplitudes of the direct wave field corresponding to the j-zone of the dispersion surface (DS), respectively. At a fixed value of the angle $\Delta\varphi$ the dependence μ_j from $\Delta\Theta$ for the slightly absorbed zones of the DS has its minimum in the centre (at $\Delta\Theta = 0$). Accordingly, $\exp(-\mu_j t)$ has its maximum in the centre, and $|E_{os}^{(j)}|^2$ has a shape of a blurred step with a centre of the transition range at $\Delta\Theta = 0$ (see Fig. 4). The shape of the curve $\varkappa(\Delta\Theta)$ is defined by compe-

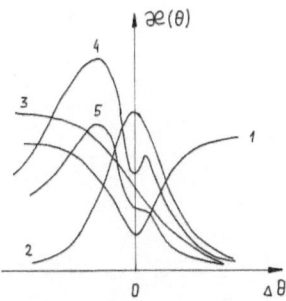

Fig. 4. Typical angular dependence curves of the absorption coefficient μ_j (curve 1), $\exp(-\mu_j t)$ (curve 2), degrees of excitation of the zone $|E_{os}^{(j)}|^2$ (curve 3) for the zone of a dispersion surface j with minimal μ_j. Curves 4 and 5 correspond to the angular dependence yield of the secondary processes in the fourwave ("gap") and twowave ("plateau") cases.

tition of these dependences. In a general case, it has a shape of the bell with a maximum, deflected from the point $\Delta\Theta = 0$ and with a "gap" or a "plateau" in its slope, which transfers through the point $\Delta\Theta = 0$ (Fig. 4). The degree of closeness of μ_{min} to zero has the decisive significance for the observation of the effect. The calculation carried out in the case of four-wave (220, 400, 220) diffraction of CuK_α-radiation in silicon at $t = 0.8$ mm confirmed this conclusion, in full. At the fourwave point $\Delta\varphi = \Delta\Theta = 0$ and μ_{min} is to a great extent smaller than at two-wave lines (the effect of intensification of anomalous transmission). Here, the "gap" appears in the curve $\varkappa$, in contrast to the two-wave case, where with the same parameters the Borrmann effect appears only as a "plateau" in one of the slopes of the maximum connected with the anomalous transmission of X-ray radiation, when Bragg's conditions are observed.

Thus, this method of X-ray standing waves happens to be quite efficient and powerful means for the study of various aspects of interaction of radiation with a crystal, and also for solving various structural problems and problems of material science in crystals with a sufficiently perfect crystal lattice.

DEVELOPMENT OF THEORETICAL AND EXPERIMENTAL INVESTIGATIONS

OF THIN SURFACE STRUCTURES BY X-RAY METHODS

I. A. Vartanyantz, A. Yu. Kazimirov, M. V. Kovalchuk

Institute of Crystallography
Academy of Sciences of the USSR
Moscow 117333, USSR

V. G. Kohn

I.V. Kurchatov Institute of Atomic Energy
Kurchatov Square 46
Moscow 123182, USSR

A new method of analysis of the curves of diffraction reflection from crystals with a disturbed surface layer has been developed. The method is based on a possibility to reconstruct the function, describing the crystal structure, with the aid of an inverse Fourier transformation. It allows to determine directly from experimental data the profile of deformation $\Delta d(z)/d$ and the degree of disorder $\exp(-W(Z))$ of the disturbed layer. The proposed method gives an opportunity to study the uniqueness of the solution thus obtained. The method is used to study a disturbed layer structure of silicon single crystals irradiated with boron ions.

INTRODUCTION

In the last years much attention has been paid to the problem of disturbances of the surface structure caused by diffusion, ionic implantation, epitaxial growth, etc. The method of X-ray diffractometry based on measurement of the angular dependence of diffracted X-rays (XR) in the Bragg geometry[1,2] is one of the attractive methods of such analysis.

The curve of the diffraction reflection (CDR) of a crystal with the disturbed surface layer has, usually, strongly distorted "tails" except the angular region of total reflection. Those "tails" contain information about the stsructure of the surface layer in the form of additional maxima or intensity oscillations. Additional reflection leads to a displacement of the Bragg angle and indicates a change of the crystal lattice parameter in the layer. If one knows the profile of the deformation, it is easy to calculate the theoretical CDR using the Takagi-Taupin set of equations[2]. However, the inverse problem, i.e. the definition of the layer structure directly from CDR, is of practical interest. However, this problem is not only quite complex, but also has not a unique solution[3,4,5,6,7,8].

For solving this problem, some methods have been suggested. The most popular of them were those of direct numerical modelling of CDR by solving the Takagi-Taupin set of equations for different model profiles. In these methods the resulting profile was selected by the best coincidence between the theoretical and experimental CDR[2,6,7,8]. A numerical realization of these methods needs, as a rule, not only a lot of computer time, but also the problem of the uniqueness remains open (the general analysis is carried out only in the works[4,5].

In this work a new method has been proposed. It would allow one to reproduce directly the structure of the surface layer from experimental CDR. This method, from our points of view, combines the relative simplicity of the mathematical procedure with the possibility of reconstruction of the complicated profile of deformation in the surface layer. Moreover, in this method the analysis of uniqueness of the solution obtained is easy to perform.

THEORETICAL ANALYSIS

By selecting the proper order of reflection for the samples under investigation, one can experimentally provide a condition, where the reflection of XR from the crystal with the disturbed surface layer is kinematic outside the angular range of total reflection. In this case, the amplitude of reflection[9] is small in comparison with unity and is defined by the following equation

$$R(q) = -i\frac{c}{L_{ex}} \int_{0}^{\infty} f(Z) \exp(-iqz)dz = -i\frac{c}{L_{ex}} F(q) \qquad (1)$$

The angular parameter q in the case of symmetric reflection is defined by

$$q = -(2\pi/\lambda) \sin 2\Theta_B \, \Delta\Theta,$$

where λ is the XR wavelength, Θ_B is the Bragg's angle and $\Delta\Theta$ is the angular deviation between the direction of the incident beam and the centre of the total reflection range. C is the polarization factor. The extinction length $L_{ex} = \lambda \cdot \sin \Theta_B/(\pi|\chi_{rh}|)$, where χ_{rh} is the Fourier transform of the real part of crystal polarizability $\chi = \chi_r + i\chi_i$.

The function f(Z) describes the crystal structure through its scattering characteristics

$$f(Z) = \begin{cases} \exp(-i\varphi(Z) - W(Z)), & 0 < Z < L \\ 1, & Z > L \end{cases} \qquad (2)$$

where $\varphi(Z) = \vec{h}.U(Z)$, $\vec{h}$ is the reciprocal lattice vector, $U(Z)$ is the displacement of the atomic layer at a depth Z from its position in a non-disturbed crystal. The function $\exp(-W(Z))$ takes into account random displacements of atoms from the mean position at a depth Z (static Debye-Waller factor), L is the thickness of the disturbed layer. The coordinate Z is directed from the surface along the normal down into the crystal.

Thus, as we have from equation (1), in the case of kinematic scattering the amplitude of the reflection R(q) is essentially the Fourier component of the function f(Z). A possibility is thus provided for reconstructing this function describing the crystal structure with the aid of an inverse Fourier transformation

$$f(Z) = \frac{1}{2\pi} \int_{\infty}^{\infty} F(q).\exp(iqZ) \, dq \, . \qquad (3)$$

However, this way of calculation of the function f(Z) cannot be per-formed directly because only the XR reflection coefficient $P_R(q) = |R(q)|^2$ is measured experimentally. Hence, only the module of the complex function F(q) is known from experimental data. It is evident that in this situation, when we have incomplete information about the function F(q), it is hard to expect a unique solution of the inverse problem. Nevertheless, we can state the problem as follows: to find out the whole class of the profiles of distortions $\Delta d(Z)/d = dU_Z/dZ$ and $\exp(-W(Z))$ that satisfy the experimental data. In other words, to find a class of profiles that entirely represent CDR in direct numerical calculations.

To carry out this program we propose the following method. In the angular range, where the module $|F(q)|$ is known from the experimental data, we define the phase $\Phi(q) = \arg(F(q))$. Now, the whole function F(q) is defined and we can use equation (3) to determine the function f(Z). It is evident that one can define the phase Φ by different methods and they would lead to different profiles of $\Delta d/d$ and $\exp(-W)$. On the other hand, all the calculated profiles satisfy the experimental data, because of the uniqueness of the inverse Fourier transformation. Now the problem is only to select physically reasonable profiles from the whole set of profiles.

This idea was carried out in the following manner. We specify some start profile of $\Delta d/d$ and $\exp(-W)$ knowing some preliminary data about the structure of the surface layer of the crystal, and also by analyzing CDR. The function f(Z) (2) thus obtained was used for calculations of F(q) (equation (1)). Knowing F(q), we can calculate the XR reflection coefficient (10)

$$P_R(q) = K|F(q)|^2$$

$$K = L_{ex}^{-2}(1+\cos^3 2\Theta_B)(1+\cos 2\Theta_B)^{-1}$$

(4)

In the next step we change the module of F(q), taking into account the experimental value of $P_R(q)$

$$|F(q)^{exp}| = ((1/K) P_R^{exp}(q))^{1/2}$$

(5)

By using this modified function, one can immediately calculate the function f(z) with the help of equation (3) and so determine the result profile that coincides exactly with the experimental CDR. The described method gives us the possibility not only to reconstruct the profile quickly, but also to study the uniqueness of the solution obtained. Indeed, by varying the phase of F(q) (for example, in our approach it is easy to do this by varying the starting profile) we obtain the class of profiles corresponding to the same experimental CDR.

EXPERIMENTAL TECHNIQUES AND RESULTS

The analysis just described shows that for the experimental implementation of the method of determination of the structure of the surface layer of single crystals described above, it is necessary to measure the CDR in the Bragg geometry. This may be done, for example, with the help of a double crystal spectrometer. However, in the disturbed layer the random displacement of atoms from a mean position takes place along with a displacement of the atomic layers. Moreover, these different defects are the source of incoherent diffusion scattering. By measuring the integrated intensity of scattered radiation, it is not possible to separate the coherent part of radiation from the incoherent (diffusion) part. That is why, usually the whole scattered radiation is considered as coherent and the profile of deformation in the layer is defined from the "fitting" of theoretical curves

(which consider only the coherent component) to the "tails" of experimental
curves. In these cases where the coherent part is of the order of magnitude
or even smaller than the diffusion part, this can lead to great mistakes.

In such a situation a triple-crystal XR - diffractometer method (TCD),
which enables one to divide the contribution of the coherent and the diffusion scattering, appears to be effective[11,12]. By analyzing the angular
dependence of intensity of the main peak, one can obtain the "tails" of the
CDR, containing only the coherent part of scattered radiation, which fully
coincides with the used theoretical model.

In this work, samples of silicon (100), implanted by boron ions (with
the energy 100 Kev, dose 10^{15} cm^{-2}) have been studied. The samples were annealed in an oxidative atmosphere at different temperatures and times: sample
1 at T=800°, t=10 min, sample 2 at T=1000°, t=4 min, sample 3 at T=1000°,
t=40 min.

The experimental results (TCD spectra, double and triple-crystal CDR
of ideal and ion-doped crystals) are presented in Figs 1-4. CuK$_\alpha$ radiation
and (400) reflection were used. A comparison of the double and triple-crystal CDR (Figs 2-4) shows that the use of the TCD method is especially top-
ical in a study of annealing of the ion-doped layers. As one can see in

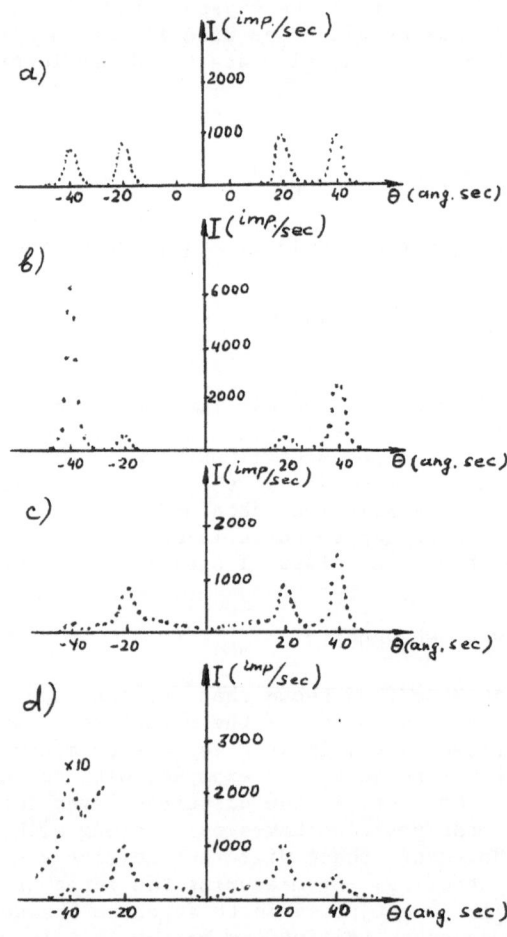

Fig. 1. Triple-crystal spectrum from investigated crystals (CuK$_\alpha$ radiation,
         (400) reflection, rotation angle α = +20"). Spectrum (a) is for
         an ideal sample, (b)-(d) for samples (1)-(3).

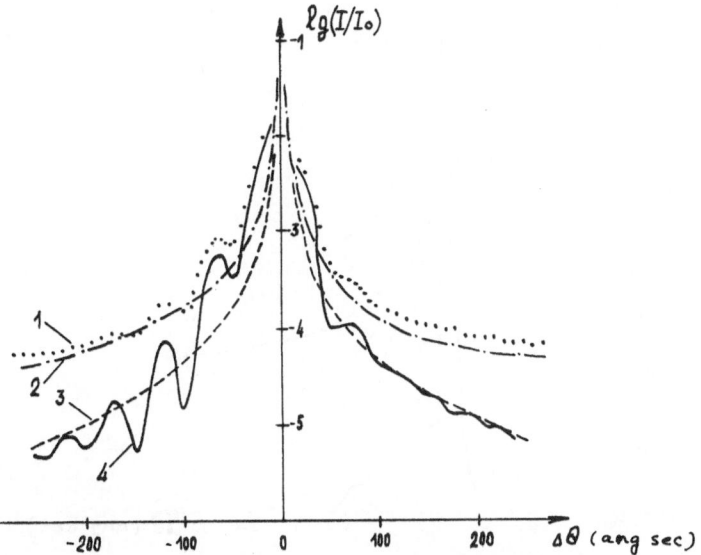

Fig. 2. Reflection curves of sample 1. Curve 1 (dotted line) is the double
crystal curve of the sample. Curve 2 (dashed and dotted line) is
the double crystal curve of the perfect crystal. Curve 3 (dashed
line) is the CDR of the perfect crystal, obtained by the TCD method.
Curve 4 (solid line) is the coherent component of the CDR of the
sample, obtained by the TCD method.

Figs 3-4, the disorder of the surface layer increases after annealing (rapid
decrease in the intensity of the coherent component of CDR (curve 4)). On
the contrary, the intensity of the double crystal CDR increases due to the
contribution of diffusion scattering (curve 1). A comparison between curve
1 and curve 4 (Figs 3-4) shows that structural information about the dis-

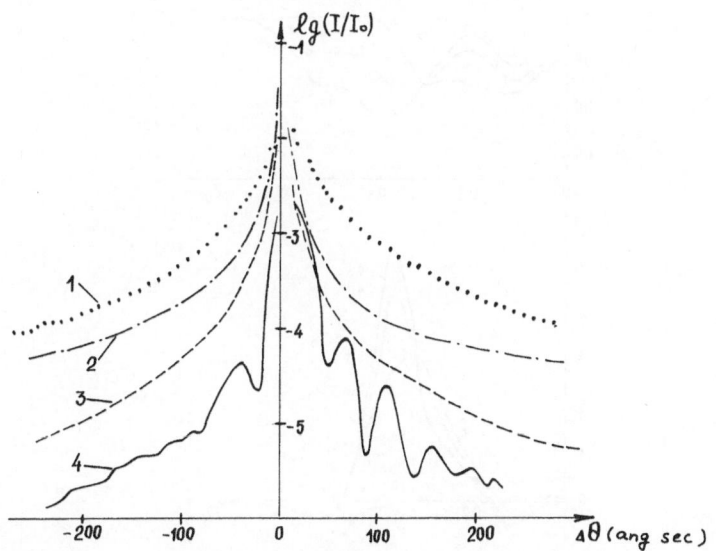

Fig. 3. Reflection curves of sample 2 (the notations are the same as in
Fig. 1).

287

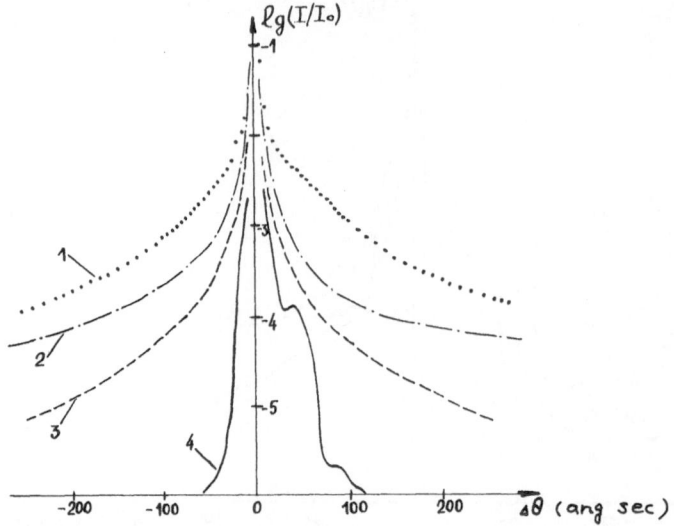

Fig. 4. Reflection curves of sample 3 (the notation is the same as in
Fig. 1).

turbed layer in the double crystal CDR is practically lost due to the increased intensity of diffusion scattering.

The profiles of $\Delta d(Z)/d$ and $\exp(-W(Z))$ for samples 1-3 reconstructed
according to the proposed method are presented in Figs 5-7. As it was stated
above, the solution to the inverse problem is not unique. For the sake of
simplicity, we show only three curves (curves 1-3 in Figs 5-6) and one curve
in Fig. 7, of the whole class of curves obtained by our method.

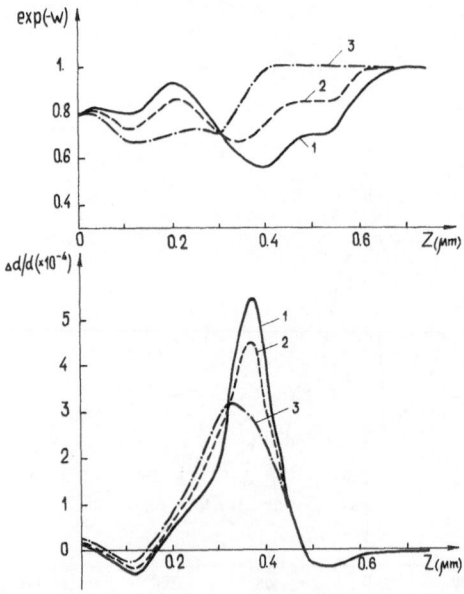

Fig. 5. Calculated profiles of deformation $\Delta d(Z)/d$ and random displacement
$\exp(-W(Z))$ for sample 1.

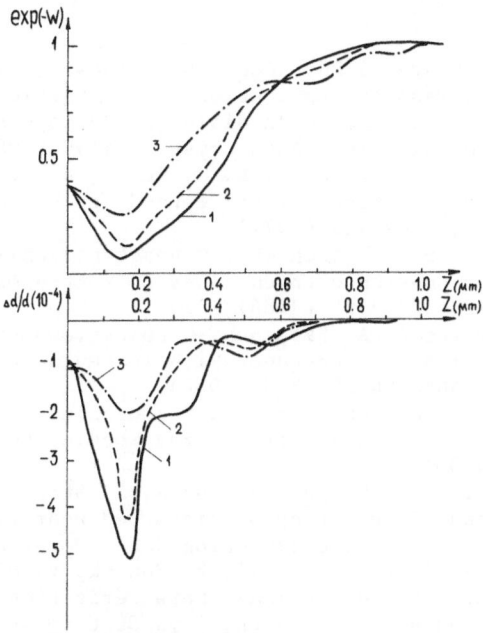

Fig. 6.   Calculated profiles of Δd(Z)/d and exp(-W(Z)) for sample 2.

This class of curves reproduces the corresponding experimental triple-crystal CDR (curve 4 in Figs 2-4) in the direct numerical calculation. Now one can select a physically reasonable profile from that class of profiles, or test them by using other methods of investigation (for example, by the standing wave technique).

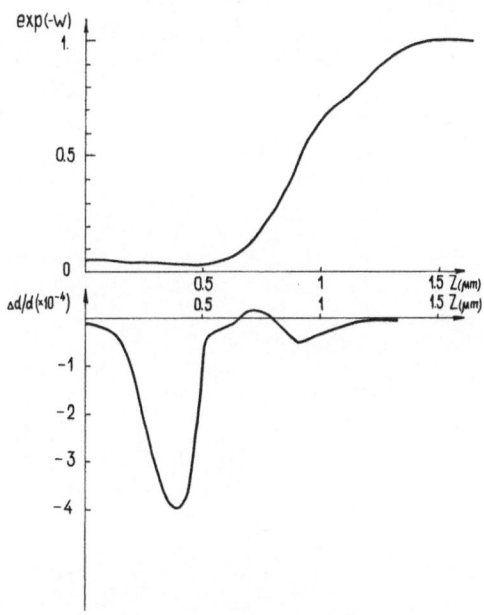

Fig. 7.   Calculated profiles of Δd(Z)/d and exp(-W(Z)) for sample 3.

REFERENCES

1. S. Kikuta and K. Kohra, <u>J</u>.<u>Phys</u>.<u>Soc</u>.<u>Jap</u>. 21:1449 (1966).

2. J. Burgeat and D. Taupin, Application de la théorie dynamique de la diffraction X à l'etude de la diffusion du bore et du phosphore dans les cristaux de silicium, <u>Acta Cryst</u>. A24:99 (1968).

3. A. M. Afanasev, M. V. Kovalchuk, E. K. Kovev, and V. G. Kohn, X-ray diffraction in a perfect crystal with distributed surface layer, <u>Phys</u>.<u>Stat</u>.<u>Sol</u>.(<u>a</u>) 42:415 (1977).

4. A. M. Afanasev and S. S. Fanchenko, O vosstanovlenii profilej naruschenij tonkich pripoverchnostnych sloev po rentgenodifrakcionym dannym, <u>Doklady</u> <u>AN</u> <u>SSSR</u> 287:1395 (1986).

5. A. V. Goncharsky and A. A. Stepanov, O edinstvennosti reschenija obratnoj zadachi difrakcii rentgenovskich luchej na tonkich monokristallach, <u>Doklady</u> <u>AN</u> <u>SSSR</u> 287:309 (1986).

6. R. N. Kyutt, P. V. Petrachen, and L. M. Sorokin, Strain profiles in ion-doped silicon obtained from X-ray rocking curves, <u>Phys</u>.<u>Stat</u>.<u>Sol</u>. (<u>a</u>) 60:381 (1980).

7. A. V. Goncharsky, A. V. Kolpackov, and A. A. Stepanov, Inverse problems of computational diagnostics of disturbed near-surface layers of crystals from the X-ray diffraction data, <u>Poverchnost</u> 12:66 (1986).

8. F. Cembali, M. Servidori, S. Solmi, Z. Šourek, V. Winter, and P. Zaumseil, Structural and electrical characterisation of boron implanted in preamorphised silicon layers, <u>Phys</u>.<u>Stat</u>.<u>Sol</u>.(<u>a</u>) 98:511 (1986).

9. A. M. Afanasev and V. G. Kohn, Vnechnij fotoeffekt pri difrakcii rentgenovskich luchej v kristallach s naruchennym poverchnostnym sloem, <u>Zh</u>.<u>Eksp</u>.<u>Teor</u>.<u>Fiz</u>. 74:300 (1978).

10. V. G. Kohn, M. V. Pvilepsky, and I. M. Sukhodreva, A simple method for the determination of the structure of a disturbed surface layer of single crystal using X-ray diffraction data, <u>Poverchnost</u> 11:122 (1984).

11. A. Iida and K. Kohra, Separate measurements of dynamical and kinematical X-ray diffractions from silicon crystals with a triple crystal diffractometer, <u>Phys</u>.<u>Stat</u>.<u>Sol</u>.(<u>a</u>) 51:533 (1979).

12. P. Zaumseil and U. Winter, Triple crystal diffractometer investigations of silicon crystals with different collimator – analyzer arrangements, <u>Phys</u>.<u>Stat</u>.<u>Sol</u>.(<u>a</u>) 70:497 (1982).

VI. CRYSTAL STRUCTURE DETERMINATION

NEW TRENDS IN DETERMINATION OF CRYSTAL STRUCTURE

Jindřich Hašek

Institute of Macromoleculaar Chemistry
Czechoslovak Academy of Sciences
162 06 Praha 6, Czechoslovakia

INTRODUCTION

Direct methods[1] have become dominant in the determination of well-ordered crystal structures. They are based*on an estimate of seminvariant values from distributions of seminvariants.* From any seminvariant one can thus form an equation

$$f_i(\varphi_1, \ldots, \varphi_n) = \Phi_i \pm d_i,$$ (1)

where Φ_i is the expected value of seminvariant and d_i is an unknown error related statistically to the distribution width. (When f_i is a periodic function, the right side has to be taken modulo.)

The functions f_i can have many different forms, e.g., for the simple triplet relation

$$\varphi_H + \varphi_K + \varphi_{-H-K} = 0 \pm d_i \mod(2\pi) ,$$

for the Sayre equation

$$\langle E_H E_K E_{-H-K} \rangle_K = |E_H|^2 . \text{const.} \pm d_i .$$

The distribution of d_i's depends on the number of summands in $\langle \rangle_K$, on the presence of atoms of different types, etc.

When the equations defining the origin are neglected, there are $m \gg n$ equations (1) for $m+n$ unknown variables $\varphi_1, \ldots, \varphi_n$, $d_1, \ldots, d_m$. To solve this set of equations, direct methods select the most reliable equations with small phase errors d_i corresponding to narrow distributions. The set of equations is then solved either under the assumption that $d_i = 0$ for $i = 1, \ldots, m$ or with additional conditions which ensure the minimum errors d_i.

However, the errors d_i increase with increasing number of atoms in the unit cell, so that the assumption that the d_i's are small cannot be fulfilled for complex structures and direct methods fail. According to ref. 3, only

---

*For exact meaning of some terms see ref. 2.

50 % of non-centrosymmetric structures are solved without difficulties and 25 % cannot be solved by these methods at all.

It is the purpose of this paper to summarize some recent attempts to suppress this drawback of current direct methods, and includes:

(a) A proper definition of the aim of the phase determining procedure as a minimization of a function of phase errors;

(b) An introduction to the theory of seminvariant graphs which describes the flow of information during the phase determining procedure, determines a hierarchy of phases and defines criteria for the best choice of both seminvariants and the starting set of phases, i.e. defines the phase determining procedure giving the optimal distribution of phase errors;

(c) A survey of distribution fitting methods which avoid entirely the formation of the set of equations (1) and allow thus the use of less reliable seminvariants and the utilization of all structure information contained in them.

THE BEST PHASE DETERMINING PROCEDURE

The aim of the phase determining procedure is to get a Fourier map with low spurious maxima and well-resolved maxima at atomic positions. It can be expressed in a mathematically convenient form[2,7] as

$$S = \int (\rho_{calc} - \rho_{exact})^2 \, d\vec{r} = minimum, \tag{2}$$

where ρ_{exact} and ρ_{calc} are exact[*] and calculated electron densities,

$$\rho_{calc} = (1/V) \sum_{H} x_H |F_H| \exp(i \varphi_H^{calc}) \exp(2\pi \, i \, \vec{H}\vec{r}), \tag{3}$$

$$\rho_{exact} = (1/V) \sum_{H} |F_H| \exp(i \varphi_H) \cdot \exp(2\pi \, i \, \vec{H}\vec{r}) . \tag{4}$$

F_H and φ_H are "exact" magnitude and phase. The weight x_H will be explained later. Upon substitution equation (2) transforms to

$$S = (1/V) \sum_{H} |F_H|^2 (x_H^2 - 2 x_H \cos \Delta\varphi_H + 1) = minimum, \tag{5}$$

where $\Delta\varphi_H = \varphi_H^{calc} - \varphi_H$. This equation can be interpreted in several ways:

(a) Let $x_H = |F_H^{calc}|/|F_H|$, $\Delta\varphi_H \neq 0$ and $|F_H^{calc}| \neq |F_H|$. Then a measure of correctness of an unweighted Fourier map (3) is

$$S = \sum_{H} (|F_H^{calc}|^2 + |F_H|^2 - 2|F_H||F_H^{calc}| \cdot \cos\Delta\varphi) = minimum. \tag{6}$$

If the calculated phases are exact (i.e. $\Delta\varphi_H = 0$) and $\Delta F = |F_H^{calc}| - |F_H|$, then

$$S = \sum |F_H|^2 (1 - x^2) = \sum (\Delta F)^2 = minimum.$$

---

[*]"Exact structure pattern" (i.e., E-map calculated with refined phases) is not isomorphous with the true electron density because it uses a small number of "sharpened" Fourier coefficients (E_H instead of F_H).

(b) Let an unweighted Fourier map be calculated with exact magnitudes $|F_H|$ (i.e. all $x_H = 1$). Then an estimate of the integral of squared errors in electron density, which should be minimized, depends on the phase errors $\Delta\varphi$ as

$$\langle S \rangle = (4/V) \, \Sigma |F_H|^2 \, \langle \sin^2(\Delta\varphi/2) \rangle = \text{minimum.} \tag{8}$$

This equation can be used as a measure of the fit between the calculated Fourier map (or E-map with $|F_H|$ replaced by $|E_H|$) and the "exact structure pattern".

For small phase errors, the estimate of S is

$$\langle S \rangle = (1/V) \, \Sigma |F_H|^2 \, \langle (\Delta\varphi_H)^2 \rangle = (1/V) \, \Sigma |F_H|^2 \, \text{var} \, (\varphi_H). \tag{9}$$

(c) Let us now suppose, that a weighted Fourier map is calculated with exact magnitudes (i.e. $\Delta F = 0$), the absolute values of phase errors are known in advance, and x_H are the optimal real coefficients minimizing the sum S. The minimum of any of summands in (5) then corresponds to $x_H = \cos\Delta\varphi_H$ and the estimate of the corresponding minimum integral of squared errors is

$$\langle S \rangle = (1/V) \, \underset{H}{\Sigma} \, |F_H|^2 \, \langle \sin^2\Delta\varphi \rangle = \text{minimum.} \tag{10}$$

For small phase errors one obtains the measure of the fit between the calculated and the exact Fourier map as

$$\langle S \rangle = (1/V) \, \Sigma |F_H|^2 \, . \, \text{var} \, \varphi_H. \tag{11}$$

This equation is valid for Fourier map weighted by $\langle \cos\Delta \, \varphi_H \rangle$. Because normalized structure factors, E_H' s, are usually used in the beginning of the structure determination instead of F_H's, we can write the following statement:

Statement 1. The best phase-determining procedure (i.e., the procedure leading to the optimal fit between "the exact structure pattern" and the E-map weighted by estimates of $\cos\Delta\varphi_H$) corresponds to variances $V_H = \langle (\sin\Delta\varphi_H)^2 \rangle \approx \langle (\Delta\varphi)^2 \rangle$ which minimize the sum

$$\langle S \rangle = (1/V) \, \Sigma |E_H|^2 \, . \, V_H = \text{minimum.} \tag{12}$$

According to (8), when no estimate of $\cos\Delta\varphi$ is available, the variances $V_H = 4 \, \langle \sin^2(\Delta\varphi_H/2) \rangle$ have to be used in (12). For small phase errors ($\Delta\varphi_H \to 0$) both criteria coincide. The next chapter will show how to find the "best" phase-determining procedure in direct methods using equation (12).

INTRODUCTION TO THE THEORY OF GRAPHS OF SEMINVARIANTS

The use of theory of graphs for solving the phase problem has already been discussed in refs 4,5,6,7. Many direct methods in current use for the determination of phases usually rely on a set of reliable seminvariants which connect all the phases together with a properly chosen starting set of phases - multisolution methods[1,8], the symbolic addition procedure[9], etc. These methods employ a well-known scheme resembling the method of elimination of

---
*Provided we knew directly the estimates of phase errors $\langle \Delta\varphi \rangle$, the weight in (3) could be taken as $x_H = \exp(-i\Delta \, \varphi_H)$ and $\langle S \rangle \to 0$.

variables for the solution of the set of linear equations. The interconnections among phases form a rather complicated structure which – without a proper mathematical apparatus – can be studied only with great difficulties. Many problems may be greatly simplified when expressed in terms of a matrix formulation of theory of graphs which gives the possibility to derive criteria for the best efficiency of the phase-determining procedure and provides a new view on the algorithms of finding "the best starting set", e.g[9] in the case of the symbolic addition procedure, the convergence procedure[1], and the greedy algorithm[10,11]. Moreover, the theory of graphs outlined here provides not only a clear and exact formulation of criteria for the best starting set, but is useful for any type of phase-determining procedure even if the starting set is not explicitly specified.

Some necessary special terms are defined in the following paragraphs. However, the set of terms given here is not closed and a simplicity of definition is preferred over mathematical exactness and uniqueness. The reader is referred to refs 13,14,15,16,30 for definitions generally used in theory of graphs. The terms having a special meaning in theory of graphs are underlined for reader's convenience when they appear first in the text.

Undirected Graphs of Seminvariants

A single H-phase seminvariant

$$\Phi = f(\varphi_1, \ldots, \varphi_N) \tag{13}$$

may be represented by a complete undirected graph $G = (\Psi, E)$, where the set of vertices $\Psi = \{\varphi_1, \ldots, \varphi_N\}$ represents N phases forming the seminvariant, and each of $N(N-1)/2$ edges of the set E represents one of the possible utilizations of phase relations following from the seminvariant Φ. Every edge must be accompanied by an identification of remaining reflections of the seminvariant and by the necessary characteristics of the theoretical distribution (e.g., the mean value and variance for acentric seminvariants or the probability of a positive sign for centric seminvariants). Thus, the presence of the edge $(\varphi_i - \varphi_j)$ in the graph G means that a knowledge of the phase φ_i suffices for the determination of the phase φ_j and vice versa, assuming that seminvariant values Φ and all the remaining phases of the seminvariant are known.

Definition 1. Graph of seminvariants (GS) is defined as a union[*] of complete subgraphs formed by all seminvariants and/or invariants that have been taken into account.

According to the definition, the graph of seminvariants is a finite undirected multigraph which summarizes the exact and "statistical" relations among phases represented here by vertices and edges of the graph. It may be used for a construction of the graphs of phase relationships corresponding directly to the phase-determining process.

The simplest graphs of seminvariants are shown in Fig. 1. A one-phase seminvariant (probability of the positive sign of the respective E_H determined by Σ_1 relation) forms a loop associated with the vertex corresponding to the phase forming the seminvariant. The loop is accompanied by some measure of probability of the positive sign.

---

[*] A union of graphs $G_1 = (V_1, E_1), \ldots, G_n = (V_n, E_n)$ is a graph $G = (V, E)$, where $V = V_1 \cup \ldots \cup V_n$, and $E = E_1 \cup \ldots \cup E_n$.

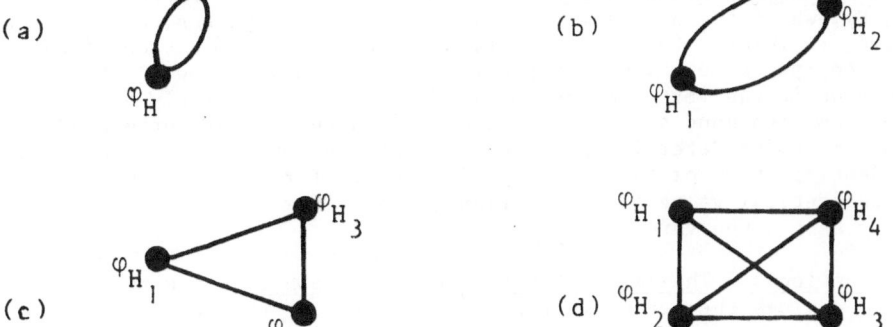

Fig. 1. Examples of graphs of the simplest seminvariants (invariants).
(a) One-phase seminvariant $\vec{H} = \vec{K}\,(R - I)$
(b) Two-phase seminvariant $\vec{H}_1 + \vec{H}_2 = \vec{K}\,(R - I)$
(c) Three-phase invariant $\vec{H}_1 + \vec{H}_2 + \vec{H}_3 = 0$
(d) Four-phase invariant $\vec{H}_1 + \vec{H}_2 + \vec{H}_3 + \vec{H}_4 = 0$

Two-phase seminvariant $\Phi = \varphi_1 + \varphi_2$ forms one edge associated with vertices corresponding to the phases φ_1, $\vec{\varphi}_2$. The edge is accompanied by information on the mean seminvariant value and its variance.

Similarly, a triplet (quartet) forms a complete graph with 3 (6) edges, each of them accompanied by the identification of one (two) remaining reflections of the seminvariant and by information on the theoretical triplet (quartet) mean value and its variance.

Directed Graphs of Seminvariants

When the phases $\varphi_1, \ldots, \varphi_{i-1},\ \varphi_{i+1}, \ldots,\ \varphi_N$ and the value of the k-th N-phase seminvariant Φ_k are already known, the phase φ_i can be calculated using the phase relation

$$\varphi_i = F(\Phi_k,\ \varphi_1, \ldots,\ \varphi_{i-1},\ \varphi_{i+1}, \ldots,\ \varphi_N). \tag{14}$$

To represent the phase relation the complete graph corresponding to the seminvariant (13) is replaced by the directed graph with N-1 arrows going from the vertices $\varphi_1, \ldots,\ \varphi_{i-1},\ \varphi_{i+1}, \ldots,\ \varphi_N$ to the vertex φ_i. They show the flow of information through the graph. The union of these directed graphs, representing all the possible phase relations, is called the directed graphs of seminvariants (DGS). Each arrow in the DGS has to be accompanied by the identification of the respective seminvariant and by its characteristics.

The variance of the phase φ_i can be approximated by

$$\operatorname{var} \varphi_i = a_k\, V_k + \sum_{j=i} b_{jk}\,\operatorname{var} \varphi_j\,, \tag{15}$$

where V_k is a variance of the k-th seminvariant, $a_k = (\partial F / \partial \Phi_k)^2$, $b_{jk} = (\partial F / \partial \varphi_j)^2$. When the seminvariant is a triplet or quartet   all the derivatives $a_k = b_{jk} = 1$. Hence,

$$\operatorname{var} \varphi_i = V_k + \sum_{j \neq i} \operatorname{var} \varphi_j.$$

297

Definition 2. Graph of phase relations (GPR) is a directed mixed graph $G = (\Psi, D)$, where the set of vertices, Ψ, is a set of phases, and the set of arrows, D, represents the phase relations used for determination of these phases. Any phase relation is represented by a single arrow in GPR. This arrow begins in the vertex corresponding to the least reliable determined phase on the righ-hand side of (14) and ends in the vertex corresponding to the phase φ_i to be determined. Of course, each arrow must be accompanied by an identification of the N-2 remaining phases forming the seminvariant, by the theoretical value of seminvariant, and by some measure of its reliability (e.g., the variance V_k).

Definition 3. The level structure of the graph of phase relations $G = (\Psi, D)$ divides the set of vertices into k levels such that:

- all vertices accessible from (i.e. reachable via one arrow from) vertices in level L_1 lie in either level L_1 or L_2,
- all vertices adjacent only from vertices in the last level, L_k, lie in the last level,
- for $1 < i < k$, all vertices adjacent from vertices in level L_i lie in either level L_i or L_{i+1}.

The set of all phases in the first level will be called the "starting set". The corresponding level structure will be called a level structure rooted at the starting set of phases. The number of levels k is called the depth of the level structure. The minimum depth of all possible level structures will be called the radius of the graph. It is evident that GPR cannot contain any cycles, while semicycles can be present.

Definition 4. Reduced graph of phase relations (RGPR) is defined as GPR in which all arrows directed to a single vertex except the arrow contained in the most permeable path progression are omitted. An information about the erased arrows should be, of course, kept along with the remaining arrow.

Thus RGPR with N phases contains maximum N arrows which show the most important features of the flow of information among phases for the actual phase determining procedure. Due to its simplicity, RGPR is suitable for graphical representation of the phase determination. The basic properties of the graph of phase relations, namely the level structure, remain unchanged in RGPR. Simple illustrations of GS, DGS, GPR and RGPR are in Fig. 2. Two quartets $\varphi_H + \varphi_K + \varphi_L + \varphi_M$ and $\varphi_H + \varphi_{K'} + \varphi_{L'} + \varphi_{M'}$ (where $\vec{H} + \vec{K} + \vec{L} + \vec{M} = 0$ and $\vec{H} + \vec{K'} + \vec{L'} + \vec{M'} = 0$) are represented by two complete subgraphs with one common vertex φ_H (articulation). Some properties of GS, DGS, GPR, RGPR are shown in Table 1.

Table 1. Maximum Number of Lines (Arrows) in the Graphs (GS, DGS, GPR, RGPR) Defined in this Chapter.

Graph	GS	DGS	GPR	RGPR
Maximum number of lines (arrows)	$s.r(r-1)/2$	$s(r-1)$	s	N
For quartets r=4, N=200, s=5000	30000	15000	5000	200

The GPR corresponds directly to an actual phase determining procedure. However, except for some trivial cases, a number of GPR corresponds to a single GS. Therefore, the best GPR, according to the criterion (12), requires the analysis of the flow of information (or of the propagation of errors) in the phase determining procedure (in GPR). The following statements simplify the analysis of GPR.

Statement 2. The level structure of GPR is not determined uniquely except in trivial cases. However, some sets of phases (vertices) exist which ensure a unique level structure when fixed in the first level.

Statement 3. The deeper is the level structure rooted in the starting set, the more difficult it is to prevent occasional significant phase errors to spread further.

Definition 5. Realizable path in GS is a path which can be actually realized in GPR.

Statement 4. Realizable path cannot contain two edges associated with the same seminvariant.

Definition 6. Excentricity of the phase φ_H (vertex in GS) is the maximum distance in GS measured from this vertex, i.e. $e(\varphi_H) = \underset{\varphi_K \in \Psi}{\mathrm{Max}}\ d(\varphi_H, \varphi_K)$, where $d(\varphi_H, \varphi_K)$ denotes a distance (i.e., the shortest realizable path) from φ_H to φ_K.

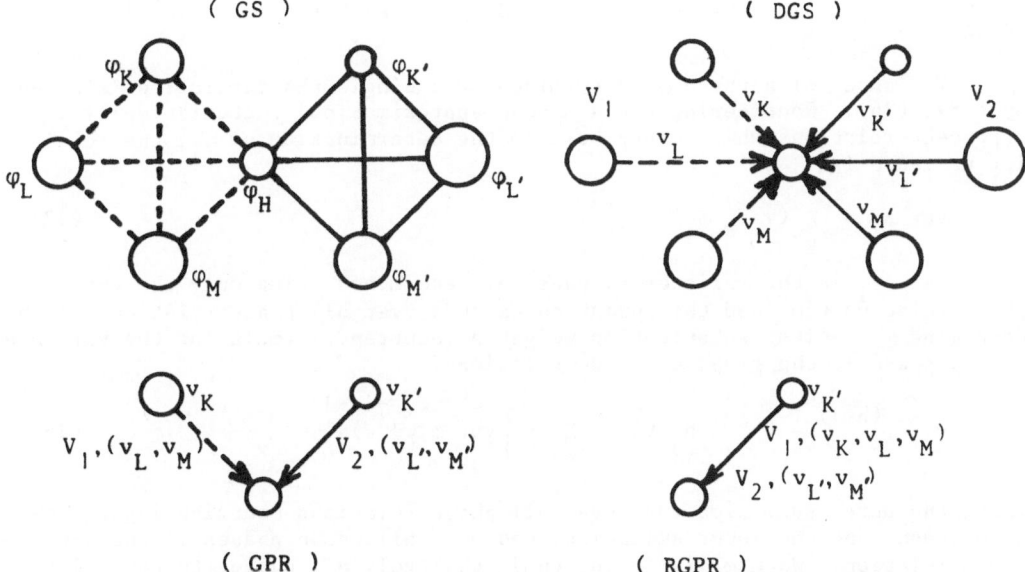

Fig. 2. Two quartets sharing one common vertex represented as Graph of Seminvariants, Directed Graph of Seminvariants, Graph of Phase Relations and Reduced Graph of Phase Relations. Theoretical variances of quartet values are $V_1 < V_2$. Variances of phases are denoted as $v_i = var\ (\varphi_i)$.

Because of their complexity, GS and GPR are handled usually in a matrix form. Most useful of them are shortly characterized in the following paragraphs:

Adjacency matrix of GS (or GPR) is a matrix $A = [a_{ij}]_{i,j=1,\ldots,N}$, where $a_{ij} = 1$ if the edge (φ_i, φ_j) exists in GS (or GPR); else $a_{ij} = 0$.

Statement 5. An element b_{ij} of the matrix $B = A^{(n)}$ (Boolean n-th power) is equal to unity if a distance φ_i and φ_j is equal to n, otherwise $b_{ij} = 0$.

Distance matrix of GPR is a matrix $D = [d_{ij}]_{i,j=1,\ldots,N}$, where d_{ij} is a length of the shortest directed path $\varphi_i,\ldots,\varphi_j$ in GPR. If no directed path $\varphi_i,\ldots,\varphi_j$ exists, then $d_{ij} = \infty$.

Capacity matrix of GPR is a matrix $V = [w_{ij}]_{i,j=1,\ldots,N}$, where w_{ij} describes the capacity of information channels (arrows) leading from φ_i to φ_j.

An exact criterion for the best phase-determining procedure (12) requires an analysis of the flow of information through GPR and the calculation of variances of all phases as shown in the next chapter.

Propagation of Errors in the Process of Phase Determination

Let us consider first methods which define the starting set of phases. The variances of phases in the starting set, i.e., in the first level of the graph of phase relations, can be determined immediately: $\text{var } \varphi_i^{(1)} = 0$ for phases fixing the origin; $\text{var } \varphi_i^{(1)} = \text{const}$ for trial (symbolic or numeric) phases in the starting set (e.g. $\text{var } \varphi_i^{(1)} = \pi^2/48$) in MULTAN.

Variance of a phase φ_i determined by a single phase relation is given by (15), (16). Considering (in a rough approximation) $\underline{r}$ statistically independent relations which contribute to the determination of φ_i, the resulting variance is

$$\text{var } \varphi_i = \left[\sum_k (\text{var}_k \varphi_i)^{-1} \right]^{-1}, \tag{17}$$

where $\text{var}_k \varphi_i$ is the variance of phase φ_i determined using only the k-th phase relation (16) and the summation extends over all phase relations which determine φ_i. After substitution we get a recurrent formula for the variance of any phase in the graph of phase relations.

$$\text{var } \varphi_k^{(n+1)} = \left(\sum_{j=1}^{r} \left[b_j V_j + \sum_{i=1}^{s} a_{ij} \text{var } \varphi_{ij} \right]^{-1} \right)^{-1}, \tag{18}$$

where the outer summation runs over all phase relations contributing to the k-th phase, and the inner summation runs over all known phases in the j-th phase relation. When φ_k is in the (n+1)-th level, all phases to the right side of (18) are in levels $1, 2, \ldots, n$. The coefficients b_j and a_{ij} are the corresponding squared values of derivatives in (15).

After multiple substitution of recurrent formula (18) the integral of squared deviation between the calculated Fourier map and the "correct" map according to the criterion (12) is

$$\langle S \rangle = \sum_{H} |E_{H}|^2 \left(\sum_{r_n} [V_i + \sum_{P_{n-1}} (\sum_{r_{n-1}} V_j^{-1} + \dots \right.$$

$$\left. + (\sum_{r_2} [V_k + \sum_{P_2} (\sum_{r_1} V_\ell^{-1} + \sum_{P_0} \text{var } \varphi_m^{(1)})^{-1}]^{-1})^{-1} \dots)^{-1}]^{-1})^{-1} = \text{minimum}. \tag{19}$$

This equation has been written for seminvariant types with all a_{ij} and $b_j = 1$ for simplicity. The symbol $\sum_{H}$ denotes summation over all normalized structure factors used for the calculation of the E-map, and $\sum_{r_n}$ designates summation over all phase relations used in the determination of the phase in the n-th level, $\sum_{P_n}$ denotes summation over all phases on the right-hand side of the respective phase relation used for determination of a phase in the (n+1)-th level; these lie in levels preceding level n+1.

The variances of phases in the starting set, var $\varphi_i^{(1)}$, are known in advance and the variances of seminvariants V_i are simple functions of only the $|E_H|$'s. Therefore, the $\langle S \rangle$ is a function of only the $|E_H|$'s and equation (19) gives an absolute measure of the fit between the calculated Fourier map and the actual electron density.

When the set of seminvariants is kept unchanged, the $\langle S \rangle$ depends only on the level structure of GPR. Conversely, a minimization of $\langle S \rangle$ with respect to the formation of level structure of GPR gives the optimal flow of information about the origin to all phases, i.e. it gives the procedure for

- optimal generation of seminvariants,
- optimal choice of the starting set of phases,
- optimal transformation of seminvariants to phase relations and the best order of their successive use.

Thus, the optimal procedure for a determination of crystal structures is transformed to already developed iterative algorithms for the maximum flow in transportation networks[30]. Some simple heuristic rules, partly already used in current direct methods, can be deduced immediately from (19) without any calculation:

- the number of <u>components</u> into which GS is partitioned after removing less reliable seminvariants should be small,
- trial phases in the starting set should be able to fix properly the origin in every component independently,
- more reliable phase relations have to be used in the lowest possible levels of the level structure of GPR,
- phase relations (arrows) in GPR should be ordered according to decreasing permeability in every path succession of GPR,
- the depth of the level structure of GPR should be kept as low as possible,
- the larger the starting set the lower the number of levels of GPR and the higher the probability of successful solution,
- the phases corresponding to lower $|E|$ values ought to form terminal vertices or should be in the highest levels of GPR,
- the higher the number of atoms, the less the number of reliable phase relations (higher d_i values in (1)), and the more "bottle-necks" (for a flow of information) can be found in GS, and finally no connected GPR can be found without increasing the number of phases in the starting set.

In classical direct methods only reliable phase relations can be used. However, their number decreases with increasing number of atoms in the unit cell and direct methods thus loose their efficiency. This drawback is not encountered in distribution fitting methods[17-27] reviewed in the next chapter.

The methods which use the fitting of empirical distributions of seminvariants to the theoretical ones are based on the hypothesis that the relative frequency of seminvariant values (the diffraction vector $\vec{H}$ is regarded as random variable) approximates for correct phases the theoretical distribution of seminvariants (based on a random variable $\vec{r}$).

The relative frequencies of seminvariant values approximate (for correct phases, random choice of diffraction vectors in seminvariants and $n \to \infty$) the theoretical distribution of seminvariants (derived for a random choice of generation of diffraction vectors $\vec{H}$ and a fixed structure). The theoretical distribution of seminvariants derived for random $\vec{H}$ and fixed structure approximates for high number of atoms $N \to \infty$ the theoretical distribution of seminvariants derived for random radius vectors of atoms $\vec{r}$ and fixed seminvariants under the assumption that the distribution of $\vec{H}$ $(\vec{r}_i - \vec{r}_j)$ modulo (1) is uniform in the interval (0.1). This distribution is non-uniform for non-uniform distribution of maxima in the Patterson map. One may therefore expect that the hypothesis is valid for structures:

- without multiple coincidences of interatomic vectors,
- with higher number of atoms in the unit cell, $N > 10$,
- with reasonably high number of measured reflections.

If any of these conditions is not fulfilled, the hypothesis need not be valid and the use of these methods is problematic if the special features of the structure are not respected in the theoretical estimate.

Numerous test of the distribution fitting have been made[17,19,20,21,23] using 5 different methods, six types of seminvariants on about 70 crystal structures.

Methods used for fitting of the distributions

1. Direct graphical comparison of the function values of both distributions (ref. 2,17).

2. The use of X^2 test as a measure of fit between corresponding distributions[2]. The seminvariants are divided according to their weights into characteristic regions and then according to their values into intervals. The measure of the fit is then the coefficient

$$K = \sum_i W_i \ (P_i^{theor} - P_i^{trial})^2 = minimum, \qquad (20)$$

where the weight $W_{ij} = N_i/(x_i^2 + N_{ij} q^2)^{1/2}$ and $x_i = min[\ 1 - c;$ $max(c; P_i^{theor})]$; N_i is a number of seminvariants in the i-th region and N_{ij} is a number of them having its value in the i-th interval; the summation runs over all regions and intervals; c,q are constants which estimate the highest possible accuracy and determine favourable properties of the weight under the extreme conditions of $N_i \to 0$ or ∞, $P_i \to 0$ or 1.

3. Minimization of a maximum difference between cumulative distributions using the Kolmogorov test,

$$\underset{i}{Max} \ \{|C_i^{theor} - C_i^{emp}|\} = minimum. \qquad (21)$$

4. Minimization of differences between some significant characteristics of theoretical and empirical distributions. Only the first and second moment of distributions were compared.

5. The criterion which determines the uniformity of the variable

$$z(\Phi) = \int_0^\Phi P(x|w)dx \; , \tag{22}$$

where $P(x|w)$ is the corresponding theoretical distribution of seminvariants[21]. This criterion makes it possible to handle all types of distributions uniformly.

Theoretical distributions used for testing

Six theoretical distributions of different types of seminvariants[1,22,26] were used for testing.

1. One-phase structure seminvariant (Σ_1 relation). The probability that $\varphi_H = 0$ is

$$P(w) = (1 + \exp(-w))^{-1} , \tag{23}$$

where $w = \sigma_2 \, \sigma_2^{-3/2} \, \sqrt{p_H} \, |E_H| \, \sum_{\vec{K}} (|E_K|^2 - 1) \, \exp(2\pi i \vec{K} \vec{t})/p_K$,

the summation runs over all available diffraction vectors $\vec{K}$ such that $\vec{K} = \vec{H}(R - I)$, $\vec{t}$ is a translation vector and R a rotation matrix. I is a unit matrix, $\sigma_s = \sum^N f_j^s$, f_j is the scattering factor of the j-th atom, N is the number of atoms in the unit cell and p_H, p_K are coefficients given by the crystal symmetry[22].

2. Three-phase universal structure invariants (triplets), $\Phi_3 = \varphi_H + \varphi_K + \varphi_{-H-K}$. For acentric triplets

$$P(\Phi_3|w) = [2\pi \, I_0(w)]^{-1} \, \exp[w \cos(\Phi_3 - q)] , \tag{24}$$

where $w = 2 \, \sigma_3 \sigma_2^{-3/2} \, |E_H E_K E_{-H-K}| \, p_{HKL}$, and q is a function of the known fragment of the structure (ref. 1). A coefficient given by the crystal symmetry was approximated by $p_{HKL} = 1$.

3. For centric triplets, the probability of $\Phi_3 = 0$ was approximated by

$$P(w) = (1 + \exp(-w))^{-1} , \text{ where the weight } w \text{ is the same as in \#2.}$$

4. Three-phase universal cosine invariants, $\Phi_3 = \cos \varphi_H + \varphi_K + \varphi_{-H-K}$

$$\langle \cos \Phi_3 \rangle = I_1(w)/I_i(w) = ((0.0106 \, w - 0.1304) \, w + 0.5658) \, w,$$

$$\text{var}(\sin \Phi_3) = I_1(w)/w(I_0) \, w = \langle \cos \Phi_3 \rangle /w , \tag{25}$$

where $w = 2 \, \sigma_3 \sigma_2^{-3/2} \, |E_H E_K E_{-H-K}|$.

5. Four-phase universal structure invariants (quartets), $\Phi_4 = \varphi_H + \varphi_K + \varphi_L + \varphi_{-H-K-L}$. For acentric quartets:

$$P(\Phi_4|R_1,\ldots,R_7) = k \cdot \exp(w_4 \cos \Phi_4) \, Q_5 \, Q_6 \, Q_7, \tag{26}$$

where $Q_i = \exp(-w_4 \cos \Phi_4) \, I_0(A_i)$ for $i = 5, 6, 7$,

$$A_5 = 2 \, R_5[(R_1 \, R_2)^2/N + (R_3 \, R_4)^2/N + w_4 \cos \Phi_4]^{1/2},$$

303

$$A_6 = 2 R_6 [(R_1 R_3)^2/N + (R_2 R_4)^2/N + w_4 \cos \Phi_4]^{1/2} ,$$

$$A_7 = 2 R_7 [(R_2 R_3)^2/N + (R_1 R_4)^2/N + w_4 \cos \Phi_4]^{1/2} ,$$

$$w = 2 R_1 R_2 R_3 R_4/N ,$$

N is an effective number of atoms in the unit cell, $N = \sigma_4/\sigma_2^2$,

$$R_1 = |E_H| , \quad R_2 = |E_K| , \quad R_3 = |E_L| , \quad R_4 = |E_{-H-K-L}| , \quad R_5 = |E_{H+K}| ,$$

$$R_6 = |E_{H+L}| , \quad R_7 = |E_{K+L}| .$$

If some of the cross reflections E_{H+K}, E_{H+L}, E_{K+L} are not measured, the corresponding Q_i are equal to unity. Thus

$$P(\Phi_4|R_1,\ldots,R_4) = k' \exp(w_4 \cos \Phi_4) ,$$

$$P(\Phi_4|R_1,\ldots,R_5) = k'' I_0(A_5) ,$$

$$P(\Phi_4|R_1,\ldots,R_6) = k''' \exp(-w_4 \cos\Phi_4) I_0(A_5) I_0(A_6) ,$$

$$P(\Phi_4|R_1,\ldots,R_7) = k'''' \exp(-2w_4 \cos \Phi_4) I_0(A_5) I_0(A_6) I_0(A_7) .$$

6. Centric quartets. The probability of $\Phi_4 = 0$ was approximated by

$$P(R_1,\ldots,R_7) = K \cdot \exp(w_4) Q_5 Q_6 Q_7 ,$$

where

$$Q_i = \exp(-w_4) \cos h(A_i) ,$$

$$A_5 = R_5(R_1 R_2 + R_3 R_4)/N ,$$

$$A_6 = R_6(R_1 R_3 + R_2 R_4)/N ,$$

$$A_7 = R_7(R_1 R_4 + R_2 R_3)/N$$

and the other variables have the same meaning as in the preceding paragraph.

Results of the Distribution Fitting

The distribution fitting methods were tested a posteriori as "figures of merit" (ref.[17,19,22]) in connection with MULTAN and SIMPEL programs and also ab initio[21] for triplets generated by a standard run of MULTAN program. They proved to be sufficiently selective in a structure solution of 40 structures included in the research program of IMC in Praha. Tests were made also with 30 other specially selected structures[3], which could not be solved by standard methods at different laboratories, but have been determined later in cooperation with specialized laboratories by more sofisticated Patterson or direct methods. Six of these "difficult cases" could not be solved by distribution fitting methods without enormous effort. The reasons for a failure can be the following:

1. Wrong estimate of theoretical distributions, e.g.
    - neglected influence of multiple coincidences of interatomic vectors,
    - neglected contribution of the heavy atom,
    - the approximation $p_{HKL}=1$ in (24) and the analogous coefficient in (26) - not specified - is too rough for special triplets (quartets) in some space groups,

- neglected members of higher order in the expansion of probability function,
- neglected super-structure effects.

2. Wrong empirical distribution because of
   - non-random choice of seminvariants (neglected correlations among seminvariants,
   - errors in E's (experiment + model),
   - partial disorder of the structure.

These effects influence more the shape of distributions than the position of their mean value. In suspicious cases it is therefore better to lower the precision in the description of distributions. In the limiting case one can lower the precision to a single parameter - mean value. The distribution fitting methods are then equivalent to the standard direct methods.

The maximum number of non-hydrogen atoms in the unit cell of tested structures was 596. No attempt has been made to apply these methods to macromolecular structures. However, Luzzati[27] presents a method where he tries to get a uniform distribution of electron density corresponding to a uniform distribution of seminvariants. Hauptman[28,29] on the other hand got sharp distributions using anomalous scattering or using pairs of isomorphous structures; this approach enabled him to use traditional direct methods. The distribution fitting methods are not restricted to these two extreme cases. They prefer of course the use of sharp distributions which carry higher information content in a single seminvariant; however, they can be used also in all other cases when only intermediate distribution types are available.

CONCLUSION

Theory of graphs and the principle of a distribution fitting have already proved to be useful in the solution of the phase problem. Both theories have been proposed for any type of seminvariant, for any type of distribution, for methods with and without definition of the starting set, and for middle-size and large structures. However, a better understanding of the behaviour of distributions of seminvariants is necessary to provide - before the structure is solved - an indication of the danger associated with "pathological" distribution. Graph of seminvariants is used for optimization of the phase determining procedure. An analysis of the graph of phase relations is recommended to determine the best starting set of phases. However, a deeper analysis of relations between the graph of seminvariants and a graph of phase relations is desirable for replacing the current iterative procedures by a straight and unique way leading to the best phase-determining sequence satisfying (12). An interesting feature is that if the description of distributions degenerates only to information about the position of the distribution maximum, the distribution-fitting methods would become identical with standard direct methods.

REFERENCES

1. M. M. Woolfson, Direct Methods - from Birth to Maturity, Acta Cryst. A43:593 (1987).
2. J. Hašek, On the phase problem solution I-IV, Acta Cryst. A40:338 (1984).
3. "Crystallography in Universities and Other Research Institutions", Full Report. Lombardy Towers, Washington (1982).
4. J. Hašek, Direct Methods. In: Methods of the Phase Problem Solution, KJT ČSVTS, Praha (in Czech) (1980).

5. J. Hašek, Theory of Graphs in Direct Methods. In: Mathematical Methods in X-Ray and Neutron Structure Analysis, KJT ČSVTS, Praha (in Czech) (1980).

6. J. Hašek, Lectures at the Univ's Utrecht, Amsterdam and York (1982-1984).

7. J. Hašek, K. Huml, H. Schenk, and J. D. Schagen, Proceedings of 8-ECM, 4.04 (1983).

8. M. M. Woolfson, Doing without Symbols - MULTAN, In: Crystallographic Computing Techniques, Munksgaard, Copenhagen (1976).

9. J. Karle and I. L. Karle, The symbolic addition procedure for phase determination for centrosymmetric and non-centrosymmetric crystals, Acta Cryst. 21:849 (1966).

10. S. Fortier, M. Fronckowiak, G. D. Smith, and H. Hauptman, An exercise in the application of a new automatic phasing procedure, Acta Cryst. B35:2062 (1979).

11. J. D. Schagen, "Some Aspects of the Determination and Elimination of Inconsistent Phase Relationships", Thesis, University of Amsterdam, (1986).

12. Y. Jia-Xing, "Program RANTAN-81, Department of Physics", University of York, England (1981).

13. J. Plesnik, "Graph Algorithms", Veda, Bratislava (in Slovak) (1983).

14. R. G. Busacker and T. L. Saaty, "Finite Graphs and Networks", New York, McGraw-Hill (1965).

15. R. P. Tewarson, "Sparse Matrices", Academic Press (1973).

16. F. Harary, R. Z. Norman, and D. Cartwright, "Structural Models", New York, Wiley (1965).

17. J. Hašek, A contribution to the determination of a correct system of signs of structure factors for centrosymmetric crystals, Acta Cryst. A30:576 (1974).

18. J. Hašek, Consistency test for the determination of a correct phase set of the structure factors, Acta Cryst. A31:818 (1975).

19. J. Hašek, H. Schenk, C. Kiers, and J. D. Schagen, Distributions fitting methods for centrosymmetric crystal structure, Acta Cryst. A41:333 (1985).

20. J. Hašek and H. Schenk, Distribution fitting methods used as figures of merit for non-centrosymmetric structures, Acta Cryst. A44, in press (1988).

21. V. Kříž, A direct method based on a fitting of distributions of seminvariants, Acta Cryst. A44, in press (1988).

22. J. Hašek, Tables of the Σ_1 relationships: triclinic, monoclinic and orthorhombic space groups, Z.f. Kristallogr. 145:263 (1977).

23. H. Hašek, Direct Methods, International Colloquium on Direct Methods, Smolenice (1973).

24. H. Schenk, On the reliability of the Σ_2 relation. I. Real structures in P2$_1$/c, Acta Cryst. A29:503 (1973).

25. W. Krieger and H. Shcenk, On the reliability of the Σ_2 relation. II. Artificial structures in P2$_1$/c, Acta Cryst. A29:720 (1973).

26. C. Giacovazzo, "Direct Methods in Crystallography", Academic Press (1980).

27. V. Luzzati, P. Mariani, and H. Delacroix, X-ray crystallography at macromolecular resolution: a solution of the phase problem, Die Makromolekulare Chemie, in press (1988).

28. H. Hauptman, On integrating the techniques of direct methods and isomorphous replacement I. The theoretical basis, Acta Cryst. A38:289 (1982).

29. H. Hauptman, On integrating the techniques of direct methods with anomalous dispersion I. The theoretical basis, Acta Cryst. A38:632 (1982).

30. S. I. Zuchoviski and I. A. Radtshik, "Matematitzeskie metody Setevovo planirovania", Nauka, Moskva (1965).

31. G. Sheldrick, Private communication, University of Cambridge, England (1982).

EASY AND UNEASY SUPERSPACE GROUPS

FOR INCOMMENSURATE CRYSTALS

P. M. de Wolff

Lab. voor Technische Natuurkunde

P.O. Box 5046, 2600 GA Delft, The Netherlands

The symmetry of an incommensurate crystal structure is usually expressed by its superspace group G. Alternatively one may examine the symmetry G' of the n-fold superstructure which results when a free coefficient of the modulation vector is replaced by a rational number p/n. It turns out that G' accounts for all symmetry elements of G only in certain cases: 'easy' superspace groups G. Even for them, this may depend on the parity of n. For many superspace groups, G' never accounts for all symmetry elements, no matter whether n is odd or even: 'uneasy' superspace groups G. The symmetry groups which so far have been reliably shown to occur point to a strong and perhaps even exclusive preference for easy groups.

INTRODUCTION

During the sixties, a new type of crystal structure came into the fore: the modulated crystal structure. It is characterized by a diffraction pattern which contains rows of satellites, one row accompanying each reflection of an otherwise normal-looking diffraction pattern. The satellites are weak and the rows they form are short (seldom extending beyond second order) so this 'satellite type' of pattern is easily recognized. Moreover, very often modulation sets in at a transition from a normal phase. By taking a series of diffraction pictures while the crystal is cooled, one then observes the appearance of satellites whereas the normal or 'main' reflections remain strong and are not much affected by the transition, either in position or in intensity.

The rows are all parallel to each other and their repetition period - the same along any row - determines the 'modulation vector' $\vec{q}$ in reciprocal space, cf. Figure 1. The satellites are explained by assuming that there is a basic structure - roughly corresponding to the main reflections - from which the actual structure is formed by a spatially periodic distortion (by displacement and/or other kinds of modification like, for instance, partial substitution of atoms) just as spectral satellites arise from optical or acoustical modulation. The modulation vector can obviously be identified with the wave vector of the periodic distortion .

In the years following this discovery, modulation was found to occur in very many crystalline substances, from pure elements up to very complex compounds. It was detected most often by the above characteristic type of

Fig. 1. Satellite type of diffraction pattern. Large dots: main reflections; small dots: satellites. The origin is encircled.

diffraction picture (other physical data, such as the nuclear resonance
spectrum, can also yield criteria for modulation). A few cases are known
where the satellites form not only a row but a two- or three-dimensional
array around each main reflection, thereby requiring two or three different
$\vec{q}$-vectors for the description of the pattern. In the present paper we shall
deal only with rows ('one-dimensional modulation') so there will appear just
one $\vec{q}$-vector.

THE POINT GROUP, AND THE CHOICE OF $\vec{q}$

For normal crystals, the conventional choice of basis vectors of the
lattice is dictated by the point group P of that lattice: they are the
shortest lattice vectors in the so-called 'symmetry directions'. As far as
directions are concerned, this assignment can be performed in reciprocal
space as well, since P is also the point group of the reciprocal lattice.
It should be recalled that in the case of centring, fractional coordinates
occur for points of the direct lattice but not for those of the reciprocal
lattice; here, the shortest lattice vector in a symmetry direction may even
have indices like 200.

For the vector $\vec{q}$, in principle any vector joining a first-order satellite with a main reflection could serve. Thus a normalizing principle is
called for. The obvious rule is to let $\vec{q}$ be parallel to a symmetry direction, which in this case means that it is invariant (apart from sign) for
all operations of the point group P of the diffraction pattern. Here we have
to remark that - apart from very artificial exceptions - both the main reflections and all satellites, taken as two separate sets of points with
intensities, each have P as their point group as well; of course, the symmetry
operations of P can image a main reflection only in another main reflection.

A second criterion for $\vec{q}$ is that it should allow all satellites to be
indexed by integers h, k, l and m, m being the order in a row of satellites
counting from the main reflection hkl which it is supposed to contain.
Calling $\vec{H}_o$ the position vector of that main reflection, the position vector
$\vec{H}$ of the satellite should be

$$\vec{H} = \vec{H}_o + m\,\vec{q} \tag{1}$$

Now it frequently happens that the vector $\vec{q}$ chosen by the above pre-
scription of invariance for P does not allow all satellites to be thus

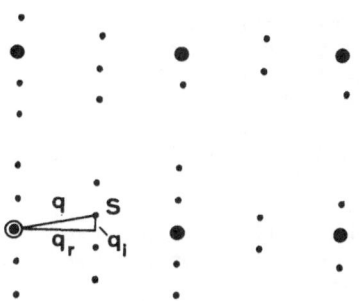

Fig. 2.   Diffraction pattern from a structure with centred modulation pat-
           tern or 'internal centring' as in Figure 3. There is a rational
           component $\vec{q}_r$ of the $\vec{q}$-vector.

indexed. An example in two dimensions is shown in Figure 2. Rows containing all of them a main reflection are obtained only by choosing the oblique
vector $\vec{q}$, which does not have the vertical symmetry direction. Conversely,
a $\vec{q}$-vector in that vertical direction, although well-adapted to the conspicu-
ous vertical rows, can index those without a main reflection only by allowing $\vec{H}_o$ in (1) to correspond to non-existing main reflections.

    Actually that is what one usually does in such a case. The 'strict'
vector $\vec{q}$ is decomposed in mutually perpendicular vectors $\vec{q}_r$ and $\vec{q}_i$, $\vec{q}_i$ being
the P-invariant part, and the indices of satellites are obtained from (1)
by substituting $\vec{q}_i$ for $\vec{q}$.

    One often finds the convenient triplet of reciprocal coordinates (used
also in the latest symbols for superspace groups) to define the 'strict'
modulation vector; in the plane case of Figure 2 the corresponding doublet
would be ($1/2\,\beta$) where β is a small number. This strict vector gives, for
the indices of satellite S in Figure 2, h k m = 001; but the actual state
of affairs is much more easily understood by using $\vec{q}_i$ instead of $\vec{q}$ in (1)
yielding h'k'm = 1/2 01, 1/2 0 being the indices h k of a non-existent main
reflection. Note that in either description the satellites of one vertical
row have orders m of the same parity: all even, or all odd. Of course, one
can obtain all-integer indices also, in this case by halving the a axis;
then the indices of S become HKm = 101 (notation as in[ ]).

    Finally it should be remarked that if the point group P belongs to the
hexagonal, tetragonal or trigonal system $\vec{q}_i$ has the direction of the 6-, 4-
or 3-fold axis. In the orthorhombic system it lies along one of the three
twofold axes. In the monoclinic system $\vec{q}_i$ can be parallel to the twofold
axis ('monoclinic axial modulation') or to the mirror plane ('monoclinic
planar modulation'); in the latter case its direction in that plane is not
fixed by symmetry, but usually the shortest vector will be chosen. A simi-
lar situation occurs when P is triclinic; then always $\vec{q}_r = \vec{0}$ and $\vec{q} = \vec{q}_i$
because there is no symmetry direction at all.

    In any case $\vec{q}_r$ is either zero or one half or one third of a reciprocal
lattice vector of the basic structure. When it is not zero, this fact can
be interpreted as a centring in the modulation pattern (see below).

RATIONAL AND IRRATIONAL COMPONENTS OF $\vec{q}$

    As stated above, the component $\vec{q}_r$ of $\vec{q}$ is an invariant simple fraction
of a reciprocal basic lattice vector (if not zero). The other component
$\vec{q}_i$, however, has at least one arbitrary coefficient, such as in the example

of Figure 2. 'Arbitrary' here means: determined not by symmetry but by the thermodynamic equilibrium of the structure, just as for instance unit cell lengths. Therefore in general these coefficients will depend upon temperature and pressure. Assuming such dependences to be continuous, one can obviously regard these 'free' coefficients as irrational, and this had led to the designation 'incommensurate' for the structures concerned: 'IC structures'.

In the early years it has often been proposed to make a sharp distinction between incommensurate structures on one hand, and 'commensurately modulated' structures on the other.

The latter category was defined as being liable to description as a modulated structure, but with the free parameter(s) equal to a rational number, preferably a simple fraction, which would be independent of pressure and temperature at least within finite ranges.

The distinction still stands. It is particularly useful in those cases where a lock-in phase transition occurs. Such a transition is observed as a continuous change of a free parameter, e.g. upon cooling, followed by a plateau where its value is constant and ostensably rational. Thioureum, for instance, has at least two such transitions with clear plateau's at 1/7 and 1/9. However, several modulated structures are known for which a free coefficient is constant but not a simple fraction, and there is no lock-in transition. From an experimental point of view, irrationality is of course meaningless. But the symmetry of incommensurate structures (next sections) has turned out to be applicable to commensurate structures as well – at least within certain limits. In the present author's opinion, these limits deserve more attention than they have received so far, so they will now be dealt with in some detail.

DESCRIPTION OF IC STRUCTURES

The IC structure in direct space is much less simple than the Fourier image in reciprocal space treated in the foregoing sections. Three different methods exist for its description:

(i) Based on that Fourier image, a four-dimensional description[1,2] has been proposed, in which the three-dimensional structure is regarded as a hyperplane section of a "super-crystal" which itself is strictly periodic in four dimensions. This superspace description is very flexible. It also covers, for instance, the composite or intergrowth structures, which do not possess a single unique basic structure and therefore do not fall under the definition of modulation given in Chapter 1. The superspace description however is difficult to visualize.

(ii) Many papers on IC structures describe them using the irreducible representation of the basic structure's space group. The advantage of this method for the analysis of phase transitions is clear, especially when the latter are regarded in the light of Landau's theory. Although essentially yielding the same symmetry groups as method (i)[3], this method when applied to structure analysis is much more complicated in use.

(iii) Finally there is the dualistic description[4]. It rests heavily on the assumption that there is a unique basic structure. (This is also the case for method (ii); therefore neither of these two methods is readily adaptable to the description of composite structures.) Through each atom centre of the basic structure, a line is drawn in the direction of $\vec{q}_i$. The periodic distortion now can be shown - or imagined - as

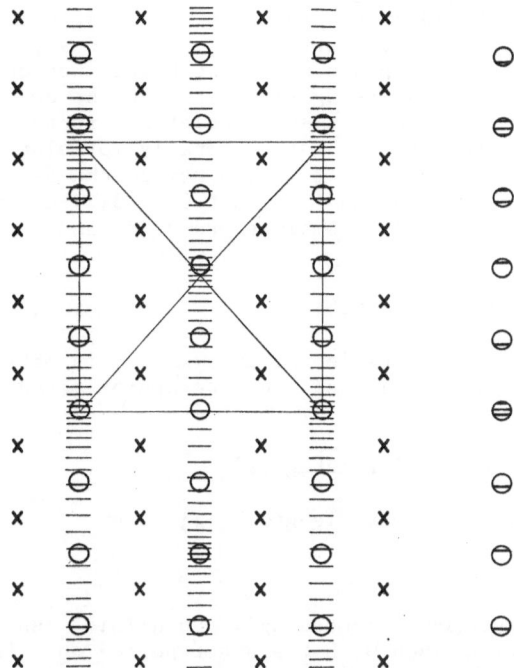

Fig. 3. Dualistic representation of a structure with a centred modulation
pattern. The scalar modulation function is shown by varying the
density of hatching. One chain of atoms parallel to $\vec{q}$ (vertical)
is shown at right so as to demonstrate the actual density for each
atom.

a periodic function displayed along this line, so that the displacement
(vector function) or substitution density (scalar function, shown in
Figure 3) of each atom lying on the line is equal to the function value
at its centre. The method therefore yields a vivid picture of the modulation in three-dimensional space. For the higher symmetries where $\vec{q}_i$
is directed along a symmetry axis, ('axial modulation'), each line
contains an infinite chain of atoms and all lines are separate. This
is not so when the modulation is planar monoclinic ($\vec{q}_i$ perpendicular
to the binary axis), or triclinic. Because of the irrational direction
of $\vec{q}_i$ the lines then merge into a stack of planes or into the entire
space, in which the modulation appears as 2- or 3-dimensional ripple
functions.

Both in the latter cases and for axial modulation the complete array of
functions displayed along lines or in planes or in space, for all atoms
of the basic structure, constitutes the 'modulation pattern' denoted
by M, which fully represents the periodic distortion. Its space group
G_M, together with that (G_B) of the basic structure B determines the
symmetry of the IC structure.

SUPERSPACE INTERPRETATION OF THE SYMMETRY OF IC STRUCTURES

Suppose a structure is axially modulated in the direction of, say, the
$\vec{b}$ axis. If the corresponding coefficient β of $\vec{q}_i$ is irrational, obviously
none of the symmetry translations of B along $\vec{b}$ are symmetry operations of
the modulated structure. However they can all be retrieved if we accept
that changing the phase of the modulation (that is, moving M with respect

to B) is equivalent to the identity operation, since it does not really change the - supposedly infinite - structure. Then to any given symmetry translation of B one can add a phase change such that not only B but also M is made to coincide with the starting situation, so the combined operation is a symmetry operation for the complete modulated structure. This reasoning, here given in dualistic terms, was originally formulated in the language of the superspace approach. It is the key to the derivation of the superspace groups which are indeed (at least for (3+1)-dimensional modulation) equivalent to the dualistic combinations of G_B and G_M.

ASYMPTOTIC SYMMETRY OF IC STRUCTURES

Quite a different interpretation of IC symmetry is obtained if one realizes that the irrational value of β can be approximated by a rational fraction, say

$$\beta' \approx p/n \text{ (p,n mutually prime integers)} \tag{2}$$

corresponding to a modulation wave length λ' given by

$$nb = p \lambda' \tag{3}$$

By changing the modulation accordingly, an n-fold 'superstructure' of the basic structure is obtained with the exact period nb. In the $\vec{b}$ direction, now only that period nb and its multiples are retrieved as symmetry translations. The actual modulated structure then appears as the asymptotic result of letting the fraction p/n approach the actual value of β.

Such approximative superstructures have often been used to solve an IC structure, usually with n below 10. In principle however, there is no limit to n so β can be approximated with any desired degree of precision. The translations b, 2b.(n-1)b are lost, but those retrieved are exact ones. Moreover, although the superstructure defines only n points per period of the modulation functions, there is a strong constraint because for increasing n these points must merge into a continuous smooth function. It should be noted that in this asymptotic interpretation, IC symmetry applies to the modulated structure as a whole, not only for the translations considered above but also for each other kind of symmetry operation. The period for the limiting case $\beta' = \beta$ is infinite. Nevertheless we can imagine the actual IC structure to be effectively identical with a superstructure for which n is high but finite, and with a normal space group G'. The question remains, then, whether G' does indeed give maximal coverage of the IC symmetry as defined by the given superspace group G. In particular, one would expect it to yield the same point group P and to account for all centrings including the "internal" ones characterized by $q_r \neq 0$; if G' does all this, we shall call it the 'rational image' of the superspace group G. The answer to the above question is:

- if it does, this may be true only for n- and p-values of a certain parity* depending on G. In spite of such parity conditions, the existence of a rational image of G greatly simplifies understanding the IC symmetry; therefore we call such a group: an _easy_ superspace group, cf. the example in the next section.

- for certain superspace groups G it never does, no matter which values of n and p are chosen. These groups will be called: _uneasy_ superspace groups, cf. last but one chapter of this paper.

---

*In higher lattice symmetries, the condition may refer to moduli of 3, 4 or 6 instead of 2. For the sake of simplicity we shall deal only with parity.

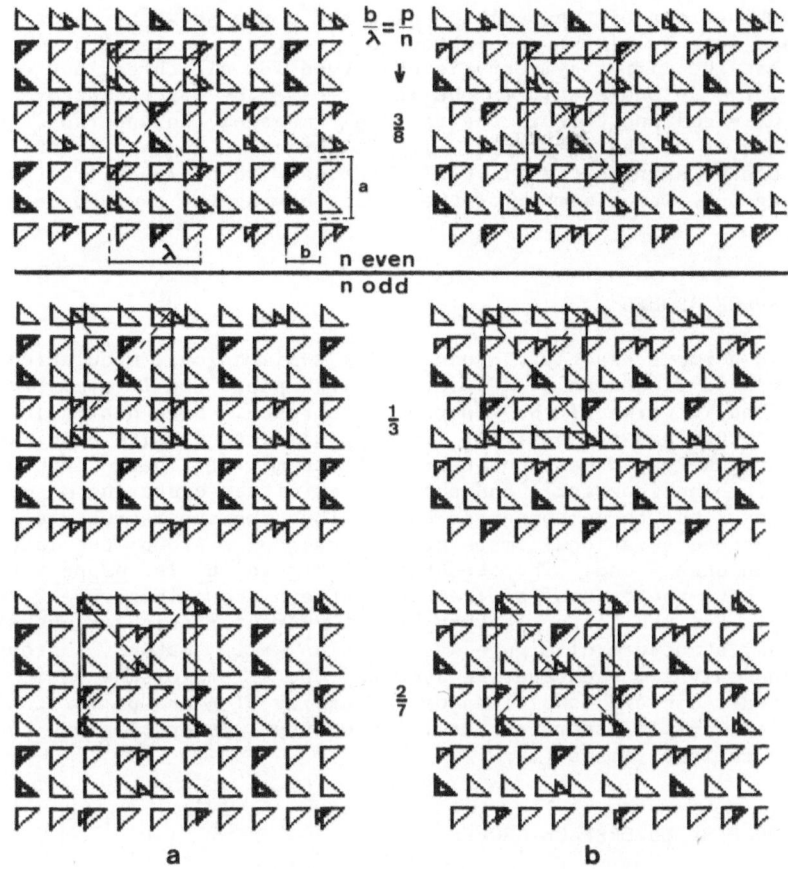

Fig. 4. Symmetry of superstructures which all of them have the same periods
b (horizontal) and a (vertical) of the basic structure. The ratio
$\beta' = p/n$ of b to the horizontal modulation wave length λ is shown
in the centre: 3/8 = odd/even, 1/3 = odd/odd and 2/7 = even/odd.
Unrealistic triangles, chosen so as to bring out the symmetry, are
used both for the basic structure B (larger triangles) and for the
modulation pattern (small triangles). The unit cell of that pattern
M has been drawn in to show that it is centred. The symmetry G'
of the superstructure as a whole is easily seen by concentrating
on the fat triangles resulting where small and large triangles
coincide.

(a) Basic structure plane group pm: Both the centering of M and the
mirrors are conserved in G' for β' = 3/8, only the mirrors for odd
n. "Easy" group G.
(b) Basic structure plane group pg: Centering occurs in G' again
for β = 3/8, the glide lines for 1/3 and 2/7, but in no case the
two occur simultaneously. (Although the glide lines are at different levels for 1/3 and 2/7, this figure also demonstrates that
the parity of the numerator p is not essential). "Uneasy" group G.

EASY SUPERSPACE GROUPS

In Fig. 4a, a (2+1)-dimensional modulated structure with an easy superspace group has been drawn. Again the dualistic representation was chosen.
The basic structure has a primitive orthogonal net, as well as mirror lines
in the direction of $\vec{q}$; M also has mirror lines but its net type is centered.

All three combinations of the parities of n and p are shown by choosing appropriate wave lengths of the modulation for identical basic structures. We observe that the mirror symmetry is conserved for all three, but centering occurs only for β' = odd/even. In the latter case the superstructure has space group G' = cm1 and this is the rational image of the superspace group which therefore is of the easy type. The IC structure for any irrational value of β can be seen as the limit of a series with odd/even fractions for β', converging to β, each member of the series having the same space group type cm1.

UNEASY SUPERSPACE GROUPS

Figure 4b depicts exactly the uneasy (2+1)-dimensional modulation symmetry which was mentioned on p. 784 in the first paper[2] on IC symmetry, where it was described in terms of the superspace approach. Seen dualistically, its basic structure has glide mirror lines parallel to $\vec{q}$, with a primitive orthogonal net, whereas M has a centered net and mirror lines. Each of the three parity combinations for n and p conserves either centering or glide mirrors, but none of the three conserves both. Thus the superstructure space group is in no case a rational image of the superspace group; the latter therefore is an uneasy one. The structure of Figure 4b also happens to be one giving rise to the type of diffraction pattern shown in Figure 2, with a non-zero q_r. But the latter feature is not characteristic for uneasiness; indeed the easy structure of Figure 4a has it as well. Rather, a group is uneasy if, and only if, the separate symmetry operations of the basic structure impose contradictory conditions on the parity of n and p* for their conservation in the n-fold superstructure. An uneasy superspace group has, therefore, no rational image.

PREFERENCE FOR EASY SUPERSPACE GROUPS

When the above characteristic of a special class of superspace groups (viz. those called uneasy here) was recognized[2], the example given could seem artificial. For real (3+1)-dimensional modulation only very few theoretically possible or actually existing symmetries were known. Now, however, we completely know all possible groups, and we also know the actual symmetries of a number of different modulated crystals. From this body of recent information we conclude that uneasy superspace groups occur very seldom, if at all. The arguments are:

(i) At least a dozen different superspace groups occur as well-established symmetries of actual crystals. With one possible exception (cf. next point) all those of which the present author is aware are of the easy type. This fact in itself indicates some bias since the number of uneasy (3+1)-dimensional superspace groups is a considerable fraction of the total (for instance, more than 30 % in the orthorhombic system).

(ii) The possible exception just mentioned is the uneasy group Pcmm (ooγ) (ss$\bar{1}$). It is uneasy because the c(s) glide plane and the m(s) mirror require n to be odd and even, respectively, for them to become glide planes of a superstructure.** It has been claimed for the symmetry of IC phases of several compounds in the beta-K_2SO_4 structure family.

---

*Strictly speaking, only the parity of n is important. If n is even, p is odd anyhow; and if n is odd, the numbers p and n-p are of different parity but they yield equivalent groups.

**The added (s) means that these planes are glide planes in M.

Indeed, some structure determinations[5] in that group have been reasonably successful.[5] However, the claim is in conflict with various experimental facts[5] originating either from very careful diffraction
work or from morphological and other physical properties. For instance, well-developed crystals of $Rb_2Zn Br_4$ have the point group symmetry 222, instead of the symmetry mmm which is expected for the uneasy space group. The latter is in accordance with Landau theory:
the high temperature form has space group Pcmn, the transition to the
IC structure is smooth, and the type of modulation is such that the
above uneasy superspace group seems almost inescapable. The fact that
this group does not actually occur for the IC phase is, therefore,
highly significant. Rather than doubting the adequacy of the experimental evidence[6], we believe that it strongly points to a tendency
in IC crystals to avoid uneasy superspace groups.

(iii) Finally, easy superspace groups provide the narrowest possible relation between IC and normal symmetry. It is at least conceivable that
the asymptotic approach reflects actual processes happening when the
IC crystal grows, or when its q vector changes under external influences. No such interpretation is afforded by an uneasy group, since
its symmetry is not the limit of any asymptotic process.

ACKNOWLEDGEMENT

Ir. A. J. van den Berg kindly assisted in designing and preparing
Figure 4.

REFERENCES

1. P. M. de Wolff, T. Janssen, and A. Janner, The superspace groups for in-
    commensurate crystal structures with a one-dimensional modulation,
    Acta Cryst. A37:625 (1981).
2. P. M. de Wolff, The pseudo-symmetry of modulated crystal structures,
    Acta Cryst. A30:777 (1974).
3. T. Janssen and A. Janner, Superspace groups and representations of ordi-
    nary space groups, Physica A126:163 (1984).
4. P. M. de Wolff, Dualistic interpretation of the symmetry of incommen-
    surate structures, Acta Cryst. A40:34 (1984).
5. A. C. R. Hogervorst, "Comparative study of the modulated structures in
    Rb_2ZnBr_4 and in related compounds". Thesis Technical University Delft
    (1986).[4]

STRUCTURE ANALYSIS OF MODULATED MOLECULAR CRYSTALS IV:

SURVEY OF OUR RECENT STUDIES

P. Coppens

Department of Chemistry
State University of New York at Buffalo
Buffalo, NY 14214, USA

V. Petříček

Institute of Physics
Czechoslovak Academy of Sciences
Na Slovance 2, 180 40 Praha 8, Czechoslovakia

INTRODUCTION

There has been a considerable upsurge in interest in modulated crystal following the development of the multidimensional description of modulated solids by DeWolff, Janner and Janssen[1,2,3]. Modulations are quite common in minerals and inorganic solids[4] which often exhibit substitutional and displacive modulations at or above ambient temperatures. However, with the greater awareness of the occurrence of modulations and the wider accessibility of low temperature diffraction equipment suitable for routine use, the number of known modulated molecular solids is rapidly increasing.[5,6,7] Among the most thoroughly studied are the low-temperature phase of biphenyl and the modulated phase of thiourea, which is stable over a narrow temperature range between 202 and 169K (ref. 8). Table 1 lists some known modulated molecular solids. In almost all studies which have been made, the atoms of a molecule are treated as individual entities, rather than as parts of a rigid covalently bonded framework. Since this can lead to unlikely distortions of the molecular geometry, we have introduced a molecular model in which the displacement of each atom is determined not by its own location, but by a point common to all atoms in a molecule or group, which is referred to as the phase reference point[9]. The molecular displacements are described in terms of rigid-body translations and rotations, thus greatly reducing the number of parameters of the model. A harmonic rigid-body description in general requires twelve parameters per molecule, which are to be determined from a usually large number of measurable satellite reflections.

We have concentrated in our work on cases where the solid state properties are strongly affected by the modulation. We will describe here results on the low-temperature organic superconductor $(BEDT-TTF)_2I_3$, in which the modulation depresses the superconducting temperature, on the modulated phase of thiourea which is intermediate between the paraelectric high-temperature and the ferroelectric low-temperature phases, and on the low-temperature modulated phase of TTF-TCNQ, which is insulating while the room-temperature phase is metallic. The very small displacements of TTF-TCNQ

Table 1. Modulations in Molecular Crystals

Compound	Temperature range (K)	q Vector	Structure analysis	Reference
Thiourea	169-202	0.13-0.11b*	yes	Shiozaki, 1971, Simonson, Denoyer and Currat, 1985 Gao, Gajhede, Mallinson, Petříček and Coppens, 1988
Phenothiazine - TCNQ	room temp.	0.23b*	yes	Kobayashi, 1974
Biphenyl	40-17	(0.5a*+ 0.46b*)	no	Baudour and Sanquer, 1983
	< 17	0.46b*	yes	Cailleau, 1986
α-bis-(N-methyl salicylaldimato) nickel(II)	room temp.	0.3a*	yes	Steurer and Adlhard, 1983
Betaine calcium chloride dihydrate	<160 (devil's staircase)	0.33-0.17	no	Brill, Schildkamp and Spilker, 1985
β-(BEDT-TTF)$_2$I$_3^-$	<195	0.076a* + 0.272b* + 0.206c*	yes	Leung et al., 1985
λ-Co(sepulchrate) (NO$_3$)$_3$	133-106 106-98 < 98	not known	no	Larsen et al., 1986
Thiophene	171-136 <136-109 <109	0.33-0.36c*	no	Andre and Szwarc, 1986
p-Azoxyphenetol	<358K	0.146a* + 0.206c*		Sciau, 1987
N,N-dimethylmor-pholinium - (TCNQ)$_2$	<207K	-0.046a* 0.04656* + 0.385c*	yes	Steurer, Visser, and DeBoer, 1987

References: Y. Shiozaki, Ferroelectrics, 2:245 (1971); T. Simonson, F. Denoyer abd R, Currat, J.Phys. 46:2187 (1985); Y. Gao, M. Gajhede, P. Mallinson, V. Petříček, and P. Coppens, Phys.Rev.B: (in press); H. Kobayashi, Acta Cryst. B30:1010 (1974); J. L. Baudour and M. Sanquer, Acta Cryst. B39:75 (1983); H. Cailleau, in: "Incommensurate Phases in Dielectrics 2, A. P. Levanyuk (Ed.), Elsevier (1986); W. Steuer and W. Adlhart, Acta Cryst. B39:349 (1983); W. Brill, W. Schildkamp, and J. Spilker, Z.f.Krist. 172:218 (1985); P. C. W. Leung, T. J. Emge, M. A. Beno, H. H. Wang, J. M. Williams, V. Petříček, and P. Coppens, J.Am.Chem.Soc. 107:6184 (1985); F. K. Larsen, P. Jorgensen, R. G. Hazell, B. Lebech, R. Thomas, R. J. Geue, and A. M. Sargeson, Private Communication (1986); D. Andre and H. Szwarc, J.Phys. 47:61 (1986); P. Sciau, Thesis, University of Montpellier, France (1987); W. Steurer, R. J. J. Visser, S. v. Smaalen, and J. L. DeBoer, Acta Cryst. B43:567 (1987).

have a dramatic effect on the transport properties, but give rise to extremely weak satellite intensities. Synchrotron radiation is required for their measurement. The phase studied is two-dimensionally modulated, which required a further generalization of the rigid-body model scattering expressions.

THE MOLECULAR DISPLACEMENT MODEL

In the molecular displacement model the displacement of an atom due to a modulation wave in the crystal is determined by a phase reference point g_ν shared by all the atoms in the rigid-body translation V and rotation W is :

$$u(r) = V(g) + W(g) \times (r - \rho) ,$$

where ρ is the center of mass of the molecule.

The translation V(g) is a sum over all n modulation waves:

$$V(g) = \sum_{i=1}^{n} V_i \sin(2\pi q \cdot g - \psi_i)$$

with phase ψ_i. This can be written as:

$$V(g) = V^x \sin(2\pi q \cdot g) - V^y \cos(2\pi q \cdot g)$$

with $V^x = \sum_i V_i \cos \psi_i$ and $V^y = \sum_i V_i \sin \psi_i$, and corresponding expressions for W.

It follows that the total number of parameters in the harmonic approximation is twelve per rigid body for a one-dimensionally modulated crystal. This number is further reduced when molecules occupy special positions in the unit cell[10]. In thiourea, for example, the molecule occupies a mirror plane in the average space group Pnma, which leads to a restsriction of $\psi = 90°$ for translations in the mirror plane and rotation around the normal to the mirror plane, and $\psi = 0°$ for the remaining rotations and translations.

MODULATED STRUCTURE OF ß-(BEDT-TTF)$_2$I$_3$[11]

Salts of bis(ethylenedithio)tetrathiofulvalene ($C_{10}S_8H_8$, BEDT-TTF) are among the few known ambient pressure organic superconductors, and the first known sulfur, rather than selenium based members of this group. The charge transfer salt ß-(BEDT-TTF)$_2$I$_3$[12] was discovered to be superconducting by Yagubskii et al.[12], with a superconducting transition temperature T_c of 1.8°. However, application of a pressure of 1.2-1.2 kbar raises T_c to 7-8 (ref. 12,13). The pressure dependence of the transport properties is thought to be related to the occurrence of an incommensurate modulated phase[11], which occurs below 200°, and gives rise to strong satellite reflections. This modulated phase is suppressed by application of a non-hydrostatic pressure of about 0.5 kbar applied at room temperature and followed by cooling[14]. Thus, the increase in the superconducting transition temperature is accompanied by a disappearance of the modulation.

The structure of ß-(BEDT-TTF)$_2$I$_3$ consists of stacks of BEDT molecules oriented along the [110] direction of the triclinic crystals (Fig. 1). Parallel stacks are connected by S--S contact interactions to form sheets in the ab planes composed of BEDT-TTF molecules. The sheets are interleaved by the triiodide anions. The q-vector of the modulation wave equals 0.076a* + 0.272b* + 0.206c*. The super space group is P:P$\bar{1}$:$\bar{1}$, with one BEDT-TTF

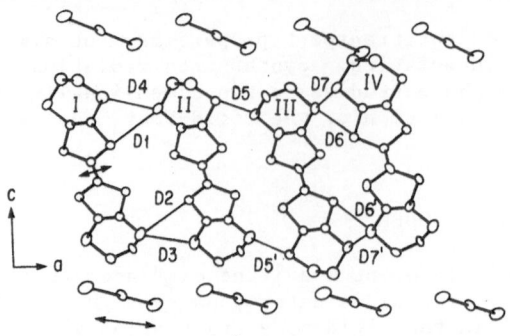

Fig. 1. View of the intermolecular contacts between adjacent BEDT-TTF molecules in ß-(BEDT-TTF)$_2$I$_3$. The arrows indicate the directions of
the major components of the displacive vectors for the BEDT-TTF
molecules and I$_3^-$ anions.

and one half triiodide ion in the asymmetric unit. Non-rigidity of the
triiodide was accounted for by allowing independent displacement of the two
iodine atoms in the asymmetric unit. In an additional refinement, which
gave a slightly better agreement factor the carbon atoms of the ethylene
group were given independent translational displacements to simulate the
occupancy of the true sites.

The major component of the molecular translation of the BEDT-TTF ions
is found to be in the molecular plane, and perpendicular to the long axis of
the molecule. Its amplitude (0.11 Å) is much smaller than the translational
displacements of the triiodide ion (0.27 Å). The rotational amplitudes are
about 1° and also contribute significantly to the displacements. A summary
of magnitudes is given in Table 2, while the directions of the translations
are shown in Figure 1.

The main contacts between the stacks of BEDT-TTF ions are between sul-
phur atoms. They are approximately perpendicular to the stack direction and
of improtance for the two-dimensional nature of the conductivity in this
solid. In the modulated structure the S••S distances in adjacent unit cells
are no longer equivalent, but vary by as much as 0.2 Å, as shown in Fig. 2.
Since the S••S contacts are of importance for the transport properties, the
relation between T$_c$ and the occurrence of the modulation is not surprising.

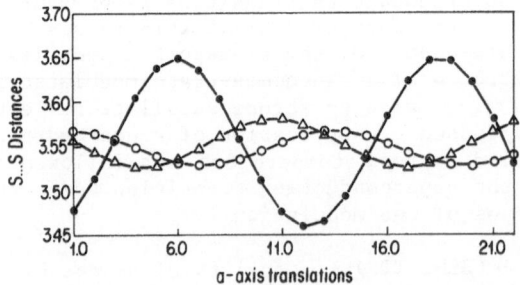

Fig. 2. A plot of the variations of interstack S••S contacts vs. unit cell
translations along the a-axis: (△) average values of D1-D4; (o)
average values of D5 and D5'; and (●) average values of D7 and
D7'.

Table 2. Magnitude of the Modulation Amplitudes in
β-$(BEDT-TTF)_2(I_3)$

	Phase (°)	Magnitude (Å, °)
Translations		
BEDT-TTF	0	0.113(1)
	90	0.041(1)
I(1)	0	0.268(2)
I(2)	0	0.277(1)
	90	0.088(1)
Rotations		
BEDT-TTF	0	0.838(1)
	90	1.034(1)

THE MODULATED PHASE OF THIOUREA[15]

Unlike urea, thiourea crystallizes at room temperature in a centrosymmetric space group, but undergoes a phase transition to a non-centrosymmetric, ferroelectric phase on cooling. Solomon found in 1953 that in an intermediate range of 202-169K a modulated structure is stable[16], the structure
of which has been the subject of several studies[8,17,18]. To test the applicability of the rigid body model we have collected two new sets of diffraction data on the intermediate phase. Since the q vector is a function
of temperature, the experimental temperatures can be estimated from comparison of the experimental magnitude of q with a q vs temperature curve given
by Denoyer and Currat[8]; they were found to be 184 and 173K for the two data
sets, respectively.

A number of low-order reflections were observed at room temperature
which violate the systematic absences of Pnma, the reported space group.
They did not behave as multiple reflections when psi scans were made. A
similar observation was made for the main reflections of the incommensurate
phase. However, since their number is not very large Pnma is at least a
reasonable approximation for the high temperature space group and the average
space group of the modulated structure. If the high temperature space group
is Pnma, the basic group, which is the three-dimensional part of the super
space group symbol, is subject to a condition formulated by Heine and Simmons[7], which states that the basic group consists of all elements of the
space group of the non-modulated phase which turn the modulation wave vector
q either into itself or its inverse. For thiourea this would imply that the
basic group is Pnma, as all orthorhombic symmetry elements fulfil this condition for a q vector along the b-axis. However, our data sets contain a
large number of satellite reflections which contradict the presence of an n
glide plane in the modulated phase. There were at least 30 such reflections
for crystal 1 and a similar number for crystal 2. Thus, either the Heine and
Simmons condition is violated, or the small deviations from Pnma in the high-
temperature phase have a dramatic effect from a symmetry point of view.

Since the thiourea molecules are located on the mirror planes of the
space group, the phases of the displacement waves are restsricted by the condition that the molecules remain rigid. This leads to phase angles of 0°
for the translation in the b-direction and the rotations around a and c, and
to phases of 90° for the translations in the a and c directions and the rotation around b. The refinement was done in both the superspace group
$P:Pnma:s\bar{1}1$, which does not allow for all the observations, and in the lower
group $P:P2_1ma:1\bar{1}1$. The latter refinement shows significantly different
translational and rotational amplitudes for the two independent molecules

Table 3. Displacement Amplitudes for Thiourea (crystal 2, T = 173 K).

	Crystal 2					
	Translations (Å)			Rotations (°)		
Molecule	a	b	c	a	b	c
Harmonic Model						
I	-0.131(4)	0.027(4)	0.103(3)	0.25(22)	-6.28(16)	0.39(35)
II	-0.112(4)	0.000(4)	0.120(3)	0.92(23)	4.74(16)	0.26(33)
Including Second Harmonic Terms						
I	-0.128(4)	0.029(3)	0.103(3)	0.20(21)	-6.21(16)	0.36(34)
second harmonic	-0.009(4)	-0.009(4)	0.015(5)	-0.25(29)	-0.38(23)	0.77(43)
II	-0.112(4)	0.002(3)	0.119(3)	0.92(22)	4.70(16)	0.22(32)
second harmonic	-0.019(4)	0.009(5)	0.006(5)	0.25(27)	0.18(23)	0.51(42)

in this space group. The results for the two data sets are in good agreement; those for the second crystal and the lower-symmetry refinement are summarized in Table 3. The main modulations are transverse translation and rotation, though small longitudinal displacements in the b-axis direction are also present. This is in agreement with earlier work, but the present results are obtained with far fewer parameters. The translation in the ac plane makes an angle of about 25° with the plane of the thiourea molecule. The amplitudes of rotation around the b-axis are at 6.3° and -4.7° intermediate between the difference in orientation of the molecules in the paraelectric and ferroelectric phases, which are 9.0° and -7.0° for the two molecules (Figure 3). This indicates that the rotational phonon mode which condenses in the transition of the modulated structure, is the same mode that plays a part in the further transformation to the ferroelectric phase.

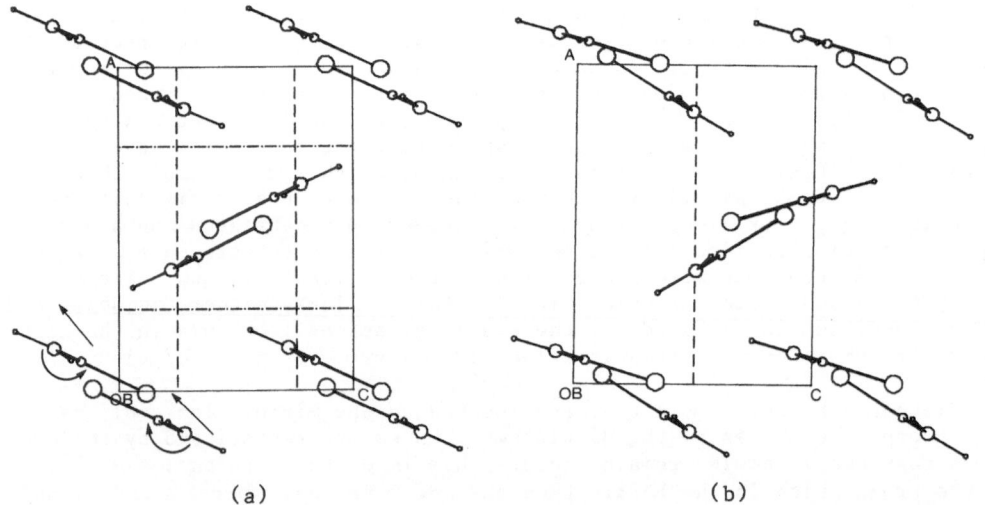

(a)   (b)

Fig. 3. Projections of the thiourea unit cell along the b-axis. The a-axis is vertical, the c-axis is horizontal. (a) Paraelectric phase, indicating the directions of the main rotational and translational modulations. (b) Ferroelectric phase. Note the shift in origin along the c-axis.

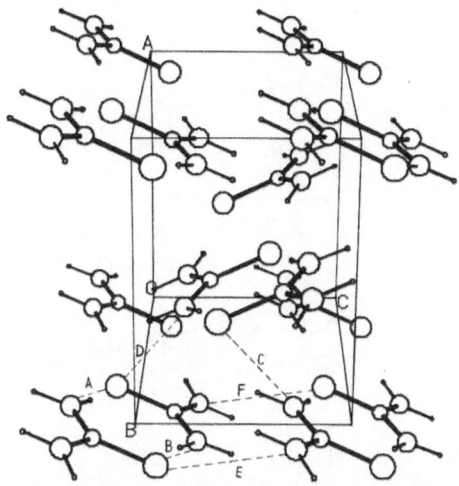

Fig. 4.   Packing diagram of the paraelectric phase of thiourea indicating
          hydrogen bonds and their labels. The b-axis is vertical in the
          drawing, the c-axis is horizontal.

There are three types of N-H··S contacts in the high-temperature phase
which can be classified as hydrogen bonds (Figure 4), though Elcombe and
Taylor note that for the longest of these the N-H··S angle is unusually
small for a hydrogen bond[19]. The deformation of the hydrogen bonds in the
modulated structure, plotted as a function of the unit cell position along
b in Figure 5, is much smaller for the stronger bond, for which it is about
0.10 Å, compared with 0.34 Å for the longest H···S contact. This is in
agreement with the force constants derived from analysis of the phonon spectrum, which are found to be 0.176, 0.037 and 0.016 mdyn/Å, in order of increasing contact distance. Clearly, the weaker bonds distort more as may
be expected. Any force field used in explaining the modulation behavior of
thiourea will have to include the restraining effect of these interactions.

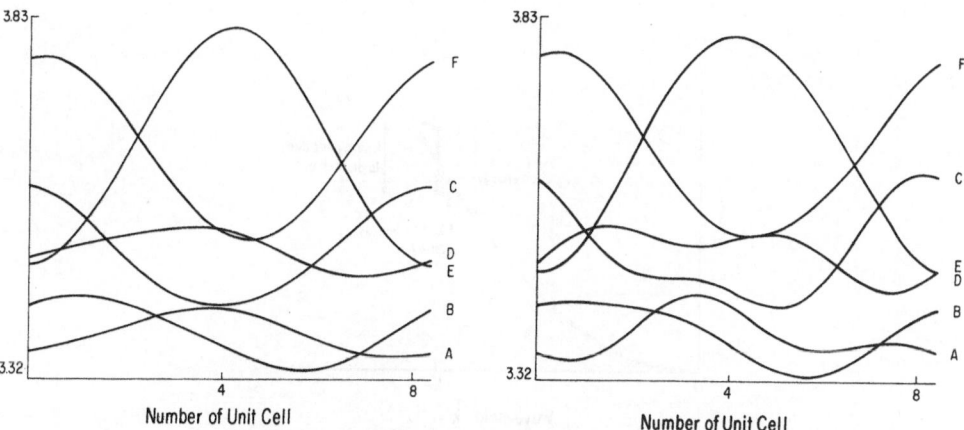

Fig. 5.   (a) Variation of hydrogen bond distance in thiourea with unit cell,
          harmonic model; (b) Variation of hydrogen bond distance with unit
          cell including second harmonics, anharmonic model.

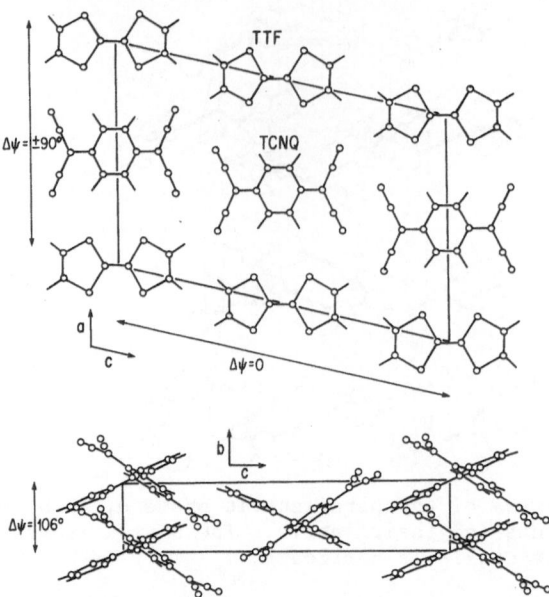

Fig. 6. Projection of the TTF-TCNQ structure along the b-axis (top) and
a-axis (bottom). Differences between the phases of the modulation
of translation-related molecules are indicated.

THE LOW-TEMPERATURE PHASE OF TTF-TCNQ[20]

The organic salt of TTF-TCNQ (tetracyanoquinodimethanide-tetrathioful-
valene) was one of the first known highly-conducting organic solids. Its
stucture consists of segregated stacks of donor and acceptor molecules, an
arrangement that favors delocalization of π electrons (Figure 6). However,
at 53K a transformation to an insulating phase occurs, and further transitions have been observed at 49 and 38K (ref. 21,22). The nature of the metal-
insulator transition and the low-temperature phases has been studied by careful measurements of diffuse scattering and satellite reflections[23,24]. As

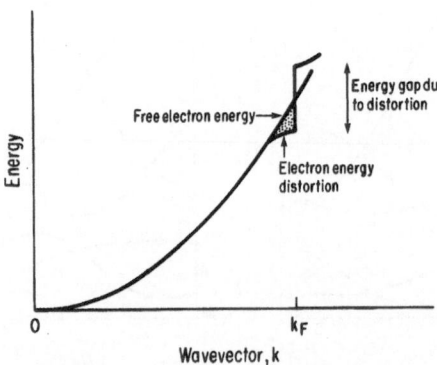

Fig. 7. Schematic representation of the Peierls distortion. The distortion
occurs at the Fermi level. It lowers the energy of the occupied
part of the band just below the Fermi energy E_F.

the Peierls metal-insulator transition corresponds to a distortion which creates a band gap at the Fermi level of the solid (Figure 7), the position of the satellite reflections is determined by the magnitude of the charge transfer between the donor TTF and the acceptor TCNQ molecules. Though this mechanism is well understood, the weakness of the satellite reflections has interfered with a structural analysis of the modulation. But indirect evidence is available. Magnetic susceptibility measurements suggest that the 53K transition involves only the TCNQ stacks, while at 49K the TTF molecules also undergo displacement[25]. A small amount of substitution of TSF (tetraselenofulvalene) into the TTF chains affects the temperature dependence of the $4k_F$ scattering much more than that of the $2k_F$ scattering, indicating that the $2k_F$ scattering is mainly due to the TCNQ chains[26]. On the other hand, analysis of the $2k_F$ diffuse scattering of isomorphous TSF-TCNQ shows the displacements to be mainly confined to the TSF stacks[27,28]. Similar evidence comes from high resolution polarized infrared spectroscopy[29].

The weakness of the satellite intensities, which are typically 10^{-4} times the intensities of the main reflections, can be overcome by the use of high intensity synchrotron radiation. Using the SUNY X21 beamline at NSLS, we have measured a total of 1342 reflections, including 1154 satellite reflections with $\sin \Theta/\lambda$ in the range of 0.2 - 0.5 Å^{-1}, using a crystal with dimensions 2.40 x 0.96 x 0.52 mm, mounted in a DISPLEX cryostat. Satellite intensities on an absolute scale were obtained by scaling each satellite intensity with the ratio of the main reflection F^2 value calculated with the parameters of the 45K structure determination[30], and the measured intensity of the main reflection. Since the orientation of the crystal is almost the same during the measurement of a main reflection and its associated satellite reflections, and the 45K data have been absorption corrected, this procedure corresponds to an absorption correction of the synchrotron intensities. After averaging over symmetry-equivalent reflections, elimination of reflections affected by instrumental instabilities, and a number of reflections for which the corresponding main reflection had not been measured, 137 unique satellite intensities with $F^2 > 3\sigma (F^2)$ were obtained which were used in the least squares refinement. No second order satellites were observed in this study.

The symmetry of the three phases of TTF-TCNQ has been discussed by Bak and Janssen[31]. The five-dimensional super space group $P:P2_1/c:cmm$ is in agreement with the observed absences. Eaquations for the displacements of symmetry-related atoms in the five-dimensional case are described in a separate publication[32], and have been incorporated in a computer program, JANA5, based on the molecular model.

The centrosymmetric site symmetry of the TTF and TCNQ molecules allows only 24 independent parameters describing the harmonic rigid body displacements:

$$u_i(q_1, q_2) = \sum_{k=1}^{2} V_\alpha^k \sin(2\pi q_k \cdot g_\alpha) + [W_\alpha^k \times (r_i - \rho_\alpha)] \cos(2\pi q_k \cdot g_\alpha) ,$$

where u_i is the displacement of ith atom, q_1, q_2 are modulation vectors related by a two-fold axis, V_α, W_α are translational and rotational molecular displacements ($\alpha = 1,2$ refer to the TTF and TCNQ molecules, respectively), r_i, g_α and ρ_α are the atomic positional vector, the phase reference point and the center of mass of the αth molecule, respectively.

The analysis shows the main displacement to be a translation of the TTF molecules with amplitude 0.0191(8) Å (Table 4, Figure 8). The polarization vector for this translational displacement lies within experimental error along the long axis of the molecule, and thus does not affect the interplanar

Table 4. Translational and Rotational Displacement Vectors of TTF and TCNQ Molecules

TTF	Phase (°)	Translations (Å)	L+	M	N
q_1	0	0.0074(9)	−0.98(12)	0.09(7)	0.16(12)
q_2	0	0.0191(8)	0.99(4)	0.05(3)	0.16(6)
		Rotations (°)			
q_1	90	0.10(3)	0.91(24)	−0.31(25)	0.26(59)
q_2	90	0.17(3)	0.96(15)	0.27(15)	0.07(30)
TCNQ		Translations (Å)			
q_1	0	0.0043(14)	0.04(3)	−0.14(31)	0.99(14)
q_2	0	0.0067(13)	−0.74(18)	0.31(20)	−0.60(25)
		Rotations (°)			
q_1	90	0.07(4)	−0.77(58)	0.62(40)	−0.15(92)
q_2	90	0.19(4)	−0.72(18)	−0.63(13)	0.28(23)

[+]Inertial axis in order of increasing moment of inertia.

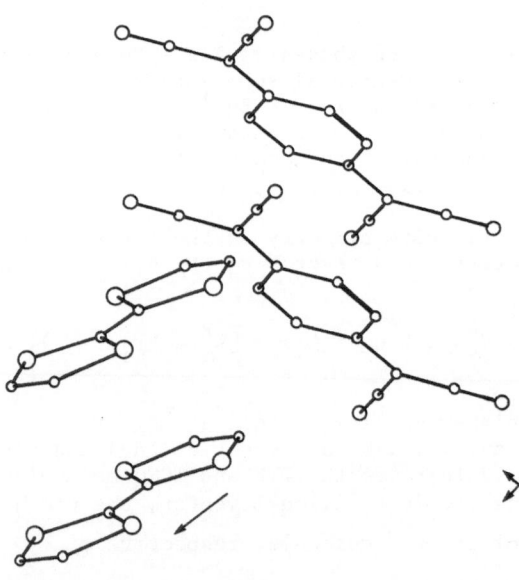

Fig. 8. Relative arrangement of TTF and TCNQ molecules. Arrows represent the amplitudes of the main modulation waves enlarged by a factor of 100. The TCNQ modulation has a component along the molecular normal, while the TTF modulation represents a slip along the long molecular axis.

spacing of molecules. As the phase difference between two adjacent molecules in a stack is 0.295 x 2π, the relative shift of two adjacent molecules in a stack is about equal to or smaller than the displacement amplitude. In some of the stacks the q_1 and q_2 displacements combine to give a relative slip of the TTF molecules which can be as large as 0.034 Å, in others the slip is at most 0.015 Å All translations of TCNQ are smaller than 0.007 Å. Though small, they have a significant component of 0.004 Å along the normal to the TCNQ planes. The combination of the two (q_1 and q_2) displacement waves leads to a shortening, or lengthening of the interplanar spacing with a maximum value of about 0.01 Å. All rotations are less than 0.2°, and about the long molecular axes. The largest atomic displacements under these rotations are about 0.007 Å. The much larger displacement of the TTF molecules, though somewhat unexpected, is in agreement with theoretical calculations of the TTF and TCNQ dimers which indicate that the energy variation with slip parameter is smaller for the dimer composed of two TTF molecules[33].

The results show that the $2k_F$ charge density wave in the low temperature phase mainly involves the TTF chains, which agrees with the conclusion drawn from the comparison with TSF-TCNQ, but not with some of the other evidence summarized above suggesting that the TCNQ stacks are mainly involved in the $2k_F$ distortion. However, a full comparison of the structural result with that obtained with other physical methods requires similar analyses of the intermediate temperature modulated phases of TTF-TCNQ.

The modulation has both transverse and longitudinal components as its main feature is a slip of the TTF molecules in their molecular plane, which is inclined to the stacking axis. The possibility that the $4k_F$ distortion is mainly localized on the TCNQ chains requires measurement of the $4k_F$ satellites. Though extremely weak, they should be accessible with synchrotron radiation.

CONCLUSION

Physical properties of molecular crystals can be strongly affected by the occurrence of modulations, which are relatively common in solids of interest. Their study is of importance for our understanding of a variety of solid state properties.

ACKNOWLEDGEMENTS

Support of this research by the National Science Foundation (CHE8711736) is gratefully acknowledged. The work reviewed here is a joint effort with our coworkers listed in references 9, 11, 15 and 20.

REFERENCES

1. P. M. De Wolff, The Pseudo-symmetry of modulated crystal structures, Acta Cryst. A30:777 (1974).
2. A. Janner and T. Janssen, Symmetry of periodically distorted crystals, Phys.Rev. B15:643 (1977).
3. P. M. De Wolff, T. Janssen, and A. Janner, The superspace groups for incommensurate crystal structures with a one-dimensional modulation, Acta Cryst. A37:625 (1981).
4. Modulated Structures, J. M. Cowley, J. B. Cohen, M. B. Salamon, and B. J. Wuensch (Eds), EIP Conference Proceedings, No. 53, American Institute of Physics, New York (1979).

5. J. L. Baudour and M. Sanquer, Structural phase transition on polyphenyls. VIII. The modulated structure of phase III of biphenyl (T 20K) from neutron diffraction data, Acta Cryst. B39:75 (1984).

6. H. Cailleau, Ch. 12 in: "Incommensurate Phases in Dielectric 2", R. Blinc and A. P. Levanyuk, eds; Elsevier (1986); H. Cailleau, J. C. Messager, F. Moussa, F. Bugaut, C. M. E. Zeyen, and C. Vettier, Main characteristic properties of incommensurate biphenyl, Ferroelectrics 67:3 (1986).

7. V. Heine and E. H. Simmons, Correct choice of superspace group for an incommensurate phase transition, Acta Cryst. A43:289 (1987).

8. F. Denoyer and R. Currat, Modulated phases in thiourea, Ch. 14 in: "Incommensurate Phases in Dielectrics 2", R. Blinc and A. P. Levanyuk (Eds), Elsevier (1986).

9. V. Petříček, P. Coppens, and P. Becker, Structure analysis of displacively modulated molecular crystals, Acta Cryst. A41:478 (1985).

10. V. Petříček and P. Coppens, Structure analysis of displacively modulated molecular crystals. V. Symmetry restrictions due to special positions, (to be published).

11. P. C. W. Leung, T. J. Emge, M. A. Beno, H. H. Wang, J. M. Williams, V. Petříček, and P. Coppens, Novel structural modulation in the ambient-pressure sulfur-based organic superconductor ß-(BEDT-TTF)$_2$I$_3$: Origin and effects on its electrical conductivity, J.Am.Chem.Soc. 107:6184 (1985).

12. N. Laukhin, E. E. Kostyuchenko, Yu. V. Sushko, I. F. Shchegolev, and E. B. Yagubskii, Effect of pressure on the superconductivity of ß-(BEDT-TTF)$_2$I$_3$, JETP Lett. 41:68 (1985).

13. K. Murata, M. Tokumoto, H. Anzai, H. Bando, G. Saito, K. Kajimura, and T. Ishguro, Superconductivity with the onset at 8K in the organic conductor ß-(BEDT-TTF)$_2$I$_3$ under pressure, J.Phys.Soc.Jap. 54:1236 (1985).

14. A. J. Schultz, M. A. Beno, H. H. Wang, and J. M. Williams, Neutron diffraction evidence for ordering in the high-T$_c$ phase of ß-di[bis(ethylenedithio)tetrathiafulvalene]triiodide, ß-(ET)$_2$I$_3$, Phys.Rev.B 33:7823 (1986).

15. Y. Gao, M. Gajhede, P. Mallinson, V. Petříček, and P. Coppens, Structure analysis of modulated molecular crystals II: The modulated phase of thiourea as described by a molecular displacement model, Phys.Rev. B: in press.

16. L. Solomon, Thiourea, a new ferroelectric, Phys.Rev. 104:1191 (1956).

17. Y. Shiozaki, Satellite X-ray scattering and structural modulation of thiourea, Ferroelectrics 2:245 (1971).

18. A. Yamamoto, Modulated structure of thiourea [SC(NH$_2$)$_2$], Phys.Rev. B22:373 (1980).

19. M. M. Elcombe and J. C. Taylor, A neutron diffraction determination of the crystal structures of thiourea and deuterated thiourea above and below the ferroelectric transition, Acta Cryst. A24:410 (1968).

20. P. Coppens, V. Petříček, D. Levendis, F. K. Larsen, A. Paturle, Y. Gao, and A. D. LeGrand, Synchrotron-radiation study of the five-dimensional modulated phase of tetrathiafulvalene-tetracyanoquinodimethanide at 15K. Phys.Rev.Lett. 59:1695 (1987).

21. J. P. Pouget, S. K. Khanna, F. Denoyer, R. Comes, A. F. Gasrito, and A. J. Heeger, X-ray observation of 2k$_F$ and 4k$_F$ scatterings in tetrathiafulvalene tetracyanoquinodimethane (TTF-TCNQ), Phys.Rev.Lett. 37:436 (1976).

22. S. Kagoshima, in: "Extended Linear Chain Compounds", J. S. Miller (Ed.), Plenum, New York (1982), Vol. 2, p. 303.

23. F. Denoyer, R. Comes, A. F. Garito, and A. J. Heeger, The X-ray diffuse scattering evidence for a phase transition in tetrathiafulvalene tetracyanoquinodimethan (TTF-TCNQ), Phys.Rev.Lett. 35:445 (1975).

24. S. Kagoshima, T. Ishiguro, and H. Anzai, X-ray scattering study of phonon anomalies and superstructures in TTF-TCNQ, J.Phys.Soc.Jap. 41:2061 (1976).

25. E. F. Rybaczewski, L. S. Smith, A. F. Garito, A. J. Heeger, and B. G. Silbernagel, Carbon-13 knight shift in TTF-TCNQ (^{13}C): Determination of the local susceptibility, Phys.Rev. B14:2746 (1976); Y. Tomkiewicz, A. R. Taranko, and J. B. Torrance, Spin susceptibility of tetrathiafulvalene tetracyanoquinodimethane, TTF-TCNQ, in the semiconducting regime: Comparison with conductivity, Phys.Rev. B15:1017 (1977).

26. L. Forro, S. Bouffard, and J. P. Pouget, X-ray diffuse scattering study of $2k_F$ and $4k_F$ anomalies in strongly irradiated TTF-TCNQ, J.Physique Lettres 45: L543 (1984).

27. S. Megtert, A. F. Garito, J. P. Pouget, and R. Comes, Lecture notes in physics, in: "Quasi One Dimensional Conductors I", S. Barisic, A. Bjelis, J. R. Cooper, and B. Leontic (Eds), Springer-Verlag, Berlin (1980).

28. K. Yamaji, S. Megtert, and R. Comes, 2D displacement pattern in TSeF-TCNQ model analysis of the $2k_F$ diffuse lines, J.Physique 42:1327 (1981).

29. F. E. Bates, J. E. Eldridge, and M. R. Bryce, High resolution polarized far-infrared vibrational spectra of semiconducting TTF-TCNQ and TSeF-TCNQ, Can.J.Phys. 59:339 (1981).

30. A. J. Shultz, G. D. Stucky, R. H. Blessing, and P. Coppens, The temperature dependence of the crystal and molecular structure of $\Delta^{2,2}$ -Bi-1,3-dithiole[TTF]7,7,8,8-tetracyano-p-quinodimethane[TCNQ], J.Am. Chem.Soc. 98:3194 (1976).

31. P. Bak and T. Janssen, Symmetry of modulated phases in tetrathiafulvalene tetracyanoquinodimethane (TTF-TCNQ): Four and five-dimensional superspace groups, Phys.Rev. B17:436 (1978).

32. V. Petříček and P. Coppens, Structure analysis of modulated molecular crystals III: Scattering formalism and symmetry considerations: Extension to higher dimensional space groups, Acta Cryst.A: in press.

33. B. D. Silverman, in: "Crystal Cohesion and Conformational Energies", R. M. Metzger (Ed.), Springer-Verlag, Berlin (1981), p. 108.

CONTROLLED GROWTH OF POLYTYPES AND THEIR IMPORTANCE

FOR SCIENCE AND TECHNOLOGY

Yu. M. Tairov and V. F. Tsvetkov

Leningrad V. I. Ulyanov (Lenin) Electrical Engineering
Institute
Leningrad, USSR

Synthesis, study and application of increasingly more complex crystalline structures are one of the most important tendencies in physics and the science of materials of solid-state electronics. The creation of A^3B^5 compounds and multi-component solid solutions based on these compounds, superlattices, complex ceramic high-temperature superconductors and other materials made it possible to enlarge substantially the possibilities of solid-state electronics, to create devices with wider ranges of operating characteristics and functional possibilities.

One of the promising trends in the science of materials is making use of polymorphism of elements and chemical compounds, polytypism of crystals being one of its varieties if the reproducible production of polymorphous modifications of a number of substances has been already mastered (graphite and diamond, cubic and hexagonal boron nitride and others) and as a result they have found wide applications in science and technology, the controlled growth of polytype modifications of a substance, their investigation and application has made its first steps as yet.

The phenomenon of polytypism as compared to that of polymorphism possesses much greater possibilities from the point of view of extending the class of materials used. If the number of polymorphous modifications for certain chemical elements and compounds does not amount to a dozen, the number of polytype modifications already discovered exceeds a hundred and is constantly increasing.[1,2]

The distinctions between substance polytypes are most markedly revealed in electrophysical properties of crystals (see Table 1), which makes their application attractive, primarily in semiconductor electronics, since having a set of polytypes of some semiconductor, one has, as a matter of fact, a group of semiconducting materials with various widths of the energy gap, mobilities of charge carriers, etc., and similar physico-chemical conditions of their production. It follows that the understanding of the nature of polytypism phenomenon and its control would make it possible to increase substantially the number of semiconducting materials, in particular, those necessary for the creation of new devices of solid-state electronics. Until recently polytypism was believed to be a specific structural feature of certain crystalline substances such as SiC, ZnS, CdI_2. As a result of thorough structural investigations of recent years this phenomenon has been discovered in much greater number of substances, including such the most

Table 1. Energy Gap Widths E_g for SiC, YnS, SnS$_2$, PbI$_2$ Polytypes

E_g	3C	18H	24H	14H	12H	8H
D, %	0	1.1	8.3	14.3	16.6	25
SiC	2.39	-	-	-	-	2.8
ZnS	3.82	-	-	-	-	3.83
SnS$_2$	-	-	0.92	-	-	-
PbI$_2$	-	1.07	-	1.44	1.65	-

E_g	24R	21R	6H	33R	15R	4H	2H
D,%	25	29	33.3	36	40	50	100
SiC	2.73	2.85	3.02	3.00	2.9	3.26	3.30
ZnS	-	-	3.84	-	-	3.85	3.89
SnS$_2$	1.15	-	1.59	-	-	1.89	2.18
PbI$_2$	-	-	-	-	-	-	2.53

important materials of electronic technology as silicon, germanium, gallium phosphide, gallium arsenide, $A^2 B^6$ compounds, triple semiconducting compounds, ferroelectrics, ferrites, etc; this phenomenon is continuously being revealed in increasing number of new materials. Apparently, under certain conditions of synthesis the polytypism phenomenon will be exhibited in a controlled way in a very wide range of solid-state electronics materials. In this connection, the studies of polytypism nature and the development of methods of controlling polytype structure become a very important problem the solution of which would permit to obtain a large spectrum of new semiconducting materials having unique electrophysical properties, for use in integrated microelectronics and power technology. The investigation of the problem must begin with a "model" material and the results obtained should be extended subsequently to other substances. Silicon carbide is of a special interest as such a "model" substance. On one hand, SiC is a classical polytype material crystallized in more then 140 polytype structures. On the other hand, having a whole set of unique electrophysical and physicochemical properties, silicon carbide is one of the most important materials of electronic technology because in the modern stage of solid-state electronics development, wide-band semi-conducting materials acquire an increasingly greater importance. Semiconducting devices based upon silicon carbide possess exceptionally stable time characteristics, high operating temperatures, resistance to high overloads of different kind during their operation, and they have already found their application in electronics. Thus, the investigation of the polytypism phenomenon and the development of physical principles of growth and polytype structure control of single crystals taking silicon carbide as an example is an urgent problem of practical significance.

The problem of polytypism control in crystals is closely connected with the general theory and practice of crystal growth. Until recently, however, the problem of controlled synthesis of crystals with specified polytype structures had not been studied. The formation of one or other polytype during the growth of crystals of various substances was accidental. Polytypism in silicon carbide and other substances was considered to be undesirable, since it often happened that several polytype structures with differing electrophysical properties co-existed in one crystal, which resulted in reducing the production of suitable crystals. The difficulties in growth control of specified polytype structures were caused primarily by the fact that the nature of the phenomenon was not fully understood. In connection with this we have carried out a detailed analysis of numerous publications on the production of polytype structures in various substances, special attention being paid to investigations which employed transmission electron microscopy. The comparative analysis of these works enabled us to draw the following conclusions:[2]

(1) The overwhelming majority of polytypes of various substances (including the substances which under normal conditions are not characteristically polytypic) are formed in the process of solid-phase martensite transformation. The moving force of phase transformation is either an excess of free energy of a similar metastable phase (metastable polymorphous structure, amorphous layers, etc.) or external shear stresses or great internal stresses in the crystalline lattice which are formed for example in the course of sectoral distribution of impurities of basic components, during a rapid cooling of crystals, etc.

(2) Such structures with excessive free energy and internal stresses are easily formed during crystallization at lower temperatures and sufficiently high growth rates when there is no equilibrium achieved between the volume of growing crystal and parent medium. This is typical for crystallization out of vapour phase, out of solution, out of solid phase as well as for super high-speed crystallization of melts, particularly viscous ones.

These conclusions formulated on the basis of analysis of experimental data agree quite well with the results of theoretical investigation conducted for establishing the connection between the crystal structure of best studied and most important group of $A^n B^{8-n}$ diamond-like semiconductors and the peculiar character of chemical bond in them. The most fruitful results were obtained in investigation using the diagram of crystalline state (DCS),[3] along the axes of which orbital coordinates of chemical bond R_π and R_σ are plotted which represent linear combinations of orbital radii of isolated atoms from which crystalline structure is formed. Orbital coordinates of chemical bond indicate the processes of redistribution of electron density in the formation of a chemical bond. All combinations of $A^n B^{8-n}$ being crystallized in wurtzite, sphalerite and NaCl structures are divided into groups on DCS and coordinates R_π and R_σ. It has been established by us that the most typical polytype compounds $A^n B^{8-n}$ lie at the boundary of the fields of compounds having the structures of wurtzite and sphalerite. The transition from the compounds with the sphalerite structure to the compounds with the wurtzite one is mainly connected with certain critical values of electron charge transfer from cation to anion $R_\sigma^{cr} \simeq 0.7$. On approaching this critical value of transfer, the stacking fault energy (SFE) of $A^n B^{8-n}$ compounds tends to a minimum value (< 10 mJ/mol), which indicates the structure instability and the ease of martensite transformation. The analysis of DCS also revealed that the further the $A^n B^{8-n}$ compound is located from the sphalerite-wurtzite boundary, i.e. the greater the value $(R_\sigma - R_\sigma^{cr})$ and hence the greater the stacking fault energy, the more non-equilibrium the synthesis conditions of this semiconductor polytype structure must be.

Proceeding from these results of theoretical and experimental studies of polytypism nature in $A^n B^{8-n}$ semiconductor crystals, X-ray structural and electron microscopic investigations of solid-phase transformations in SiC polytypes were performed with the view of elucidating the conditions of their formation. The relationships between the transformation extent versus crystal annealing time and temperature were obtained and processed applying the theory of transformations created by A. N. Kolmogorov.[4] The activation energy of transformation equal to 222 kJ/mol found out correlates well with the well-known energy of dislocation motion activation in SiC and confirms the dislocation mechanism of martensite solid-phase transformation observed.

Disordered layers of similar structure were observed and studied by us in α-SiC crystals grown by sublimation method during spontaneous seed formation and mass crystallization as well as when growing silicon carbide ingots.[5]

The formation of such disordered layers consisting of thin inter-lamels of different polytype structures and 3C-SiC initial structure is bound with solid-phase transformations in 3C-SiC layers formed in the initial non-equilibrium stages of silicon carbide crystal nucleation.

From the experimental investigation data stated on solid-phase transformations in silicon carbide single crystals it follows that the 3C structure is the most unstable one and serves as a source of the existing variety of numerous polytype structures. This fundamental conclusion was confirmed by electron microscopic studies of ceramics based on silicon carbide.[6] Thermodymanic calculations and crystallo-chemical investigations cited by us confirmed that at normal pressure the 3C-structure is unstable and represents a high-pressure phase which is stable in the 10^9-10^{10} Pa range. This results in its thermodynamic stability in nucleus and its metastability in bulk state. Solid-phase transformations occurring in 3C-SiC during nucleation and subsequent crystal growth result in the formation of dislocations with a screw component in Burgers vector (Fig. 1). These dislocations moving onto the surface of growth front become growth centres, i.e. the source of growth steps succession of polytype structure whose crystal cell parameter C corresponds to the Burgers vector value of the dislocation formed.

From investigations carried out it follows that the dislocation theory of formation and growth of polytype structures is the most adequate and realistic one; according to this theory polytype structures are formed and grown from matrix defect which in the case of silicon carbide is represented by disordered layers formed in result of solid-phase transformations of a 3C-SiC metastable structure. The effect of point defects, vacancies, impurities is directed mainly to the energy of stacking fault formation which in-

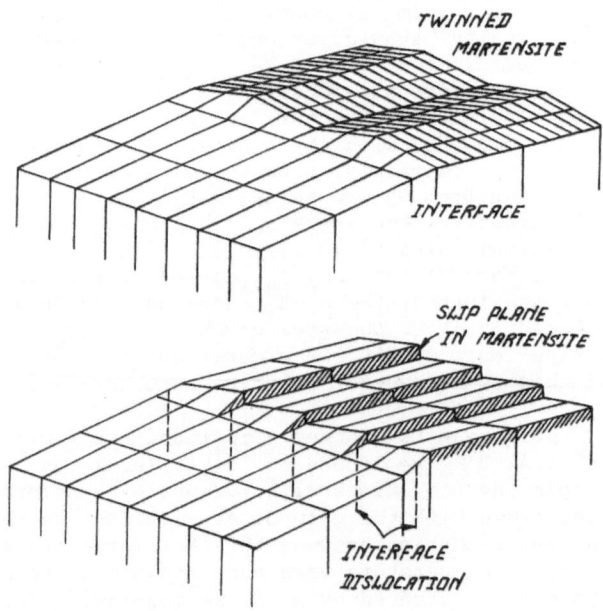

Fig. 1. Schematic of formation of screw-component dislocations by means of solid-phase transformation of a martensite type.[19]

itiates phase transformations and to the propagation rate of dislocations that define the phase transformation kinetics.

The studies performed of the polytypism nature in crystals of $A^n B^{8-n}$ semiconductor group enabled us to formulate the direction of investigations in the controlled production of polytypes of these substances. The main of these directions are the control of stacking fault energy by means of point defects, controlled annealing of materials in amorphous state, crystallization on the specified facets of the seed crystals of polytype structure required, crystallization kinetics control, the formation of homovalent and hetero-valent solid solutions.

Let us consider special features of these methods of control.

To obtain massive silicon carbide crystals with specified polytype structure a bank of seed crystals of various polytype structures was created. It has been found out that the best reproduction of the seed polytype structure occurs when growing crystals on the (0001)Si facet as well as on facets deflected from (0001). To obtain a good reproduction of the seed polytype structure the crystal growth must take place under conditions approaching the "vapour-crystal volume" equilibrium. This can be achieved when crystal growth rate V is much slower than self-diffusion rate (V_D) of crystal own components. The worst reproduction of polytype structure is characteristic for the (0001)C facet.

When it is required to obtain heteropolytype structures, i.e. to grow a different polytype structure from the seed of a certain polytype structure or to change the crystal polytype structure during its growth, the most effective way is to control the structure of the crystal being grown with growth kinetics. In this case it is advisable to grow crystal on the (0001)C facet – which is the least stable from the point of view of passing on the structural information from the seed. Crystal growth must be performed under conditions approaching the "vapour adsorbed layer" equilibrium ($V \geq V_D$) under which the dependence of the basic components in silicon carbide crystals on their growth rate exists. The variations in the growth rate, composition gradient and hence the value of stresses in the growing layer performed in accordance with a certain law permitted to plot kinetic structural diagrams

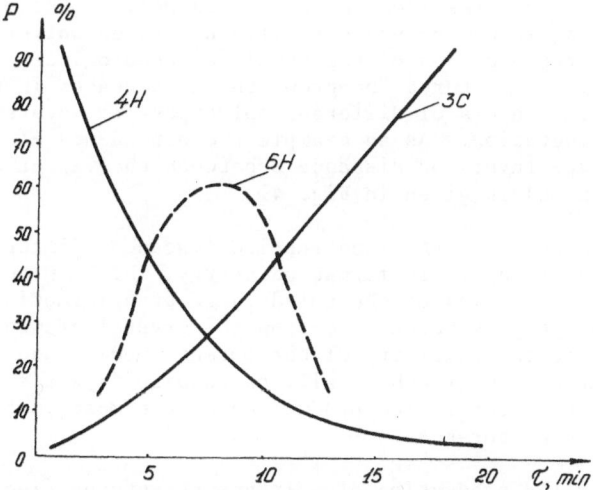

Fig. 2. Dependence of probability of 3C, 6H and 4H polytype structure formation on time constant of exponential increase in the rate of silicon
carbide vapour deposition in the initial growth stage.

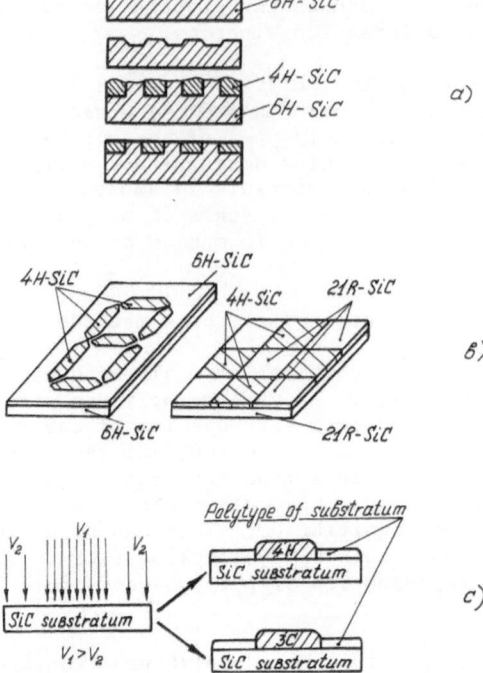

Fig. 3. Ways of local growth control of polytype structures:
   (a) The substrate surface is locally evaporated or etched in accord-
       ance with the topological pattern specified, then the growth of
       epitaxial layers is carried on under the conditions of 4H–SiC
       formation (see Fig. 2);
   (b) the variants of local growth of SiC polytypes according to the
       schematic of Fig. 3a;
   (c) local growing of polytype structures by controlling SiC vapour
       flow density along the coordinate.

(Fig. 2) and elaborate methods of growing massive single crystals of the most
important polytype structures from any seeds, as well as the method of local
growing the epitaxial layers of various silicon carbide polytypes (Fig. 3)[2].
To obtain an exact reproduction of the specified topological pattern when
local hetero-polytype structures "co-grow" the differences in the growth
kinetics of epitaxial layers of different polytypes from oversaturation must
be taken into consideration. As an example the dependence of growth rates V
of 6H and 3C polytype layers on distance H between the vapour source and substrate (oversaturation) is given in Fig. 4.

    The point defects in crystalline lattice (vacances, impurities) substan-
tially affect the stacking fault formation energy, which in turn, determines
the nature and the directions of the solid-phase transformation resulting in
the formation of polytype structures during the crystal growth. On this
basis by changing the stoichiometry of the parent medium and injecting certain impurities during the growth of silicon carbide crystals and epitaxial
layers the processes of controlled production of the most important polytype
structures have been elaborated.[8,9]

    For the controlled production of polytype structures (including the
local one) the employment of proportioned annealing of amorphous layers of
various materials obtained both by condensation from the vapour phase with

336

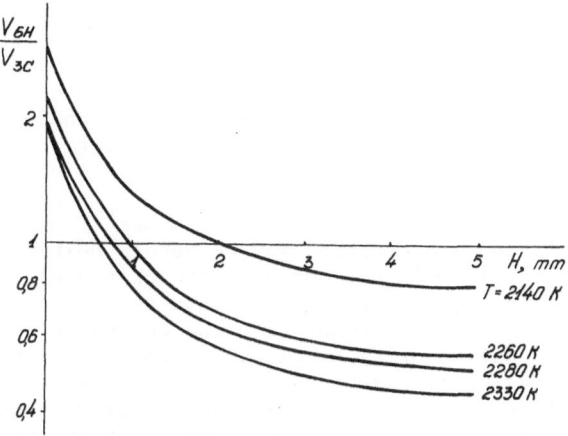

Fig. 4. Dependence of growth rates ratio for 6H and 3C polytypes on temperature and distance H between vapour source and substrate.

super high-speed cooling of melts and by ion bombardment is of great interest. The amorphous state of a substance is characterized with a high non-equilibrium accompanied by the formation of fragments of numerous structures. The proportioned annealing of amorphous layers performed in accordance with the Rule of steps creates conditions for the successive transformation of a number of metastable structures into one another. In this case in structures formed there arise considerable stresses including the sheer ones as a result of interactions with the surrounding matrix,[10] leading to the formation of growth centres of polytype structures. The annealing of amorphous state has already allowed polytype structure to be produced even in materials for which the phenomenon of polytypism is not characteristic under conventional conditions of production. In such a way, e.g. the following silicon polytypes have been accidentally obtained so far: 2H, 6H, 27R, 51R, 141R.[11] The stresses arising in grains when producing polycrystalline silicon also result in the formation of polytypes.[12]

The variations in the composition of homovalent or heterovalent solid solutions also present a promising way of crystal polytypism control. So, for example, by varying the proportion of X components in solid solutions $(SiC)_x(AlN)_{1-x}$[13] the polytype structure of crystals grown can be successfully controlled.

Let us now dwell upon the significance of polytype structures for science and practical applications.

Since in the polytype structures of the same material the nearest surrounding is similar and the differences are observed only in the further surrounding, the discovery of the relationships between the polytype structure and the crystal properties is of great significance for establishing the role of the nearest and further surroundings in the formation of solid--state electrophysical characteristics. In addition, as a result of intensive studies in the field of artificial superlattices in which superperiodicity is caused by the variations in the chemical composition of a substance the investigation of polytypes as natural structural superlattices in which superperiodicity arises due to the variations in the crystalline structure with the same chemical composition becomes a fundamental problem. The creation of heterotransitions on the basis of different polytype crystals is also of great significance.

Polytypes may serve as a model specimen for studying the specific nature of physical processes in high-temperature superconductors[15] since these materials are also layered ones. It is relevant to note here due to the layered nature of materials for high-temperature superconductors, polytypism is also observed in them.[16] Therefore for deep understanding of the phenomenon of high-temperature superconductance it is necessary to study its specific features in different polytype structures.

Many crystals in certain temperature and pressure range spontaneously reveal periodic modulation whose period may be both a multiple (a commensurate phase) and a non-multiple (a non-commensurate phase) of the period in the initial lattice. The problem of non-commensurate phases and phase transitions of commensurate - non-commensurate phase constitute the subject of intensive studies. In the majority of cases the commensurate phase - non-commensurate phase transitions occur due to the formation of a regular series of domain walls (solitons) of small widths separating nearly commensurate areas. In recent years it has been found out that there is a close connection between two interesting phenomena - polytypism and non-commensurability. A rather rare phenomenon has been established - polytypism in tetragonal and monoclinic crystals.[17]

The investigation of intercalation processes in laminated crystals, which permits considerably to modify the properties of these crystals and substantially enlarge the nomenclature of new materials, presents a great interest. On the other hand, it has been shown that polytypism is characteristic for intercalated crystals.[18] Thus, the use of intercalation and the polytypism control in intercalated crystals make it possible, on one hand, to investigate new physical phenomena and on the other hand, still more to increase the spectrum of materials used.

We shall illustrate the importance of producing new materials by making use of the polytypism phenomenon and phenomena accompanying it taking silicon carbide polytypes as an example. In Fig. 5 one can see luminescence spectrum for different silicon carbide polytypes.[2] It can be seen that the employment of the polytypism phenomenon permits to produce light-emitting diodes quite easily which cover all visible range of wave lengths spectrum. On the other hand, it allows multicoloured matrix screens to be produced on a single crystal due to local variations of polytype structure.[2]

In conclusion we should like to notice that polytypes have been observed both in crystals forming meteorites and in protein crystals. This permits to consider them as tests with the help of which it is possible to study processes occurring both in universe and in human organism.

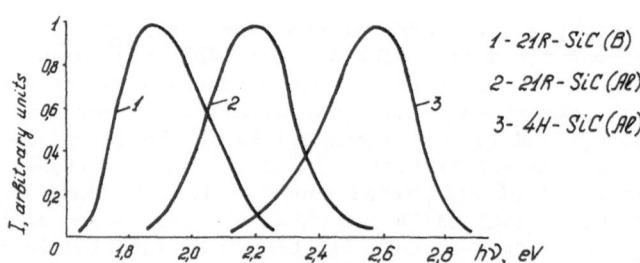

1 - 21R - SiC (B)
2 - 21R - SiC (Al)
3 - 4H - SiC (Al)

Fig. 5. Electroluminescence spectra for light-emitting diodes on the basis of local epitaxical structures of 4H and 21R polytypes (T = 300 K).

REFERENCES

1. "Crystal Growth and Characterization of Polytype Structures" ed. by P. Krishna, Pergamon Press (1983).
2. Yu. M. Tairov, V. F. Tsvetkov, "Progress in controlling the growth of polytype crystals", in: "Crystal Growth and Characterizations of Polytype Structures", P. Krishna, ed., Pergamon Press (1983).
3. G. Sukhanek, V. F. Tsvetkov, Crystal growth and polytypism, Izvestiya LETI, 17:16 (1980).
4. Yu. M. Tairov, V. F. Tsvetkov, M. A. Chernov, and V. A. Taranets, Investigation of phase transformations and polytype stability of β-SiC, Phys. Stat. Sol.(a) 43:363 (1977).
5. B. I. Levchuk, Yu. M. Tairov, V. F. Tsvetkov, and M. A. Chernov, X-ray diffractometric analysis of silicon carbide polytype structures, Kristallografia 27:392 (1982).
6. L. U. Ogbuji, T.E. Mitchell, and A. H. Heuer, The β→α transformations in polycrystalline SiC, J. Amer. Ceram. Soc. 64:91 (1981).
7. N. D. Sorokin, F. Raikhel, and Yu. M. Tairov, Silicon carbide - variable component compositions, Lett. J. Tekhn. Fiz. 8:101 (1982).
8. H. Vakhner, Yu. M. Tairov, Studies of silicon carbide doped with scandium, Fiz. Tverd. Tela 12:1543 (1970).
9. Yu. A. Vodakov, G. A. Lomakina, and E. N. Mokhov, Non-stoichiometry and polytypism of silicon carbide, Fiz. Tverd. Tela 24:1377 (1982).
10. A. A. Sokol, V. M. Kosevich, Growth of crystals in amorphous films, in: "Growth of Crystals", Nauka, Moscow (1983).
11. Y. Miyamoto, M. Hirata, Polytypism and amorphousness in silicon carbide, J. Phys. Soc. Japan 44:181 (1978).
12. H. Menso, Additional X-ray and electron diffraction peaks of polycrystalline silicon films, Thin Solid Films 113:53 (1984).
13. Sh. A. Nurmagamedov, G. K. Safaraliev, Yu. M. Tairov, N. D. Sorokin, and V. F. Tsvetkov, Specific features of forming epitaxial layers in $(SiC)_{1-x}(AlN)_x$ solid solutions, Neorganicheskiye Materialy 22:1672 (1986).
14. G. B. Dubrovsky, Natural superlattices in crystals, in: "Problems of Physics and Technology in Wide-band Semiconductors," Leningr. Inst. Yadernnoi Fiziki, Leningrad (1980).
15. J. D. Jorgensen, H. B. Schüttler, D. G. Hinks, D. W. Capone, H. K. Zhang, N. B. Brodsky, and D. J. Scalapino, Lattice instability in high-Tc superconductivity in $La_{2-x}Ba_xCuO_4$, Phys. Rev. Lett. 58:1024 (1987).
16. L. N. Bulaevsky, Superconductivity and electron properties in layered compounds, Uspekhi Fiz. Nauki 116:449 (1975).
17. A. U. Sheleg, V. V. Zaretsky, Polytypism and non-commensurability in zinc diphospides, Fiz. Tverd. Tela 28:935 (1986).
18. V. P. Mushinsky, M. I. Karaman, "Optical Properties of Gallium and Indium Chalcogenides," Shtiintsa, Kishinev (1973).
19. C. M. Wayman, Martensitic transformations: electron microscopy and diffraction studies, in: "Diffraction and Imaging Techniques in Material Science," S. Amelinckx, R. Gevers, and J. van Landuyt, eds. North Holland Publishing Co. (1978).

SYMMETRY AND DIFFRACTION PATTERNS OF POLYTYPES

Konrad Fichtner

Central Institute of Physical Chemistry
Academy of Sciences of the G.D.R.
1199 Berlin-Adlershof, G.D.R.

POLYTYPISM AND OD STRUCTURES

The notion of polytypism - coined by Baumhauer[1] in the case of SiC - describes a special kind of polymorphism, exhibited by a large number of materials. The following definition has been approved by the IUCr[2]:

(i) An element or compound is polytypic, if it occurs in several structural modifications, each of which may be regarded as built up by stacking layers of (nearly) identical structure and composition, and if the modifications differ only in their stacking sequence.

For the stacking of layers certain conditions have to be fulfilled[3]:

(ii) The two lattice parameters parallel to the layers are the same for all the polytype modifications.

(iii) The nearest neighbor relationships for the constituent atoms are the same in the different polytypes.

Because of (iii), disorder in the stacking of layers should be expected - and is observed with most polytypic substances. The property to be able to form periodic ("ordered") as well as non-periodic ("disordered") stacking sequences, was considered as a most characteristic attribute, when Dornberger-Schiff[4] developed her theory of Order-Disorder structures (OD theory) in the fifties. "OD theory is actually the symmetry theory of polytypic substances".[5] Most of the material presented in this paper is taken from OD theory. Thus, it is necessary to characterize the relationship between the two notions, OD structures and polytypes. OD structures consisting of layers are polytypes. The difference is that for OD structures, condition (iii) is more strict:

(iii)$_{OD}$ Let L' and L" be two next layer positions of a layer L. Then the pairs of layers (L, L') and (L, L") are geometrically equivalent, i.e. may be transformed one into another by a motion of space.

Summarizing the experience with many polytypic substances, the difference between the two notions, polytypes and OD structures consisting of layers, may also be described as follows: often there are two or even more ways to choose the layers in order to describe the polytypism of a substance (Fig. 1). One of the choices conforms to the definition of OD structures, the others not ("non-OD layers"). From the point of view of symmetry, the

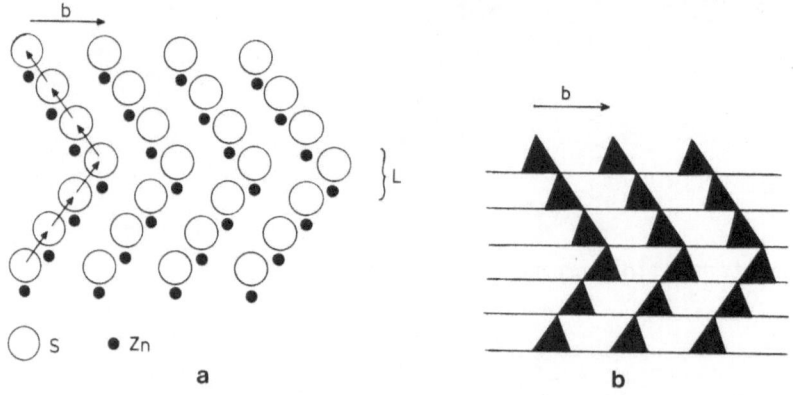

Fig. 1. Two ways for the choice of layers in a polytypic compound.
(a) Pairs of atomic planes S-Zn (OD layer); (b) sheets of tetrahedra with Zn in the centre and S at the corners (non-OD layer).

divison into OD layers is the most appropriate one[6]. Non-OD layers may,
however, be the most convenient layers for some other reason. The chemical
building units, for instance, may be non-OD layers.

We mention that there is a tendency to define the notion of polytypism
in a more general way[7], including families of structures consisting of rods
or even finite blocks, but also families of structures with essentially different chemical composition as well as mixed layer structures.

SYMMETRY THEORY OF POLYTYPES

There are many indications that the symmetry of polytypes is incomplete-
ly reflected by their space groups. Some of these indications are discussed
further below (e.g. systematic absences not resulting from space group symmetries, diffraction enhancement of symmetry). OD theory contains an adequate symmetry theory that may be applied to any set of polytypes, even if
non-OD layers are used. This symmetry theory is built up in analogy to the
symmetry theory of normal crystal structures: based on a mathematical funda-
ment, it contains several classifications of polytypic substances and permits
a complete characterization of the common symmetry of the infinite number of
stacking possibilities by a finite number of data. This latter point is the
most important one for our purposes.

The two basic notions of this theory are: <u>partial coincidence operation</u>
or <u>partial symmetry</u> and <u>symmetry groupoid</u>. A partial symmetry of a polytype
is a motion of space transforming a layer into itself or into another layer
of the polytype. The set of all such partial symmetries of a polytype forms
a groupoid[8], a notion coined by the German mathematician Heinrich Brandt[9] in
1926. For a polytypic substance, the symmetry groupoids of the individual
polytypes are closely related to each other as can easily be seen. The information contained in the symmetry groupoid may be split into three parts:
(i) the symmetry of the layer, described by the symbol of the layer group
and the net constants; (ii) the possible positions of any next layer, given,
for instance, by the transformations of a layer into the positions of the
next one; (iii) the actual stacking sequence, usually characterized by a
stacking symbol. Whereas the third part contains individual features of a
polytype, parts (i) and (ii) describe the symmetry features of the whole
family of polytypes.

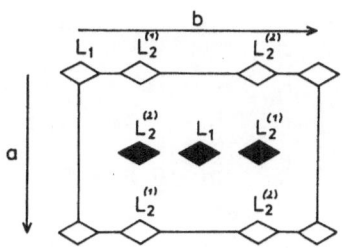

Fig. 2. Schematic representation of the two possible positions of layers L_2 relative to L_1, if the symmetry of the polytype family is described by the symbol $Pmm(n)| 1, x=0, y=1/4$.

Any individual periodic polytype has a space group that is uniquely determined by the symmetry groupoid and may be derived from the three parts characterizing the groupoid.* For any polytypic substance, rules may be given, how the space group may be derived from the stacking symbol. Such rules have been worked out, for isntance, in the case of close packings of equal spheres[10], CdI_2-type structures[11], FeB-CrB stacking variants[12,13].

The characterization of the symmetry of polytypes may be illustrated by the schematic example of Fig. 2. The layer group is $Pmm(n)$. Any next layer has two possible positions, and the layer L_1 is transformed into the next one by a translational vector either $\vec{c}_o + \vec{b}/4$ or $\vec{c}_o - \vec{b}/4$. Thus the symmetry features of the family may be characterized by the symbol $Pmm(n)| 1, x=0$ $y=\pm 0.25$[14]. For the characterization of the stacking, the Zhdanov notation may be used. Any polytype can have one of 5 possible space groups: $Pm11$, $P2_1/m11$, $Pm2_1n$, $Pmc2_1$, and $Pmcn$. Examples are the polytypes 4213, 31, 221111, 321321, and 22, respectively.

DIFFRACTION PATTERNS OF POLYTYPIC SUBSTANCES

Basic Properties

The intensities in a diffraction pattern of a crystal with a slightly mosaic structure fulfil the following relation

$$I(\vec{r}) \sim \left| \int \rho(\vec{r}) \cdot \exp(2\pi i \vec{r}\vec{r}^*) d\vec{r} \right|^2 = |F(\vec{r})|^2 .$$

For a structure consisting of layers, the Fourier transform of the electron density is the sum of the contributions of the layers

$$F(\vec{r}) = \sum_p F_p(\vec{r}) .$$

The Fourier transform may be different from zero only for points (h,k,ζ) in reciprocal space as all layers are supposed to be periodic with two common translational vectors $\vec{a}, \vec{b}$.

The stacking of the layers may or may not be periodic. Thus the Fourier transform may be different from zero either only for discrete values of ζ or for (almost) all values of ζ. Accordingly, for a polytypic substance, diffraction patterns with only sharp reflections and such containing

*) There are some exceptional cases. Example: the space group operation $\bar{4}$ of ß-SiC (polytype 3c) is not contained in the symmetry groupoid as it does not transform the layers into each other.

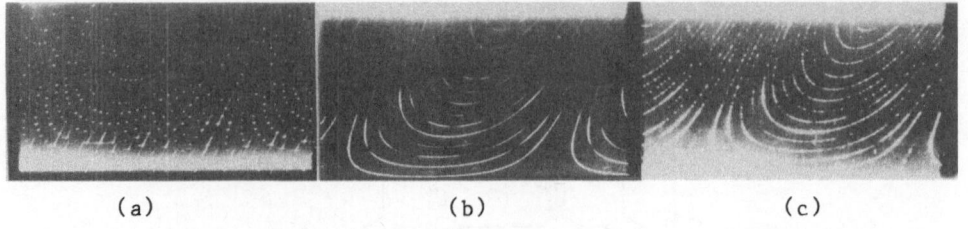

(a)   (b)   (c)

Fig. 3.   Diffraction patterns of a polytypic substance may contain:
          (a) sharp reflections; (b) diffuse streaks; (c) streaks with maxima,
          (after ref. 15).

"streaks" are possible. As a disordered polytype may contain ordered regions, maxima on the streaks are also possible (Fig. 3)[15].

Family Reflections and Family Structure

For a correct interpretation of the diffraction patterns of a polytypic
substance, reflections that are independent on the stacking ("family reflections") play a key role. The 001 reflections, for instance, are family reflections for any polytypic substance. This is true, because the projection
of a layer on a line perpendicular to the layer is the same for all possible
positions of this layer. There may exist other family reflections, if
certain additional conditions are fulfilled.

Let us suppose that every layer L has Q possible positions,

$$L_p^{(j)} = L_o + t_p^{(j)}, \quad t_p^{(j)} = (j/Q,0,p) \quad j=0,\ 1,\ 2,\dots,\ Q-1.$$

Then the layer structure factor,

$$F_p^{(j)}(h,k,\zeta) = F_o(h,k,\zeta).\exp(2\pi\ i.j.h/Q).\exp(2\pi\ i.p\zeta),$$

is independent on j, if h=nQ. For non-integral values of ζ, the contributions of the layers compensate each other. Thus, for fixed n and k, the
(nQ,k,ζ) rod in reciprocal space is a rod of sharp family reflections. The
intensity distribution for Q=4 is characterized in Fig. 4. Any (h,k) rod
of family reflections corresponds to Q-1 non-family rods in reciprocal
space. Thus, for a disordered polytype, the number of possible positions of
a layer may be concluded from the number of streaks between rods of sharp
reflections. The values of the Fourier transform at the points of the family
reflections, taken by themselves, are the structure factors of a fictitious

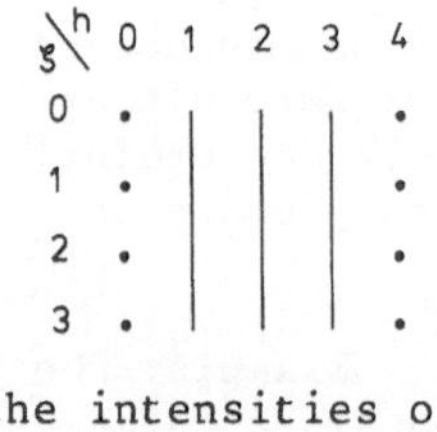

Fig. 4.   Distribution of the intensities of a polytype with disordered
          stacking sequence (schematically).

structure periodic in three dimensions, the family structure (superposition structure, average structure). It is the structure, for which every possible position of a layer is occupied with weight 1/Q.

Rather general sufficient conditions for the existence of a family structure may be formulated: the structures of the family may be described as sequences of layers such that:

(i) all vectors between translationally equivalent layers of the structures belong to a lattice $m\vec{a}+n\vec{b}+p\vec{c}$ (m,n,p=0,$\pm$1,$\pm$2,...),

(ii) the possible positions for any next layer differ only by translational vectors.

Structures of Maximum Degree of Order (MDO Structures)

It is useful to single out in a family of polytypes a small number of important or potentially important polytypes. There are several concepts to define such "privileged" polytypes. In a pragmatic approach, the most frequently observed stackings are taken and called <u>basic polytypes</u>.

OD theory contains the concept of <u>MDO structures</u>[16], a geometrical concept based on energetical considerations. Polytypes are singled out, which contain the smallest number of kinds of layer triples. The concept of <u>simple</u> or <u>regular polytypes</u>[17,18] pays special attention to those stackings, for which a (total) symmetry operation transforms any layer into the next one. Thus this concept is based on symmetry properties. The term simple refers to the stacking, but is justified with respect to the diffraction pattern as well. The knowledge of the possible positions of the family reflections and of the reflections of the simple polytypes is a solid base for understanding any diffraction pattern of the substance under consideration.

For most polytypic substances, simple polytypes are MDO structures, and vice versa. In the schematic example of Fig. 5, all layers are translationally equivalent, and any next layer has one of two possible positions, which may be described by translations of the starting layer in $\vec{c}$-direction with sideway displacement either to the left or to the right. Obviously, two kinds of layer triples exist: stretched and bent triples leading to the two MDO structures.

For a polytypic substance, a multiple twinning is frequently observed. The reason for that may be understood from Fig. 5. Let us suppose that during crystallization stretched triples are strongly preferred. Because of stacking faults, the polytype may be a mixture of MDO_1 and MDO_1' regions.

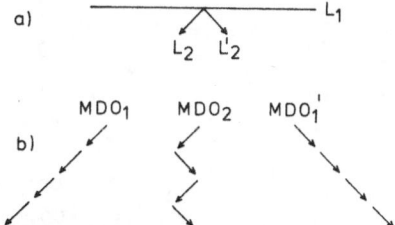

Fig. 5. Schematic example of MDO structures. Any next layer has two possible positions. The layers are translationally equivalent.

If the proportions are approximately equal, the diffraction pattern has a higher symmetry than expected from pure MDO_1 or MDO_1' stacking. If MDO_1 is monoclinic or trigonal, the diffraction pattern may have orthorhombic or hexagonal symmetry, respectively.

Systematic Absences Caused by Partial Symmetries

Crystal symmetries with glide components cause systematic absences, and these are used for the determination of the symmetries. The following two examples illustrate that partial symmetries may also cause absences.

(i) If the layers are translationally equivalent, the structure factor is the product of the layer structure factor and a term depending on the stacking[19]

$$F(h,k,\zeta) = F_o(h,k,\zeta).S(h,k,\zeta).$$

Partial symmetries of the layer with glide components lead to absences, independent on the stacking, e.g. a layer symmetry 11(a) causes absence of reflections hk0 for h=2n+1.

(ii) Partial symmetries transforming different layers into each other may also cause extinctions. Let us suppose that any layer is transformed into the next one by a twofold screw rotation with translational component $\vec{a}/4$. Then reflections h00 for h≠4n are missing as the projection of the structure on the $\vec{x}$-axis is periodic with a/4.

In the following, the classical example of calcium monoborate dihydrate, is used to demonstrate, how the symmetry of an OD structure may be determined[20]. An investigation of the diffraction pattern (Fig. 6) revealed:

sharp reflections	hkl for	k + l = 2n
diffuse streaks	ζkl for	k + l = 2n + 1
systematic absences	ζ0l for	l = 2n + 1.

From the existence of diffuse streaks in $\vec{a}*$-direction, layers perpendicular to a with translational vectors $\vec{b}$ and $\vec{c}$ have been concluded. Because of the systematic absences the symmetry of the layer is at least P(1)c1. From the distribution of sharp and diffuse intensities, two possible positions for any layer, differing by a vector $(\vec{b}+\vec{c})/2$, were concluded.

An intensity statistics of the sharp reflections showed the presence of a centre of symmetry for the family structure. Its space group was determined to be A12/m1. The two layer groups compatible with this space

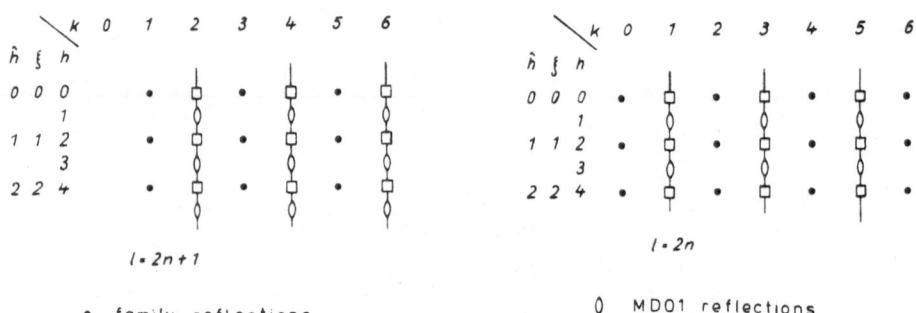

Fig. 6. Distribution of intensities in reciprocal space of $Ca[B(OH)_4]_2.2H_2O$, schematically (after [20]).

group are $P(1)2_1/c1$ and $P(1)2/c1$. The first one was excluded because reflections 0k0 for k=2n+1 exist.

The different structures may indeed be considered as consisting of one kind of layer with layer group $P(1)2/c1$. For such a sequence of layers, pairs of layers are not always equivalent. Therefore, from the point of view of OD theory, this structure has to be considered as consisting of two kinds of layers. The additional layer has higher symmetry, layer group $A(1)2/m1$.

SPECIAL FEATURES

Powder Patterns of Polytypic Substances

The possible powder patterns of a polytypic substance may be deduced from the discussion of the corresponding single crystal patterns, taking into account that

(i) For a powder pattern, reciprocal space is transformed into one dimension.

(ii) A great number of crystallites contribute to the diffraction pattern. Whether a line is visible or not, depends on the percentage of crystallites contributing to this line.

From this, the following conclusions may be drawn[21]: The streaks in single crystal patterns, caused by a thoroughly disordered crystal contribute in a powder pattern only to the background. Whenever the Ewald sphere starts to cut a diffuse streak, the background increases sharply, and then decreases with a long tail.

Long-periodic polytypes will not produce specific lines as the number of crystallites with the same long-periodic stacking sequence will be small. Characteristic powder patterns of a polytypic substance are (Fig. 7):

- the powder pattern of the family structure, corresponding to a thoroughly disordered sample;
- the powder patterns of pure basic polytypes (MDO structures).
  The pattern of any concrete sample is the weighted superposition of the quoted characteristic patterns.

For powder patterns, some special possibilities have to be taken into account:

- A non-family reflection may have the same Θ-value as a family reflection. Then, the relative intensities of this line may also change from sample to sample.
- A disordered sample that in spite of many stacking faults contains ordered regions with the same MDO structure, produces a powder pattern, which contains additionally to the sharp family reflections broadened lines. The maxima of these lines correspond to the positions of the strongest lines of that MDO structure.
- Sometimes a distinction of certain MDO structures by its powder patterns is practically impossible (e.g. in case of layer silicates). Then, for identification purposes, the set of all MDO polytypes is divided into groups of structures with the same powder pattern.
- Samples may be textured.

Diffraction Enhancement of Symmetry

In 1951, Ramsdell and Kohn[22] reported on the structure of a 10H SiC polytype, stacking sequence 3223. The space group was P3m1, but the diffraction pattern strictly hexagonal, point group 6/mmm. 15 years later, Ross

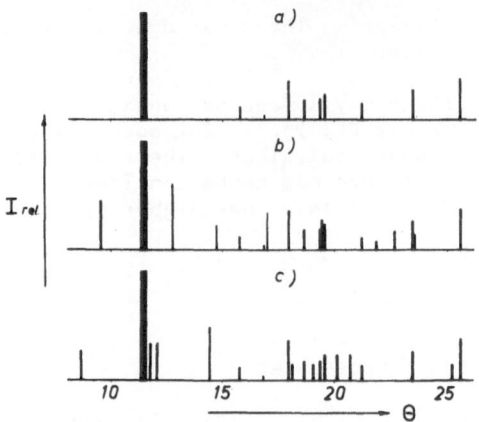

Fig. 7. Calculated powder patterns of three samples of copper oxalate.
(a) Family structure; (b) MDO_1 structure; (c) MDO_2 structure,
(after[21]).

et al.[23] stated that a triclinic mica polytype had a monoclinic diffraction
pattern.

The phenomenon that the symmetry of the diffraction pattern is higher
than expected from the point group of the crystal and Friedel's law was
given the name "diffraction enhancement of symmetry"[24]. This typical phenomenon of polytypes may be understood – at least in practical cases – as the
result of the presence of partial symmetries of the layer on the one side
and of regularities in the stacking on the other side. Let us check this
explanation in the classical case of 10H-SiC.

The single layer has the symmetry P(6)mm. For any stacking in SiC, the
3-fold axes and the mirror plane perpendicular to the main axes are total.
It follows that the diffraction symmetry is at least 3ml, taking Friedel's
law into account. Diffraction enhancement of symmetry means therefore that
the set of interatomic vectors has a 2-fold axis along $\vec{z}$. This axis results
from the 2-fold axis in the layer and from the mirror plane in the sequence
of origins of layers (compare Fig. 8). Thus, this kind of diffraction en-

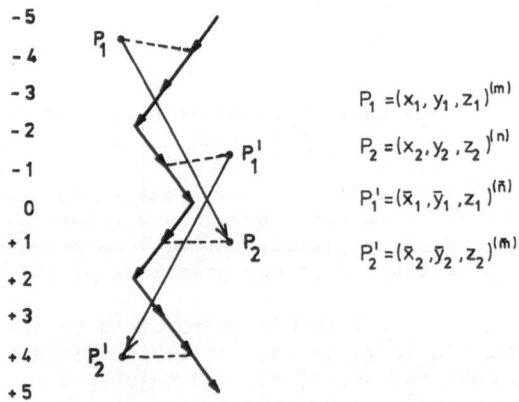

$$P_1 = (x_1, y_1, z_1)^{(m)}$$
$$P_2 = (x_2, y_2, z_2)^{(n)}$$
$$P_1' = (\bar{x}_1, \bar{y}_1, z_1)^{(\bar{n})}$$
$$P_2' = (\bar{x}_2, \bar{y}_2, z_2)^{(\bar{m})}$$

Fig. 8. Stacking sequence in the SiC polytype 10H. The structure is strictly trigonal, but the symmetry of the diffraction pattern is 6/mmm.
By a 2-fold (screw) rotation, any interatomic vector $\overrightarrow{P_1P_2}$ is transformed into another one $(\overrightarrow{P_1'P_2'})$. The coordinates of the points P_1,
P_2, P_1', P_2' refer to the origin of the respective layer, indicated
by an upper index..

hancement is not restricted to MX structures. For MX structures, all polytypes, whose Zhdanov sequences consist of two parts related to each other reflection - symmetrically, have a hexagonal diffraction pattern. There are,
however, other stacking sequences that have also diffraction enhancement of
symmetry, e.g. 24H:44422422 [25].

Homometric Structures among Polytypes

Homometrics are two or more structures that are neither congruent nor en-
antiomorphous, but would give identical diffraction patterns. The possibility
of this phenomenon was detected by Pauling and Shappell[26] in 1930 and later
investigated by Patterson, Hosemann and Bagchi, Buerger and others. Until
1970, however, no actual structure with a homometric mate was known; only
numerous theoretical examples had been constructed. Polytypes changed this
situation.

In 1970, Dornberger-Schiff and Farkas-Jahnke[27] reported that any ZnS
polytype with a non-symmetrical Zhdanov sequence has a homometric mate. This
means, most of the long-periodic MX polytypes have homometric mates. The
simplest examples are 9H:4221 and 4122, 24R:$(3221)_3$ and $(3122)_3$. Homometrics
have also been reported for MX_2 (CdI_2-type) structures[28-31]. Miklos[32] has
found homometrics among mesooctahedral micas.

Three general theorems on the existence of homometric polytypes have
been proved[27,31], which refer to polytypes consisting of parallel non-centrosymmetric layers or of two kinds of parallel layers. By means of the general
theorems, it has been shown that the known criteria for MX and MX_2 structures
do not cover all possible cases of homometrics in these compounds. It has
also been concluded that homometrics are to be expected for any polytypic
compound. Up to now, there is, however, no procedure to deduce all possibilities for homometrics of a certain compound, and there is no general procedure for a quick checking, whether a certain polytype has homometric mates
or not.

Desymmetrization in Polytypic Substances

Symmetry of real structures includes certain idealizations, i.e. deviations of the symmetry supposed. It is well known that physical properties
are correlated to both, symmetries and asymmetries. The crystal structures
of $K_4(Si_8O_{18})$ and $\gamma-H9_3S_2Cl_2$ show a special kind of violation of symmetry,
which is typical for polytypic substances and has been named desymmetrization[33,34].

A layer in a polytype has (true) partial symmetries, i.e. the environment of the layer does not have these symmetries. As a consequence, the environment has a disturbing influence that is different for different stackings. The disturbing effect is most drastic in pure MDO structures because
of the monotonous stacking of layers. In thoroughly disordered stackings,
the disturbing effects are neutralized, at least statistically. The phenomenon of desymmetrization may be expressed as the rule: the more disordered
a polytype is, the more symmetric it appears.

CONCLUDING REMARKS

The symmetry theory, provided by OD theory and based on the notions
partial symmetry and symmetry groupoid, describes the symmetry of the structural building units (layers) and their arrangement. For a family of polytypes, symmetry features of all polytypes as well as the properties of individual members are reflected.

It is well known that abstraction can lead to a simplification and thus to a deeper insight. With crystals, symmetry abstracts from the concrete arrangement of atoms and is considered as a most outstanding property, helpful for the solution of many problems. The symmetry theory described in this paper has proved to fascilitate the understanding of the phenomenon of polytypism and to be of inspiring value in the interpretation of the diffraction patterns of polytypes. It is advisable to start the investigation of a polytypic substance with the study of its symmetry.

REFERENCES

1. H. Baumhauer, Über die verschiedenen Modifikationen des Carborundums und die Erscheinung der Polytypie, Z.Kristallogr. 55:249 (1915).
2. A. Guinier, G. B. Bokij, K. Boll-Dornberger, J. M. Cowley, S. Ďurovič, H. Jagodzinski, P. Krishna, P. M. de Wolff, B. B. Zvyagin, D. E. Cox, P. Goodman, Th. Hahn, K. Kuchitsu, and S. C. Abrahams, Nomenclature of polytype structures. Report of the International Union of Crystallography Ad-hoc committee on the nomenclature of disordered, modulated and polytype structures, Acta Cryst. A40:399 (1984).
3. P. Krishna, ed., "Crystal growth and characterization of polytype structures", Pergamon Press, Oxford (1983).
4. K. Dornberger-Schiff, On Order-Disorder structures (OD structures), Acta Cryst. 9:593 (1956); OD structures – a game and a bit more, Krist.Tech. 14:1027 (1979).
5. S. Ďurovič and Z. Weiss, OD structures and polytypes, Bull.Minéral. 109:15 (1986).
6. H. Grell, How to choose OD layers, Acta Cryst. A40:95 (1984).
7. R. J. Angel, Polytypes and polytypism, Z.Kristallogr. 176:193 (1986).
8. K. Fichtner, On groupoids in crystallography, Comm.Math.Chem. 9:21 (1980).
9. H. Brandt, Über eine Verallgemeinerung des Gruppenbegriffs, Math.Ann. 96:360 (1926).
10. A. L. Patterson and J. S. Kasper, Close packing, in: International Tables for X-ray Crystallography, vol. II, pp. 342-354. The Kynoch Press, Birmingham (1959).
11. K. Fichtner, A new polytype notation for CdI_2 type structures. The thr symbols, Cryst.Res.Technol. 18:77 (1983).
12. K. Klepp and E. Parthe, FeB-CrB stacking variants in the system GdNi-TbNi, Acta Cryst. B37:495 (1981).
13. K. Fichtner, Generalizations of the hc notation, Z.Kristallogr. 167:261 (1984).
14. K. Fichtner, Short symbols for OD groupoid families, in: Symposium Special topics of X-ray crystal structure analysis, Hohengrün, Coll. contrib., p. 22 (1980).
15. K. Szulzewsky, Röntgenkristallstrukturanalyse der fehlgeordneten Kupfersalicylatmodifikation, in: Diss.Akad.Wiss.DDR, Berlin (1976).
16. K. Dornberger-Schiff, Geometrical properties of MDO polytypes and procedures for their derivation. I. General concept and applications to polytype families consisting of OD layers all of the same kind, Acta Cryst. A38:483 (1982).
17. B. B. Zvyagin, "Electron Diffraction Analysis of Clay Mineral Structures", Plenum Press, New York (1967).
18. K. Fichtner, On special ordered OD structures. A generalization of the concept of regular polytypes, in: Symposium Special topics of X-ray crystal structure analysis, Hohengrün, Coll.contrib., pp. 75-78 (1980).
19. K. Dornberger-Schiff, "Lehrgang über OD-Strukturen", Akademie-Verlag, Berlin (1966).
20. P. Sedlaček, Methoden zur Strukturanalyse OD-fehlgeordneter Substanzen, in: Diss. B, Akad.Wiss.DDR, Berlin (1979).

21. H. Fichtner-Schmittler, On some features of X-ray powder patterns of OD structures, Krist.Tech. 14:1079 (1979).

22. L. S. Ramsdell and J. A. Kohn, Disagreement between crystal symmetry and X-ray diffraction data as shown by a new type of silicon carbide, 10H, Acta Cryst. 4:111 (1951).

23. M. Ross, H. Takeda, and D. R. Wones, Mica polytypes: systematic description and identification, Science 151:191 (1966).

24. R. Sadanaga and H. Takeda, Monoclinic diffraction patterns produced by certain triclinic crystals and diffraction enhancement of symmetry, Acta Cryst. B24:144 (1968).

25. R. Sadanaga, K. Ohsumi, and T. Matsumoto, On the diffraction enhancement of symmetry in SiC, Proc.Japan Acad. 49:816 (1973).

26. L. Pauling and M. D. Shappell, The crystal structure of bixbyite and the c-modification of the sesquioxides, Z.Kristallogr. 75:128 (1930).

27. K. Dornberger-Schiff and M. Farkas-Jahnke, A direct method for the determination of polytype structures. I. Theoretical basis, Acta Cryst. A26:24 (1970).

28. P. C. Jain and G. C. Trigunayat, Resolution of ambiguities in Zhdanov notation: actual examples of homometric structures, Acta Cryst. A33:257 (1977).

29. G. K. Chadha, Homometric structures in MX_2-type compounds, Acta Cryst. A37:843 (1981).

30. K. Ohsumi and W. Nowacki, On enhanced vector symmetry and homometry in polytype, Bull.Minéral. 104:211 (1981).

31. K. Fichtner, Homometric polytypes in cadmium iodide, Acta Cryst. A42:98 (1986).

32. D. Mikloš, Personal communication (1981).

33. S. Ďurovič, Die Kristallstruktur des $K_4(Si_8O_{18})$: eine desymmetrisierte OD-Struktur, Acta Cryst. B30:2214 (1974).

34. S. Ďurovič, Desymmetrization of OD structures, Krist.Tech. 14:1047 (1979).

STRUCTURAL CHARACTERISTICS OF POLYTYPES

DELIVERED BY TEXTURE DIFFRACTION PATTERNS

B. B. Zvyagin

Institute of Ore Mineralogy (IGEM)
Academy of Sciences of the USSR
109017 Moscow, USSR

INTRODUCTION

The formation of textures in specimens is a natural consequence of the tendency of crystals highly anisotropic in shape to deposit on the substrate surface in a preferred orientation. The corresponding diffraction patterns have special advantages in the solution of problems of phase and structural analysis. Lamellar textures composed of crystals with the most developed face parallel to the substrate plane but randomly rotated about its normal are especially important. Such specimens satisfy the experimental requirements of transmission electron diffraction (TED) and patterns obtained under oblique position of the textures to the incident beam (oblique texture patterns - OTP's - Fig. 1) are valuable property of the electron diffraction method. They contain regular hkl-reflection sets distributed in two dimensions permitting to perform structural studies for a great variety of objects sometimes with location of hydrogen positions[1-4]. In this respect, they are the most effective means in application to fine-grained materials and imperfect structures. In view of the great efficiency of the texture patterns (TP's) they were recently introduced also in the X-ray diffraction practice[5-7].

In case of layer structures the TP's distinctly separate diffractional features relating to single layers and to layer sequences, characterizing families, groups and particular structures. Therefore, the TP's are most suitable for the structural study and identification of polytypes.

The Texture Reciprocal Lattice and Polytype Texture Patterns

Polytypes built of layers usually form textures with a basic coinciding with the layer cell a, b, γ. Therefore, it is convenient to divide the imagination of the direct lattice in two parts considering the spatial lattice parameters c, α, β which characterize the sequence of layers at the next step. It has been shown[4] that it is better to replace the angular parameters α, β by linear parameters, components of the normal projection c_n of the axis c on the plane ab:

$$x_n = (c/a)(\cos \beta - \cos\alpha \, \cos\gamma)/\sin^2 \gamma,$$

$$y_n = (c/b)(\cos \alpha - \cos\beta \, \cos \gamma)/\sin^2 \gamma.$$

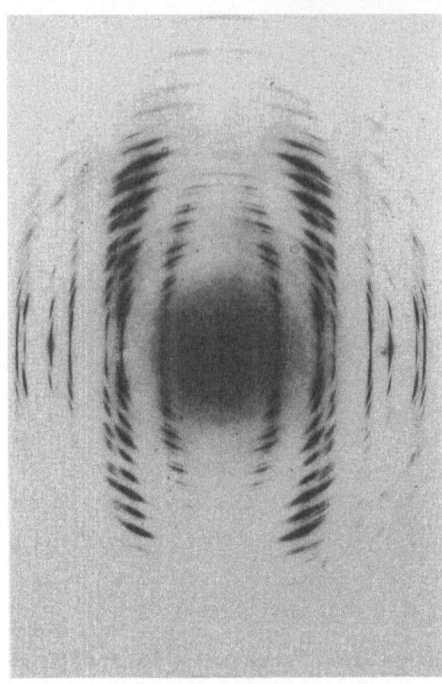

Fig. 1. Oblique-texture electron diffraction pattern of the phyllosilicate
         nacrite.

The set of x_n, y_n values differing by integers characterize all the c-vec-
tors differing by translations parallel to the plane ab permitting to make
the best choice of c according to the shortest projection c_n. The reciprocal lattice is then presented (in two respective steps) by a two-dimensional
net of intersection points of the hk reciprocal lattice rows parallel to
the c*-axis with the plane ab (1) and by the sequence of the reciprocal lattice points hkℓ along the hk-rows (2). The hk-net has periods $1/a \sin \gamma$,
$1/b \sin \gamma$ and an angle $\gamma' = \pi - \gamma$ between them depending thus only on layer
cell and independent of the layer stacking. The position of the hkℓ-points
along the hk-rows are defined by their distances from the plane ab

$$D_{hk\ell} = (ha^* \cos \beta^*/c^* + kb^* \cos \alpha^*/c^* + \ell)c^* = (-hx_n - ky_n + \ell)/d_{001}.$$

They are repeating at intervals $c^* = 1/d_{001}$ but their real values depend on
the projections of the axes a* and b* on the axis c*

$$a^* \cos \beta^*/c^* = -x_n, \qquad b^* \cos \alpha^*/c^* = -y_n$$

which are presenting additional relations connecting both lattices.

     The reciprocal lattice of a texture is a rotation body around the c*-
axis of the reciprocal lattice of a single crystal. The hk-rows describe
coaxial cylinders with radii

$$B_{hk} = (h^2/a^2 + k^2/b^2 - 2 hk \cos \gamma /ab)^{1/2}/\sin \gamma$$

and the points hkℓ describe rings lying on these cylinders at distance
$D_{hk\ell}$ from the plane ab.

At small wave length λ the respective diffraction patterns are approximated by plane cross-sections of the reciprocal lattice passing through the origin normal to the incident beam. This is fully justified for high-energy electron diffraction (HEED). For X-ray diffraction the curvature of Ewalds sphere (radius $1/\lambda$) is to be taken in account[5-7]. At angles φ between the pattern plane and ab-plane the reflections (ring sections) are distributed along ellipses (oblique cylinder sections) having small axes B_{hk} and great axes $A_{hk} = B_{hk}/\cos \varphi$. It is convenient to consider axial plane sections at $\varphi = \pi/2$ where the ellipses degenerate into pairs of directs parallel to c^*. The reflection positions are characterized by the B_{hk} and $D_{hk\ell}$ values which are the base for the geometry analysis of the TP's resulting in reflection indexing and cell determination.

Each reciprocal lattice point $hk\ell$ is supplied with the weight F or $|F|^2$ defining the reflection intensity. In case of layer structures in general and polytypes in particular the structure amplitude $F(hk\ell)$ may be expressed through the layer amplitudes Φ $(hk\ell)$ as

$$F(hk\ell) = \sum_{m=0}^{N-1} \Phi_m(hk\ell) \exp 2\pi i(hx_{0m} + ky_{0m} + \ell m/N) =$$

$$= \sum_{m=0}^{N-1} \Phi(h_m k_m \ell) \exp 2\pi i(hx_{0m} + ky_{0m} + \ell m/N).$$

Here, N is the number of layers per repeat, m – the index of successive layers, x_{0m}, y_{0m} and m/N are their origin coordinates, Φ are layer amplitudes for a fixed orientation[4]. Such expressions indicate which reflections are sensible or, on the contrary, insensible to layer stacking variations and stacking faults depending on the values of x_{0m}, y_{0m} characteristic for the polytypes of a family or group.

With the use of the B_{hk}, $D_{hk\ell}$ and $F(hk\ell)$-values it is easy to construct schemes of TP's imaging the ellipses as direct lines hk parallel to the 00ℓ-line at distances B_{hk} from it and the reflections $hk\ell$ – as direct pieces of a length proportional to F or $|F|^2$ intersecting the line hk normal to it at distances $D_{hk\ell}$ from the reference zero line belonging to the plane ab.

In case of phyllosilicate polytypes information of different kind is presented by "ellipses" I, II and V (numbered in the order of increasing B_{hk} values). The "ellipsis" V, for which both indices hk= 06,33 (referred to an orthogonal centred basis with $b=a\sqrt{3}$) are multiples of 3, characterizes the interlayer distances distinguishing for example layers 1:1 and 2:1. The reflections of the "ellipsis" II (hk=13, 20, only k=3n) are distinguishing polytype groups differing by their structural projections on the plane ac, revealing deviations of the monoclinic angle from ideal value defined by the relation $-c/a.\cos\beta = 1/3$. The reflections of the "ellipsis" I (hk=02, 11, k≠3n) are in general characteristic for particular polytypes (Fig. 2).

Polytypism Problems the Solution of which was Inspired and Realized by the Use of Texture Patterns

The theoretical derivation of polytype structures supplied with calculated diffractional characteristics[3,4,8,9] provides an effective scientific cognition tool possessing a prediction power. In particular it is a highest form of the trial and error method.

Thus, the need for the interpretation of the TP's of the kaolin mineral nacrite in order to establish the real stacking of the 1:1 layers in its structure was the reason for a derivation of all 36 kaolinite polytypes. In the course of this work the main features of the symbolism describing all phyllosilicate structures were elaborated. The diffractional characteristics

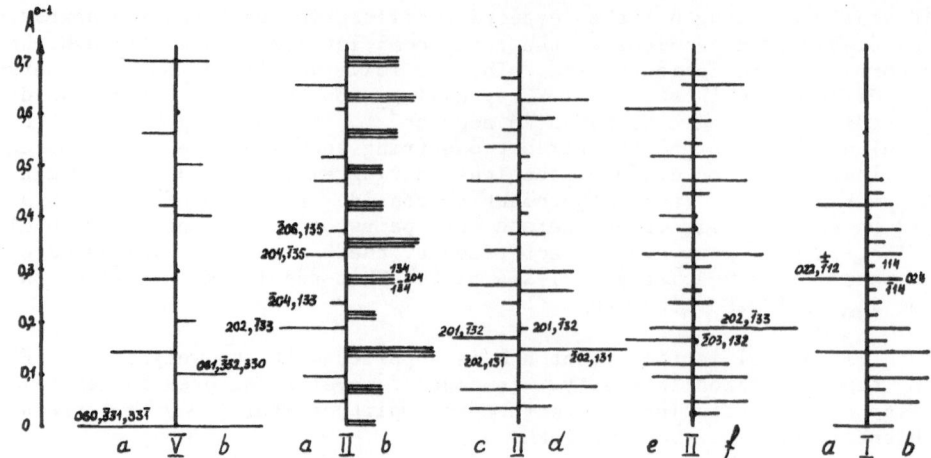

Fig. 2. Schemes of "ellipses" V, II and I characterizing different features
of phyllosilicate structures; sequences of 1:1 and 2:1 layers with
7Å and 10Å interlayer distances (Va and Vb);
projections on the plane ac for the polytype groups B and D, at an
ideal and increased angle ß examplified by the kaolin minerals
dickite and nacrite (IIa and IIb, respectively);
the same projections for mica and pyrophyllite structures with the
same parity of layer orientations (IIc and IId);
the same projections for chlorites composed of packets (layer combinations 2:1+0:1) A and F[4] (IIe and IIf), note the reverse intensity relations for some reflection pairs in IIc and IId, IIe and
IIf;
the real stacking of 1:1 layers in the strictly periodic homogeneous
structures of lizardite 1T and dickite 2M (Ia and Ib).

calculated for kaolinite polytypes enabled to resolve the ambiguity concerning this polytype structure and to refine it in result of detailed analysis of intensities[3,4]. Furthermore ·they served later as a base for the subsequent identification of halloysite as an original kaolinite polytype ideal-
ly having a peculiar two-layer monoclinic structure[4].

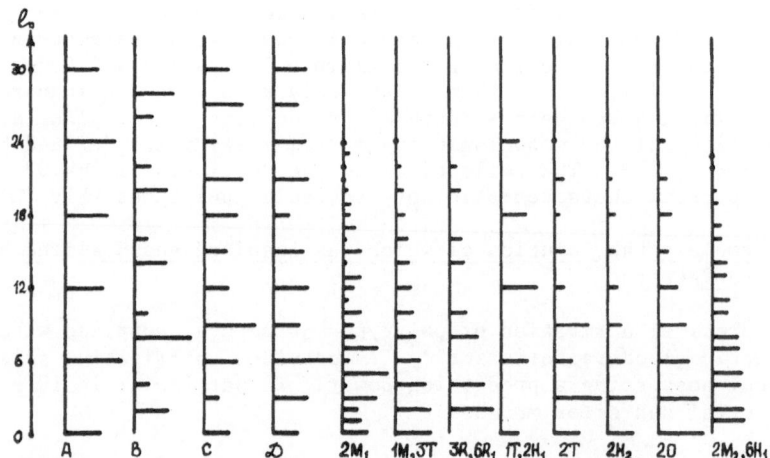

Fig. 3. Distinguishing diffraction features of the serpentine polytype
groups A-D presented by the ellipsis II and of particular regular
polytypes $2M_1$ - $6H_1$ presented by the ellipsis I.

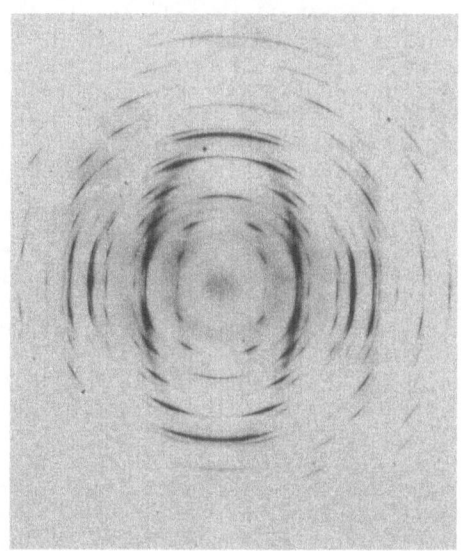

Fig. 4. Oblique-texture electron diffraction pattern of a lizardite polytype combination $1T+2H_1$.

The use of TP's has accompanied the development of the polytypism theory
for trioctahedral phyllosilicates built of 1:1 layers (serpentines). If the
reflections of ellipsis II clearly define the serpentine polytype groups
(A – D), those of the ellipsis I sometimes do not distinguish particular
polytypes (Fig. 3). This is the case for polytypes 1T and $2H_1$ (Fig. 4),
although they belong to different polytype groups (A and D). The comparison
of the reflection intensities of the ellipses I and II has permitted not only
to discover a new lizardite (Mg-serpentine) polytype $2H_1$ but to identify also
mixtures of lizardites 1T and $2H_1$ (ref. 10). It has been also established
by means of TP's that Zn bearing phyllosilicates may have serpentine like
structures[4].

The TP's have revealed the principle differences of the stackings of
the 2:1 layers which form the mica and pyrophyllite structures. These phyllosilicates present thus two essentially different polytype families and the

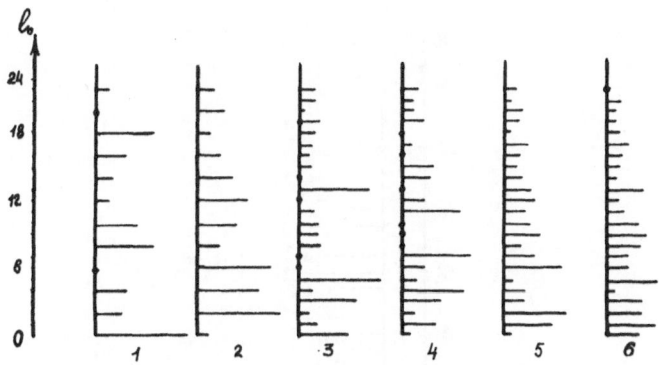

Fig. 5. Scheme of the ellipsis I of pyrophyllite polytypes, the scheme 3
being characteristic for the polytype 2M identified in case of
ferripyrophyllite, too.

derivation of the original pyrophyllite-talc family is completely a merit of the electron diffraction practice with the use of TP's[4]. The diffraction features displayed by the ellipsis I (Fig. 5) of one of the 2M polytypes are so peculiar that could be fixed in case of a much less perfect structure giving poor patterns with few diffuse reflections. In such a way a new mineral with unexpected composition has been discovered. It was ferripyrophyllite isostructural with pyrophyllite but containing Fe cations instead of Al in the octahedral positions[4].

Even more surprising was the discovery of the minerals chapmanite and bismutoferrite composed of phyllosilicate 1:1 layers similar to those of kaolinite. The complete replacement of Fe for Al is there accompanied by the elimination of hydrogenes of the layer surfaces and introduction of interlayer cations Sb or Bi. Under such conditions the layer stacking of the polytype 1M is realized which is improbable at the kaolinite composition. The EDTP's were sufficient for a further structural refinement of these phyllosilicates[4].

The EDTP's have revealed the existence of peculiar dioctahedral 2:1 layers where the tetrahedral sheets are related by a two-fold axis of the octahedral sheet or, in other terms, layers with vacant cis-octahedra. Before this only centrosymmetrical 2:1 layers (with vacant trans-octahedra) were known for the phyllosilicates although the possibility of non-centrosymmetrical 2:1 layers has been theoretically considered and polytypes containing such layers were deduced[4]. In a more wide sense dioctahedral layers are attributed to the so-called meso-octahedral layers where two octahedral sites have the same, while the third - another cation occupation[11], and 14 MDO-polytypes (6 centrosymmetrical and 8 non-centrosymmetrical) were deduced for mica stackings of such layers[12]. By means of EDTP's one of these non-centrosymmetric polytypes has been identified among smectites brought in an ordered state after potassium treatment and subsequent cycle of hydration-dehydration procedures[13]. This opened a new era in the study of these crystallochemically important and complicated expandable clay minerals. The same polytype [150.. (ref.[4]) or |5.1| (ref.[12])] has been found for unique mica according to the peculiar features of geometry and intensities of the EDTP's (Fig. 6).

The EDTP's proved their efficiency in the study of polytypism of astrophyllite, molybdenite and graphite. In the latter case evidence has been obtained on the specific nature of the polytypes. The EDTP's always contained reflections of two graphite polytypes 2H and 3R as an indication of the reversible transformation 2H - 3R at the sample grinding for the prep-

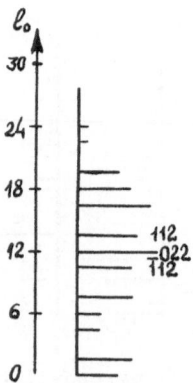

Fig. 6. Scheme of the ellipsis I of the mica polytype 150... (ref.[14]).

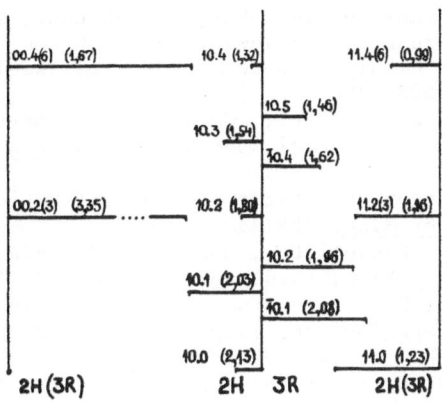

Fig. 7. Scheme of the texture pattern of a combination of graphite polytypes 2H and 3R.

aration of specimens. The absence of an intermediate background between
reflections and intensity relations for the reflections 2H- and 3R- with
indices h-k ≠ 3n and h-k = 3n (Fig. 7) have shown that both polytypes are
present as zones in one and the same crystal. The stacking faults are thus
boundaries between segregations of layers forming areas of coherent scattering of diffracted waves. The absence of other polytypes (of the in principle infinite number of them) together with the absence of random stacking
faults may be interpreted as an indication that the layer pairs of polytypes
2H and 3R are physically not equivalent although pure geometrically they are
fairly equivalent. Such a feature may be expected under conditions when
single atomic planes serve as polytype layeers. The physical state of a layer
depends then on the arrangement of adjacent layers at both sides of it. When
different polytypes alternate in the same crystals the relative polytype contents has to be evaluated according not to the intensities but to the square
roots of the intensities of the respective reflections[15].

It should be noted that the general conclusions on the layer kinds and
their stacking are usually obtained by visual inspection of TP's. The more
detailed analysis of the TP's may result in structural refinements delivering
exact atomic positions and revealing structural distortions characteristic
for particular polytypes both as preconditions and as consequences of their
realization.

REFERENCES

1. Z. G. Pinsker, "Electron Diffraction", Butterworth, London (1953).
2. B. K. Vainshtein, "Structure Analysis by Electron Diffraction", Pergamon
   Press, London (1964).
3. B. B. Zvyagin, "Electron Diffraction Analysis of Clay Mineral Structures",
   Plenum Press, New York (1967).
4. B. B. Zvyagin, Z. V. Vrublevskaya, A. P. Zhukhlistov, O. V. Sidorenko,
   S. V. Soboleva, and A. F. Fedotov, "High-voltage Electron Diffraction
   in the Study of Layered Minerals", Nauka Press, Moscow (1979).
5. G. A. Krinary, On the possibilities to use oriented specimens for recording of non-basal X-ray reflections of fine-grained layer silicates,
   in: Crystal Chemistry of Minerals and Geological Problems, Nauka Press,
   Moscow (1975).
6. J. Mamy and J.-P.Gaultier, Evolution structurale de la montmorillonite
   associée au phénomène de fixation irréversible du potassium, Ann.
   Agron. 27:1 (1976).

7. A. Plancon, F. Rousseaux, D. Tchoubar, C. Tchoubar, G. Krinari, and V. A. Drits, Recording and calculation of hk rod intensities in case of diffraction by highly oriented powders of lamellar samples, J.Appl.Cryst. 15:509 (1982).

8. K. Dornberger-Schiff and H. Grell, Geometrical properties of MDO polytypes and procedures for their derivation, Acta Cryst. A38:483 (1982).

9. S. Ďurovič, OD-Charakter, Polytypie und Identifikationen von Schichtsilikaten, Fortschr.Miner. 59:191 (1981).

10. M. A. Litsarev, A. P. Zhukhlistov, and B. B. Zvyagin, The first find of lizardite $2H_1$, Doklady Akad.Nauk 277:188 (1984).

11. K. Dornberger-Schiff, K.-O. Backhaus, and S. Ďurovič, Polytypism of micas: OD-interpretation, stacking symbols, symmetry relations, Clays and Clay Min. 30:364 (1982).

12. K. O. Backhaus and S. Ďurovič, Polytypism of micas. I. MDO polytypes and their derivation, Clays and Clay Min. 32:453 (1984).

13. S. I. Tsipursky and V. A. Drits, The distribution of octahedral cations in the 2:1 layers of dioctahedral smectites studied by oblique-texture electron diffraction, Clay Minerals 19:177 (1984).

14. B. B. Zvyagin, V. T. Rabotnov, O. V. Sidorenko, and D. D. Kotelnikov, A unique mica composed of non-centrosymmetrical layers, Izvestiya AN SSSR (ser. geol.) 5:121 (1985).

15. F. V. Chukhrov, B. B. Zvyagin, A. P. Zhukhlistov, N. I. Organova, and L. P. Ermilova, Characterization of structural features of natural graphites, Izvestiya AN SSSR (ser. geol.) 7:3 (1986).

THE X-RAY METHOD FOR DETERMINING CRYSTALLINE MULTILAYER

POLYTYPE STRUCTURES IN METAL ALLOYS

B. I. Nikolin

Institute of Metal Physics
Academy of Sciences of the Ukrainian SSR
Kiev, USSR

During many years (1912-1963) polytypism was known only for semiconductors, minerals, but not for metals. Recently, in some alloys with a low stacking-fault energy ($\gamma < 30$ mJ/m^2) martensitic phases with multilayer (long-period) crystal structures unusual for metals and alloys were discovered. Thus, e.g., in Fe-based alloys (Fe-Mn-C, Fe-Mn-N, Fe-Mn-Cu) the ε'-phase with the 18R, structure appears, in the Cu-based alloys (Cu-Al, Cu-Al-Ni, Cu-Si) the martensite phases with predominantly 9R structure are formed and in Co binary alloys (Co-Cu, Co-Ti, Co-C, Co-Ta, Co-Nb, Co-Sn, Co-W) $7T_1$ and other multilayer structures exist. It is impossible to determine polytype structures formed in metal alloys by martensite transformation by classical methods, by X-ray Weissenberg, rotation or oscillation methods. This difficulty is due to the impossibility to obtain single crystal samples without residual matrix phase and size influence on crystal structure of polytypes. To solve this problem we used X-ray rotation and oscillation methods with arbitrary orientation of samples ($\emptyset \approx 0.8 - 1.0$ mm)[1,2]. The application of this method allows to obtain X-ray diffraction such as for equi-inclination Weissenberg photographs and using X-ray maxima of residual matrix phase as standard points to calculate the number of close-packed layers N in the unit cell, and to determine the crystal structure of martensite polytypes. The main advantage of the proposed method is that according Wilson theory the reflections with h-k=3n $\pm$ 1 are broadened due to stacking faults and between them diffuse streaks arise. As a result, such reflections have constant two indices $\underline{h}$ and $\underline{k}$ and variable $\underline{\ell}$. In reciprocal lattice there are corresponding nodes rows with constant $\underline{h}$ and $\underline{k}$ indices. Such reflections with diffusion streaks represent fastoon of equi-inclination Weissenberg X-ray photograph. The value N was determined by two methods:

1. From the number of polytype reflections $\underline{n}$ located between the neighbour reflections of the residual f.c.c. phase along one streak on an X-ray photograph using the relations N = 3n for the rhombohedral (R) and N=n for the hexagonal (H) and trigonal (T) lattices.

2. From the distance between the neighbour polytype nodes (Ln) located along the row 10.ℓ of the reciprocal lattice using the formulas $N = 3L_o/Ln$ for the R-lattice and $N = L_o/Ln$ for the H, T lattices. The distance between the neighbour f.c.c. lattice nodes along the 10.ℓ row of the reciprocal lattice is given by the quantity $L_o = 1/d_{111fcc}$.

It is necessary to calculate $d_{10.\ell}$ and $d_{10.\ell+1}$ the interplanar distances for the neighbour polytype reflections located along one diffuse streak and using the relation

$$Ln = \left| (\frac{1}{d_{10.\ell+1}})^2_{NR} - (\frac{1}{d_{112}})^2_{fcc} \right|^{-1/2} - \left| (\frac{1}{d_{10.\ell}})^2_{NR} - (\frac{1}{d_{112}})^2_{fcc} \right|^{-1/2}$$

to find Ln and consequently N and parameter c.

Two other parameters are equal $a=b=2d_{220\ fcc}$. R-rhombohedral, H-hexagonal and T-trigonal symmetry are determined from the location of $10.\ell$ nodes row of the reciprocal lattice with $+\ell$ and $-\ell$. The nodes with $+\ell$ and $-\ell$ are located symmetrically for H and T-lattices and asymmetrically for R-lattices. It should be noted that polytype martensite structures are not perfect but contain randomly distributed stacking faults with density $\alpha = 0.10 - 0.15$.

REFERENCES

1. B. I. Nikolin, N. N. Shevchenko, Scripta Met. 14:467 (1980).
2. B. I. Nikolin, "Multilayer Structures and Polytypism in Metallic Alloys", Naukova Dumka, Kiev (1984).

THE ANALYSIS OF DISORDER POLYTYPE STRUCTURES

BY STATISTICAL PARAMETERS

M. J. Kozielski and A. Tomaszewicz

Institute of Physics
Warsaw Technical University
ul. Koszykowa, 75, 00-662 Warsaw, Poland

The method of investigation of the polytype disorder of cp structures is presented. The curves of intensity of X-ray of ($10.\ell$) reflections obtained from the oscillation crystal film are analytically elaborated. The statistical parameters derived by the Farkas method are plotted in special graphs.

The positions of experimental points in the graphs give immediate information on the layer orderings in the general structure of the investigated crystal.

INTRODUCTION

A^2B^6 compounds usually crystallize in the basic close-packed structures, i.e. sphaleritte 3C and wurtzitte 2H, but in case of ZnS other cp polytypes can also be obtained. Various ZnS polytypes are grown under special conditions of the vapour crystal growth[1-6]. Most of the crystals have one dimensional disorder structure[6]. The doped ZnS crystals and also some of A^2B^6 solid solutions show usually a complex disorder structure with the specific orderings of the layers as short-period LH polytype unit cells (where L=2,4...)[7-10]. Other solid solutions, such as $Zn_{1-x}Mn_xS$ obtained from the melt show, in general, equal structure type (ref.[11,12]). Also during thermal treatment the polytype transformations run through the complex disorder structures[13-17].

It is evident that the one-dimensional disorder is a very common phenomenon in polytype crystals. But there are not many publications in this field. The early works dealt with the problem how typical stacking faults, e.g. growth faults or intrinsic faults, can influence X-ray diffraction patterns[18-20].

Jagodzinski[20] has introduced the probability parameters for the description of these structures. Then, this method of description was developed by Patterson[23] and by Kakinoki and Komura[24].

Farkas-Jahnke[6,24] showed that the definitions of intrinsic, or growth faults as well as other faults of this type have essential limitations. The same fault may correspond to quite different local stacking in different initial basic structures. But in some cases of studying the nearly-perfect structures it is useful to single the deformation, displacement, or other faults of that type (ref. 14,25,26).

Farkas-Jahnke[6] has introduced other structural parameters, π (m,p), the relative rate of different configurations occurrence of short sequences of the layers; m and p being the stacking vector coordinates (m =-1,0,+1 and p = 1,2,3...).

For the investigation of the phenomenon of the polytype ordering of the layers in general disorder structure an interesting question is how many layers are ordered in specific polytype units. For that purpose other method of disorder structure description was introduced by Kozielski and co-workers (ref. 7,8,9,27,28). There are distinguished from all possible configurations of sequences of the layers, only the special ones, which are characteristic for LH polytypes, e.g. 2H, 4H, 6H, 8H, and 10H. The structural parameters: P_{LH} - the probability of 1H polytype occurrence and α - the "hexagonality" which is the ratio of the number of hexagonal stacked layers to the total number of the layers are used.

The automatisation of computing of the structural parameters due to Farkas method is an important advantage: however the information on the polytype orderings of layers is hard to obtain by the method. Therefore, the authors of this work have made the attempt to propose a unified method, which includes the advantage of Farkas method and to provide us with informations on polytypic orderings of the layers.

π_ℓ Parameters

The layer labels: "e" (equal) and "u" (unequal) as first introduced by Hirth-Lotte[29], Tiwari et al.[30] and Panday and Krishna[31] will be used.

From all possible π (m,p) Farkas parameters we will use π (0,p) only. For simplicity we label:

$$\pi_\ell = \pi (0,p), \tag{1}$$

where ℓ=p, layer spacing in a direction as measured in terms of interlayer distance taken as a unit of measure. π_ℓ is equal to the relative number of "e" layer pairs (A-A, B-B or C-C), the layers being ℓ interlayer distances away. It is assumed that interlayer distances of adjacent layers are equal although the stacking of cp layers is different in a given material.

It can easily be seen that $1-\pi_\ell$ means the relative number of layer pairs that are ℓ interlayer distances away, and in the "u" position. π_1 is equal to zero in all cp structures. π_2 is equal to the relative number of hexagonally "h" stacked layers, but on the other hand, π_2 is also equal to the value of α, $\pi_2 = \alpha$.

The Dependence of π_3 on Hexagonality

Parameter π_3 is equal to the relative number of pairs of neighbouring cubic layers, cc. Simultaneously, it also means the probability, P(cc), that orbitrary two neighbouring layers are in cubic stacking.

$$\pi_3 = P(cc). \tag{2}$$

The simple rules of the probability theory lead to the conclusion, that the possible values of π_3 must be limited as follows: $1 - 2 \alpha < \pi_3 < 1 - \alpha$ for $\alpha < 1/2$, and $0 < \pi_3 < 1 - \alpha$, for $\alpha > 1/2$.

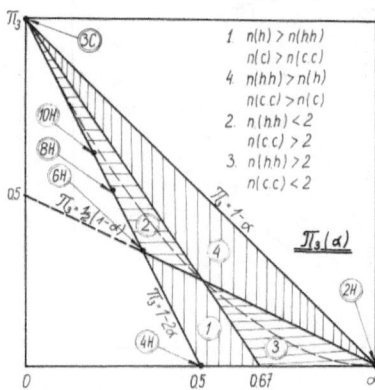

Fig. 1. The characteristic areas of $\pi_3(\alpha)$ and possible structures in the areas.

Fig. 1 represents the graph of π_3 vs. α. All possible values of π_3 are within the triangle limited by lines: $\pi_3 = 1-2\alpha$. $\pi_3 = 0$ and $\pi_3 = 1-\alpha$.

The fully disorder structure "DS" occurs when the h and c layers are fully randomly distributed in general structure (the model of DS structure is presented by Kozielski and Tomaszewicz[32]). In this case, the probability $P(cc)$ is $(1-\alpha)^2$, or

$$\pi_3 = (1-\alpha)^2. \tag{3}$$

Then the full disorder manifests itself in the upper curve in Fig. 1.

On the other hand, fully ordered polytypes of LH type (L=2,4,6...) have values that are on $\pi_3 = 1 - 2\alpha$ line. The line represents structures with each layer h surrounded by c layers. The area of possible π_3 values can be divided into the following areas in Fig. 1:

1. Individual h layers are in predominance over the blocked ones.

2. In blocks; on the average, h layers are less than 2 and c layers are more than 2.

3. As in 2., but reversely.

4. Layers aggregated in blocks cccc..., and hhhh... prevail over the individual ones; $\pi_3 = 1-\alpha$ shows that all c layers form a block and the same is valid for h layers. It means that the structures are dissociated into 3C and 2H polytypes.

The Dependence of π_4 on Hexagonality

The parameter π_4 is the sum of probabilities of occurrence of the following layer sequences

$$\pi_4 = P(chc) + P(hch) + P(hhh). \tag{4}$$

It is easy to prove that all possible values of π_4 are inside of the parallelogram of (π_4,α): (0.0), (1,1/2), (1,1) and (0,1/2), Fig. 2. Now, the interpretation of the particular ranges in the graph is more difficult. The formula (4) is more complex then respective formula (3) for π_3. But the particular lines and points can give good interpretation.

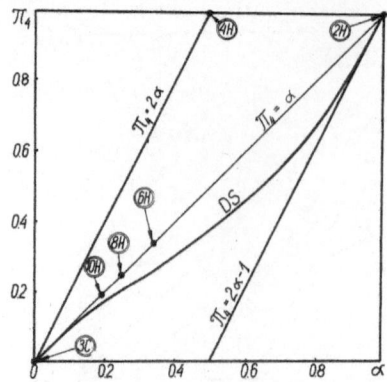

Fig. 2. π_4 vs. α and related structures.

The polytypes: 3C, 10H, 8H, 6H and 2H are distributed on the line $\pi_4 = \alpha$, and the 4H polytype is represented by the point $(1, 1/2)$. The mixed structures 3C+4H are distributed on the line $\pi_4 = 2\alpha$, and the 4H+2H structures are placed in the line $\pi_4 = 1$.

The DS structures are shown by the curve given by the equation:

$$\pi_4 = \alpha \, [\, 1 - \alpha \, (\, 1 - \alpha \,) \,]. \tag{5}$$

The Dependence of π_5 on Hexagonality

The formula becomes more complicated when interlayer distances are increasing, eg..

$$\pi_5 = P(ccch) + P(chhc) + P(cchh) + P(hhcc) + P(hccc) \ , \tag{6}$$

Our analysis will be limited to $\ell = 5$, because the difficulties in the interpretation of $\pi_\ell (\alpha)$ graphs become greater with the increase of ℓ-value; also there is higher influence of experimental errors.

Experimental Graphs of $\pi_\ell \, (\alpha)$

The π_ℓ parameters for $A^2 B^6$ solid solutions and for ZnS:Al crystals are computed by the Farkas method (ref.6). All the crystals are analysed pre-

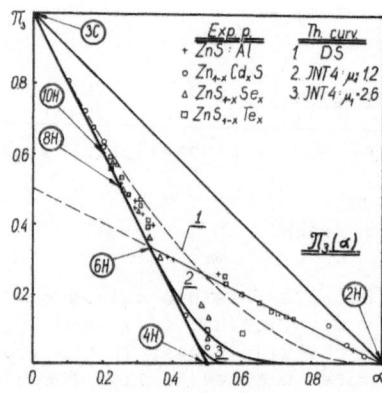

Fig. 3. Experimental points of π_3 parameter and theoretical curves.

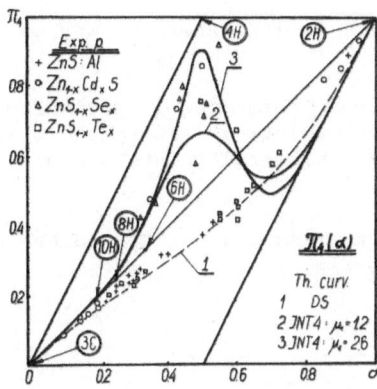

Fig. 4. π_4 vs. α; experimental points and the model curves.

viously by the P_{LH} parameters method[7-9] and described in frame of the statistical model[28]. The π_ℓ parameters of these crystals are plotted in Figs 3, 4 and 5.

As shown in Fig. 3, the expxerimental points of π_3 are distributed inside the area 1) in the range of α : $(0,1/2)$. Then hexagonal layers are mostly individually distributed and are separated by the cubic layers. The experimental points are adjacent to the "order polytype line" $\pi_3 = 1-\alpha$, and become more distant from the "disorder curve" $\pi_3 = (1-\alpha)^2$. It means that the investigated structures show high tendency to ordering into LH polytypes which is in agreement with the previous analysis[7-9].

This conclusion is confirmed by the statistical model ("INT-4" model)[28]. But the agreement with theoretical curves appears in the range of $\alpha:(0.1/2)$. The distinct deviation of the theoretical curves occurs for $\alpha < 1/2$. This fact was indicated by authors of the previous work[28]. Now it is possible to interpret this deviation.

The experimental points are close to the line $\pi_3 = 1/2(1-\alpha)$. But this line corresponds to the structures in which c layers appear in blocks, and the average number of layers in a block is 2. The assumed type of configurational interaction of the layers in the model keep away such blocking of

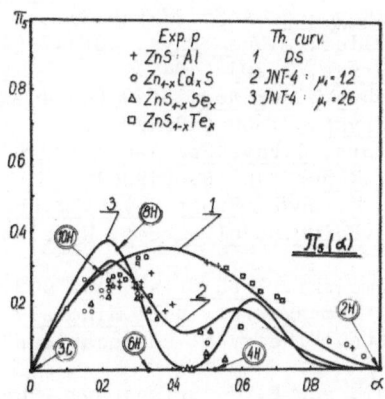

Fig. 5. π_5 vs. α : experimental points and the model curves.

c layers in this range of α. It indicates the way of the improvement of the structural models.

The dependence of π_4 on α parameter is presented in Fig. 4. The distribution of the experimental points in the graph confirms the above interferences. Generally, the experimental points are distributed along the model curves.

The agreement of the experimental points and the model curves is also observed for π_5 parameters, Fig. 5.

CONCLUSION

The method of construction of $\pi_\ell(\alpha)$ graphs is useful for the analysis of complex disorder polytype structures. The positions of experimental points in the graphs give immediate information of the type of the layer orderings in general disorder structures. The maximum information gives the $\pi_3(\alpha)$ graph. The other graphs give the additional information on the investigated structures. The analysis of the positions of experimental points in the graphs allow to verify structural models.

REFERENCES

1. K. Pátek, Czech.J.Phys. B11:686 (1961).
2. S. Madrix and J. T. Steinberger, J.Appl.Phys. 41:5339 (1970).
3. s. Madrix, J.Appl.Cryst. 17:328 (1984).
4. S. Madrix, Bull.Mineralog. 109:131 (1986).
5. D. C. Reynolds and S. J. Czyzak, Phys.Rev. 79:543 (1950).
6. M. Farkas-Jahnke, Acta Cryst. B29:407 (1973).
7. M. J. Kozielski, J.Cryst.Growth 30:86 (1975).
8. M. J. Kozielski, Bull.Acad.Polon.Sci. XXVI:193 (1978).
9. M. J. Kozielski, Bull.Acad.Polon.Sci. XXIV:367 (1976).
10. W. Palosz, M. J. Kozielski, and B. Palosz, J.Cryst.Growth 58:185 (1982).
11. S. Kaczmarek, E. Michalski, M. Demianiuk, and J. Zmija, Acta Phys.Polon. A59:723 (1981).
12. E. Michalski, M. Demianiuk, and S. Kaczmarek, Electron Technol. 13:3 (1980).
13. H. Gobrecht, H. Nelkowski, J. W. Baars, and G. Brandt, Proc.Int.Conf. Lumin. Budapest 1086 (1966).
14. D. Pandey, Acta Cryst. B40:567 (1984).
15. D. Pandey and P. Krishna, Progr.Cryst.Gr.Charact. 7:213 (1983).
16. D. Pandey, V. K. Kabra, and S. Lele, Bull.Mineral. 109:49 (1986).
17. A. Krol, M. J. Kozielski, and W. Nazarewicz, Bull.Mineral. 109:81 (1986)
18. A. J. C. Wilson, Proc.Roy.Soc. A180:77 (1942).
19. S. Hendricks and E. Teller, J.Chem.Phys. 10:147 (1942).
20. H. Jagodzinski, Acta Cryst. 2:201 (1949).
21. H. Jagodzinski, Neu.Jahrb.Miner.Mon. 3:49 (1954).
22. H. Jagodzinski, Acta Cryst. 7:300 (1954).
23. J. Kakinoki and Y. Komura, J.Phys.Soc.Jap. 7:30·(1952).
24. M. Farkas-Jahnke, Bull.Miner. 109:69 (1986).
25. M. T. Sebastian and P. Krishna, Progr.Cryst.Gr.Charact. 14:103 (1987).
26. E. Michalski, Int.Conf.Polyt.Modul.Struct. Wroclaw (1986); abstract only.
27. B. Palosz and J. Przedmojski, Acta Cryst. A32:409 (1976).
28. M. J. Kozielski and A. Tomaszewicz, Bull.Mineral. 109:89 (1986).
29. J. P. Hirth and J. Lotte, "Theory of Dislocations", McGraw Hill, Inc. New York (1968).
30. R. S. Tiwari, A. K. Rai, and P. N. Srivastava, Phys.Stat.Sol.(a) 31:419 (1975).
31. D. Pandey and P. Krishna, Phys.Lett. 51A:209 (1975).

VII. DATA ACQUISITION

BRAGG REFLECTION ANALYSIS FOR POWDERED SAMPLES

G. Stanisz, J. Holender, and J. Sołtys

Institute of Physics
Jagellonian University
ul. Reymonta 4, Cracow, Poland

INTRODUCTION

Precise quantitative X-ray analysis often requires an analytical description of the Bragg reflection profile. In the recent papers[1-4] the
authors were studying the usefulness of some selected functions, which could
have approximated the X-ray diffraction patterns. It seems very important
to further analysis which of those functions is the most efficient one.
The main question of the present paper is to try to solve this problem.

EXPERIMENTAL

The whole diffraction pattern of Si (purity: 99.999%, grain size:
0.08-0.1 mm) was taken using DRON X-ray diffractometer with Fe $K\alpha_1$, $K\alpha_2$,
Kß radiation. The intensity was measured using step scanning regime
(0.01° in 2 theta scale).

PROFILE ANALYSIS

Ten functions were taken into account: Gauss(G), Lorentz(L), Modified
Lorentz(ML), Intermediate Lorentz(IL), Pearson VII(PVII), Pseudo-Voigt(PV),
Asymmetric Pseudo-Voigt(PVA), Rational (RF), Asymmetric Rational (RFA) and
Mignot-Rondot(MR) functions. They were fitted to each Kα and Kß peak. Two
different procedures were followed while fitting the double reflections
(α_1, α_2):

- independent parameters for both doublet profiles were assumed (two single
peaks)

- for α_2 peak the same shape, intensity 0.5 of α_1's were taken on and doublet
separation was evaluated on the basis of the Bragg equation[5-6].

The background was assumed linear and was added to each function. It means,
that two more parameters have been calculated. Each function was fitted
using standard SIMPLEX procedure with the following criteria:

$$FCN = [\Sigma(y_i(exp) - y_i(calc))^2/\sigma_i^2]/(N - NP - 1) ,$$

where y_i - intensity at the i-th step, σ_i - standard deviation of i-th intensity, N - number of experimental points, NP - number of fitted parameters.

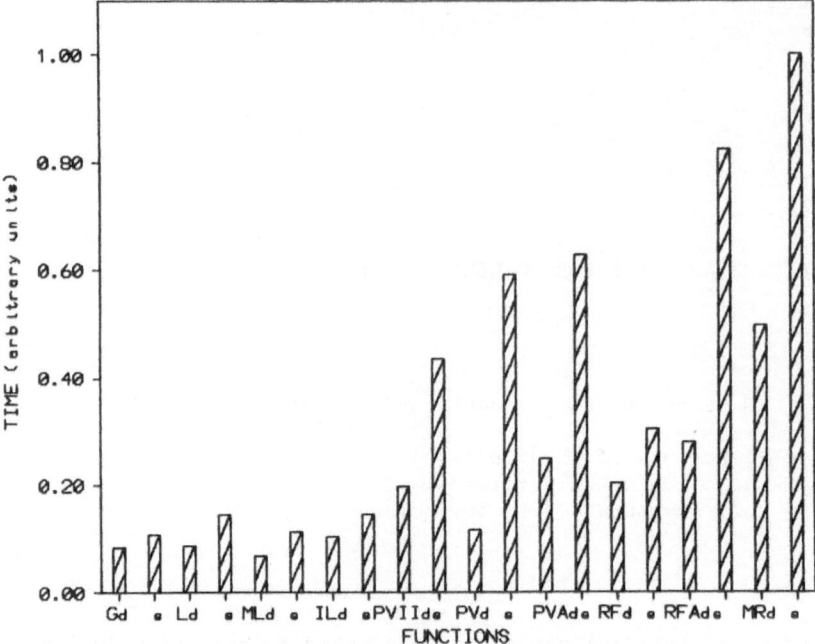

Fig. 1.   Time of fitting the whole spectrum.

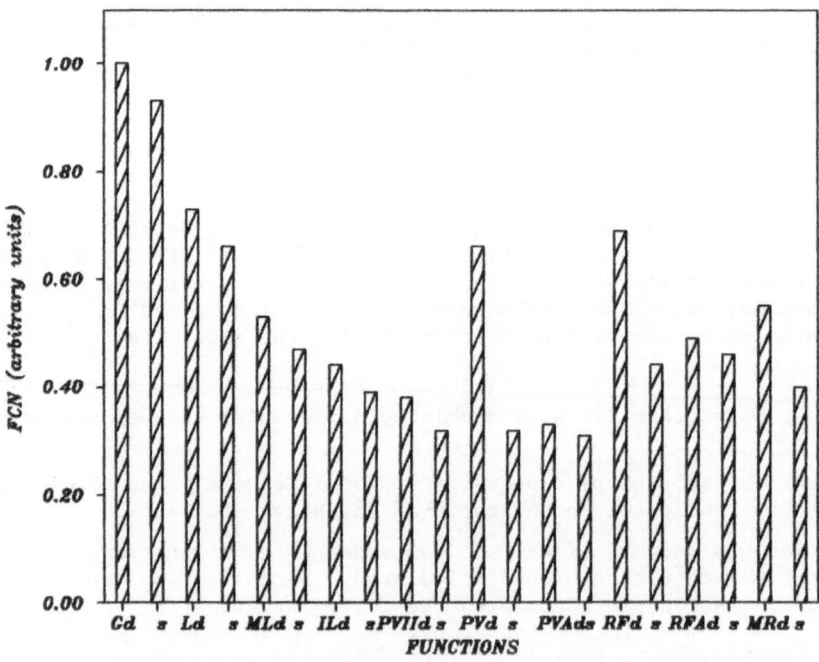

Fig. 2.   Values of FCN function for whole spectrum.

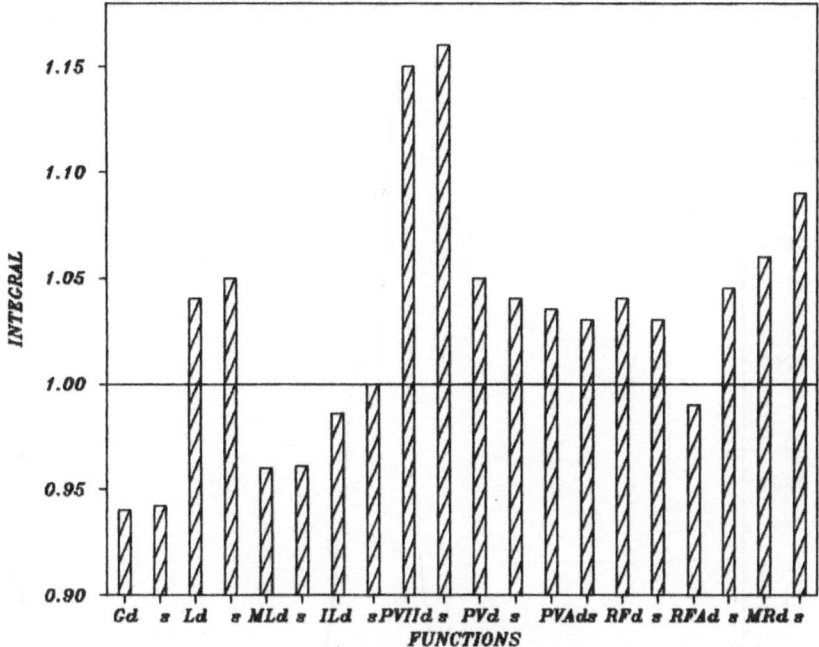

Fig. 3. Calculated integral intensities for 311α reflection.
Solid line - experimental integral intensity.

RESULTS

Three factors were taken into account in the analysis of applicability
of given function:

- the final value of FCN function,

- period of time taken by the calculations,

- difference between the calculated and experimental integral intensities.

For all symmetrical functions calculated positions of the peaks are the same
within accuracy 0.001°. Among simple 3 parameters functions (G, L, ML, IL)
Intermediate Lorentzian has the smallest value of FCN function (Fig. 1) and
almost the same period of time taken by the calculations (Fig. 2). Among
4 parameters functions PVII and PV give almost the same fit (FCN value)
and similar time, though the integrals calculated (Fig. 3) using PVII seem
to be too big (because of the poor result of fitting near the maximum of
fit). RF gives the worse fit and in much longer time. PVA gives the same
fit as PV but time is longer and the integrals are too small. Rational
functions (RF, RFA) and MR give rather poor fits and time is very long.
The assumption of the same shape of α_1 and α_2 profiles is not fulfilled for
low angle peaks, though the intensity ratio is close to 0.5.

CONCLUSION

As long as the position of a peak is important L function is sufficient
because of the shortest time of fitting. For more complicated analysis
Pseudo-Voigt function is strongly recommended. Unfortunately, effects of
asymmetry were not measured, so in this case no essential conclusion can be
drawn about the asymmetrical functions. The α_2 elimination does not seem
as exact approximation.

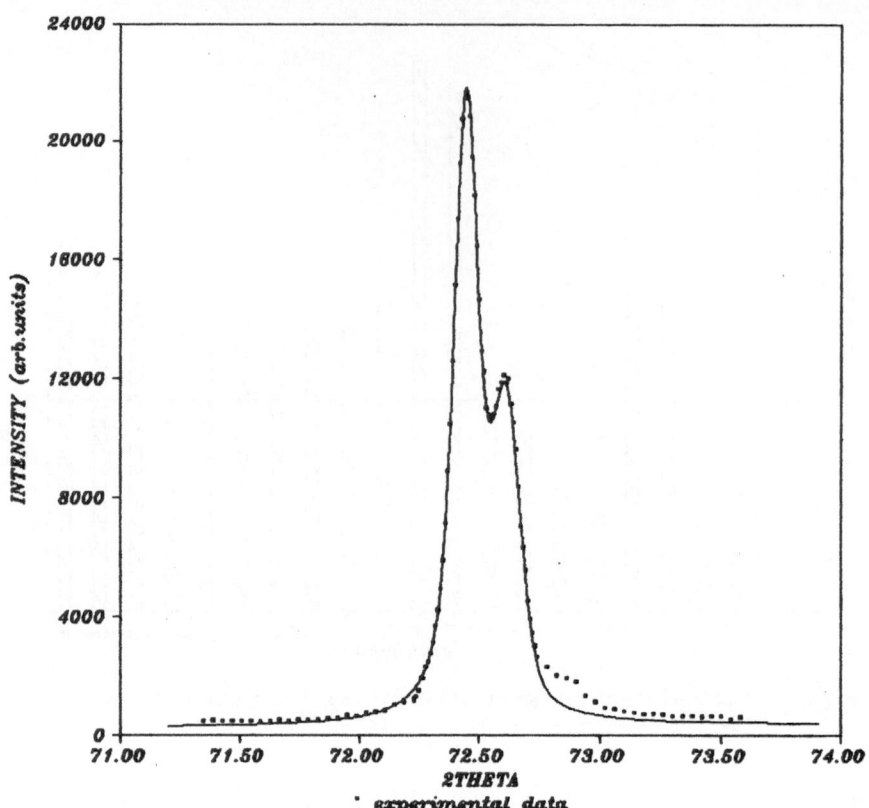

Fig. 4. Pseudo-Voigt single function fitted to the experimental data
Solid line - fitted function.

REFERENCES

1. N. P. Pyros, J.Appl.Cryst. 16:289 (1983).
2. E. J. Sonneveld and J. W. Visser, J.Appl.Cryst. 8:1 (1975).
3. M. M. Hall, V. G. Veeraraghavan, H. Rubin, and P. G. Winchell, J.Appl.
    Cryst. 10:66 (1977).
4. M. Hecq, J.Appl.Cryst. 14:60 (1981).
5. R. Delhez and E. J. Mittenmeijer, J.Appl.Cryst. 8:609 (1975).
6. F. Pompa and S. Zirilli, J.Appl.Cryst. 15:602 (1982).

RIETVELD REFINEMENT OF Y_2O_3 - COMPARISON OF TWO PROFILE FUNCTIONS

Ľ. Smrčok

Institute of Inorganic Chemistry CCHR
Slovak Academy of Sciences
842 36 Bratislava, Czechoslovakia

P. Ďuriš

Computing Center of Slovak Academy of Sciences
842 35 Bratislava, Czechoslovakia

This paper is an extension of the communication[1] where all experimental and computational details are given, but the attention is paid only to the Pearson VII type profile shape function (PVII). Because the refinement described in the paper[1] was initiated with the modified Lorentz function (ML), a brief comparison of the results obtained with both these functions is given here.

Table 1. Review of Refinement Results
Atomic positions ($\times 10^4$), isotropic temperature factors $B(\text{Å}^2)$, R-factors, goodness-of-fit-index S, Durbin-Watson $\underline{d}$ statistic with the significance points Q and the quasi-external indicator U_α. The e.s.d.'s given in parentheses refer to the last decimal place. The results of previous single crystal refinements are given for comparison. PVII - Pearson VII profile function, ML - modified Lorentz function, N - neutron data[4], X - X-ray data[5]

		PVII*	ML**	N	X
O(48e)	x	3911(5)	3915(8)	3907(3)	3890(10)
	y	1516(5)	1517(8)	1520(3)	1540(10)
	z	3827(6)	3832(9)	3804(3)	3780(10)
Y(24d)	x	-320(1)	-320(1)	-327(3)	-328(3)
Y(8b)	x=y=z=1/4				

*B(Y in 8b) = 0.63(7), B(Y in 24d) = 0.47(5), B(O) = 0.41(11), R_{wp} = 0.13, R_B = 0.02, S = 1.51, $\underline{d}$ = 1.425, Q(0.1 %; 1 %; 5 %) = 1.868, 1.904, 1.935, $U_{0.01}$ = (1.454 - 1.569).

**B(Y in 8b) = 0.60(8), B(Y in 24d) = 0.40(5), B(O) = 0.36(17), R_{wp} = 0.20, R_B = 0.05, S = 2.34, $\underline{d}$ = 0.968, Q(0.1 %, 1 %, 5 %) = 1.868, 1.904, 1.935), $U_{0.01}$ = (2.254 - 2.432).

The fractional coordinates (Table 1) are in a perfect agreement despite the significantly different values of R_{wp} and R_B. On the other hand, smaller standard deviations achieved in the PVII case unequivocally give the preferrence to this model: the e.s.d.'s of the coordinates of the O atom have been reduced by a factor of 1.5. Very similar results have been reported by Eddy et al.[2] who studied the cubic Na-ZK-4 zeolite. In that case a Voigt profile function instead of a Gaussian in addition to the new treatment for asymmetry has reduced the R_{wp} from 19.7 to 12.6 per cent. Despite this striking improvement, the atomic coordinates have remained essentially unchanged but the precision has also been improved. This, however, may not be a rule of thumb especially for more complicated structures and the patterns with the higher degree of overlap of individual Bragg peaks.

The values of the Durbin-Watson d statistic (less than the expected if no serial correlation in residuals is present) are due to the misfit either in profiles, background or in both. Because the background in both models has been evaluated in the same way, e.g. by linear interpolation between ten mean intensities, $d_{ML} \ll d_{PVII} < Q_\alpha$ is due to bad behavior of the ML mainly in the tails (Fig. 1). Note, that the plot of sign (Δy) is an useful tool, because any lack-of-fit either in the profile or background approximation is clearly indicated by a longer sequence of residuals having the same sign.

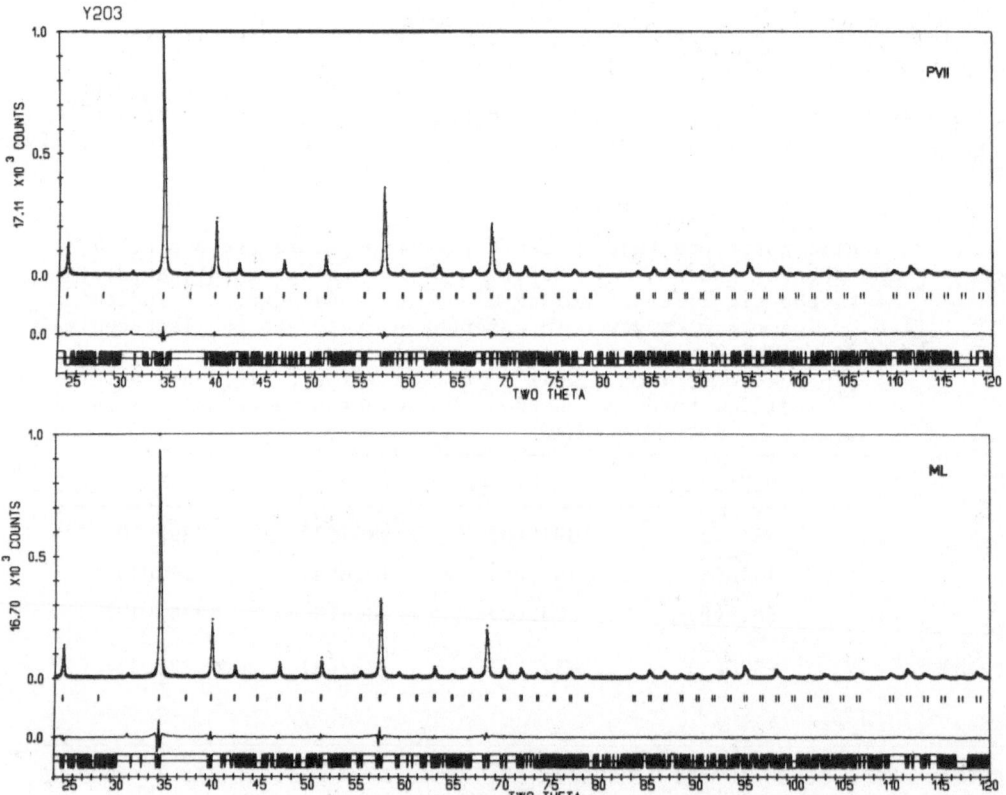

Fig. 1. Observed, calculated and difference data. Position of possible Bragg diffractions are marked by vertical bars. The lowermost oscillating line represents the function sign ($y_{Oi}-y_{ci}$). The peaks were fitted with (a) the PVII and (b) the ML profile function.

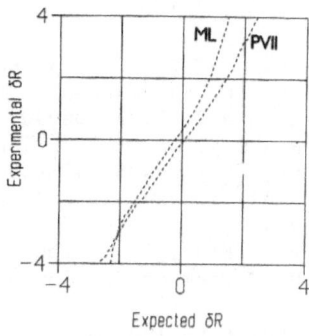

Fig. 2. Normal probability plots based on δR_i. After (1)

The slopes and intercepts of the normal probability plots (Fig. 2) are as follows: PVII - 1.46 and 0.06, ML - 1.75 and 0.5, respectively. The ML curve neither passes through the origin nor is linear, so that this model cannot be considered to be very good. The PVII on the other hand, is very close to linear and passes through the origin so that the conclusion can be drawn, that δR_i have a normal distribution and that there are no large systematic errors inabsorbable in the model. The slope greater than one indicates, that either Δy are too large, the $\sigma(y_o)$ too small or both conditions are applicable. The inspection of the Δy curve (Fig. 1) and of the Δy vers y_o plots (not given here) has not revealed any large discrepancies. The alternative hypothesis that $\sigma(y_o)$ are underestimated can be confirmed by the quasi-external indicator U_α (ref.[3]). It comes out from the interval (Table 1) assigned to the PVII model that the error in $\sigma(y_o)$ lies between 45 and 57 per cent with 98 per cent probability. The slope of 1.46 means that the $\sigma(y)$ are too small by 46 per cent which is in good agreement with the values above.

Fig. 3 gives the simple analysis of the weighting scheme ($w_i = 1/y_{oi}$) applied during the calculations. The 1884 values of $w_i \Delta_i^2$ were divided into 19 overlapping intervals with respect to y_o. While in the ML case there is a significant but undesirable trend, in the more correct PVII case the weighting scheme has ensured reasonable constancy of $\langle w \Delta^2 \rangle$ with respect to y_o.

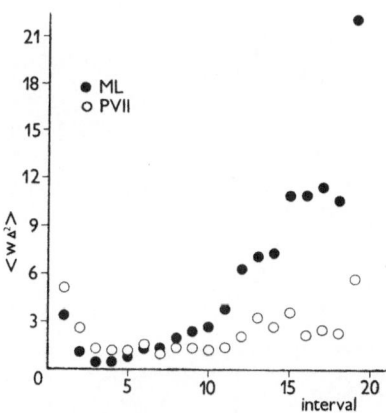

Fig. 3. Variation of the individual means $\langle w \Delta^2 \rangle$ with the intervals of y_o.

REFERENCES

1. Ľ. Smrčok, Rietveld refinement of Y_2O_3 using the Pearson VII profile shape function, J.Appl.Cryst. (submitted) (1987).
2. M. M. Eddy, A. K. Cheetham, and W. I. F. David, Powder neutron diffraction study of zeolite Na-Zk-4; an application of new functions for peak shape and asymmetry, Zeolites 6:449 (1986).
3. S. C. Abrahams, Indicators of accuracy in structure factor measurement, Acta Cryst. A25:165 (1969).
4. B. H. O'Connor and T. M. Valentine, A neutron diffraction study of the crystal structure of the C-form of yttrium sesquioxide, Acta Cryst. B25:2140 (1969).
5. M. G. Paton and E. N. Maslen, A refinement of the crystal structure of yttria, Acta Cryst. 19:307 (1965).

THREE-DIMENSIONAL MULTIPLE X-RAY DIFFRACTION

D. Korytár

Institute of Physics CEPR
Slovak Academy of Sciences
Bratislava, Czechoslovakia

In the theory of multiple crystal X-ray diffraction, coplanar arrangement has usually been used. That means incident, as well as all diffracted beams used, and unit vectors of diffracting planes, all lie in the same plane. The aim of this contribution is to discuss conditions for three-dimensional (coplanar and non-coplanar) multiple diffraction. For the sake of illustration of the theory outlined, experimental results on monolithic X-ray magnifier will be given.

Let us consider Bragg's Law in the vector form[1]

$$\vec{S} - \vec{S}_o = 2.\sin \Theta_B \cdot \vec{n} = \frac{\lambda . \vec{n}}{d_{hk\ell}} \ , \tag{1}$$

where $\vec{S}_o$, $\vec{S}$ are unit vectors of incident and diffracted beams, respectively, Θ_B - Bragg angle, $d_{hk\ell}$ interplanar spacing of diffracting planes $hk\ell$, λ wavelength of the X-rays used, and $\vec{n}$ unit vector perpendicular to the diffracting planes. For double successive diffraction, beam $\vec{S}^I$ diffracted from the first lattice planes (I) must fulfil equation (1) for the second lattice planes (II), which means that $\vec{S}$ must be equal to $\vec{S}^{II}$. Lattice planes I and II need not be from two different crystals necessarily. Multiplying (1) by $\vec{S}_o$ and $\vec{S}$, and after some simple mathematics, we can get

$$\sin \Theta_B^I = \vec{n}^I \cdot \vec{S}^I \tag{2}$$

$$\sin \Theta_B^{II} = -\vec{n}^{II} \cdot \vec{S}^I \tag{3}$$

$$(\vec{n}^I + \vec{n}^{II}) \cdot \vec{S}^I = \sin \Theta_B^I - \sin \Theta_B^{II} = \frac{\lambda}{2} \cdot (1/d^I - 1/d^{II}) \tag{4}$$

$$\vec{S}^{II} - \vec{S}_o = \frac{\lambda}{a} (\vec{H}^I + \vec{H}^{II}) \ , \tag{5}$$

where

$$\vec{n}_{hk\ell}/d_{hk\ell} = \vec{H}_{hk\ell}/a \ , \tag{6}$$

a being lattice parameter. Equation (5) reflects translation symmetry of reciprocal lattice and says that the double diffraction in question can be regarded as a single diffraction from lattice planes $\vec{L} = \vec{H}^I + \vec{H}^{II}$ and equations (2)-(4) in connection with the unit magnitude of S represent a rule of choice for the beams. Solving these equations for a given experimental set-up $[\vec{n}^I, \vec{n}^{II}, \Theta_B]$ leads to a quadratic equation with either two or one, or no real solution $\vec{S}^I$. In the case of triple successive diffraction, similar equations can be derived.

In the next, let us concentrate on the double diffraction. Unit vector $\vec{S}^I$ can be expressed in the form

$$\vec{S}^I = p\vec{n}^I + q\vec{n}^{II} + r\frac{\vec{n}^I \times \vec{n}^{II}}{|\vec{n}^I \times \vec{n}^{II}|} , \tag{7}$$

where p,q,r are real numbers. Combining (4) and (7) leads to

$$(p + q)(1 + \vec{n}^I . \vec{n}^{II}) = \sin \Theta_B^I - \sin \Theta_B^{II} = 0 , \tag{8}$$

which is equal to zero for $\Theta_B^I = \Theta_B^{II}$. In this case, we can distinguish three possibilities:

1. Let $\vec{n}^{II} = -\vec{n}^I$. Then equation (8) holds for every $\vec{S}^I$, there is no restriction upon $\vec{S}_o$ to be diffracted on $\vec{L} = \vec{H}^I + \vec{H}^{II}$, and we have [n,-n] non-dispersive arrangement.

2. Let $q = -p$, then for every angular setting there exists just one vector in the plane of $\vec{n}^I, \vec{n}^{II}$ (r=0), it is equal to $\vec{S}^I = p(\vec{n}^I - \vec{n}^{II})$ and we have [n,n] dispersion arrangement.

3. Non-coplanar arrangement. $\vec{S}^I$ does not lie in the same plane as $\vec{n}^I, \vec{n}^{II}$. For equation (4) to hold, $\vec{S}^I$ must be perpendicular to $\vec{m} = \vec{n}^I + \vec{n}^{II}$. That means, $\vec{S}^I$ must all lie in the plane perpendicular to $\vec{m}$.

All these possibilities can be easily demonstrated by means of the model depicted in Fig. 1. Well-known [n,m] or [n,-m] arrangements can be. deduced analogically.

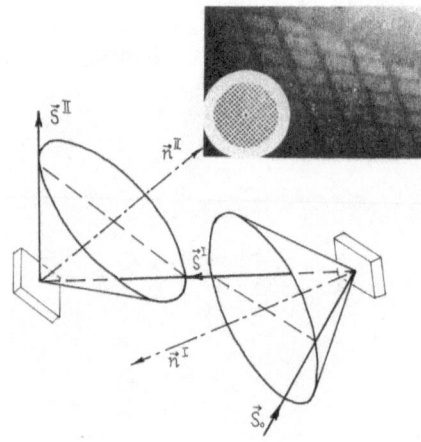

Fig. 1. Non-planar double diffraction. Inlet: microscopic mesh (ϕ 2,3 mm) and its X-ray magnified image.

Using two asymmetrically cut crystals, non-planar arrangement was used as X-ray magnifier[2]. On the basis of Eqs (1)-(5) we designed a monolithic magnifier. It consists of a silicon block shaped in such a way that $\bar{S}_o$ was incident onto $0\bar{1}0$, $\bar{S}^I$ onto 001 surface, $\vec{n}^I = [1\bar{1}0]/\sqrt{2}$, $\vec{n}^{II} = [011]/\sqrt{2}$, $\Theta_B = 53.3°$. Equations (1)-(5) will give $\vec{S}_1^I = [0.16; -0.974; -0.16]$, $\vec{S}_{o1} = [-0.974; 0.16; -0.16]$, $[\vec{S}_1^{II} = 0.16; 0.16; 0.974]$. X-ray magnification, based on asymmetric diffraction is equal to $Z = \sin(\Theta_B + \alpha)/\sin(\Theta_B - \alpha)$, where α is the angle between surface and diffracting planes. In our case, $Z = 6$ in both directions. The inlet in Fig. 1 shows a microscopic mesh and its X-ray magnified image demonstrating possibilities of Equations (1)-(5) in X-ray diffraction optics. To make progress, measures must be and are being, taken to have the same Z in both directions, to lower chromatic aberration, as well as to test another arrangements.

REFERENCES

1. L. H. Schwartz and J. B. Cohen, "Diffraction from Materials", Academic Press, New York (1977).
2. W. J. Boettinger, H. E. Burdette, M. Kuriyama, X-ray magnifier, Rev.Sci. Instrum. 50:26 (1979).

REDUCTION OF THE ELECTRONICS OF A 1D POSITION

SENSITIVE DETECTOR*

J. H. Duijn, C. W. E. van Eijk, R. W. Hollander,
and G. W. Sloof

Department of Applied Physics, Radiation Technology Group
Delft University of Technology
Delft, The Netherlands

A readout method has been developed for a 1D position sensitive detector which reduces the readout electronics dramatically. The method combines features of both cathode-strip and cathode-wedge readout and has been applied to a 36-strip detector. The detector electronics is reduced to 8 charge amplifiers without loss of spatial resolution.

INTRODUCTION

The readout system for our position sensitive detector described in[1] makes use of 36 independent signal processors, each processor containing a charge amplifier and an ADC. Before making a measurement many things had to be adjusted, e.g. the offset and the amplification factor of each processor. Interconnecting the strips using the grouped wire method[2] reduces the number of channels but makes the adjustment of the amplification factors quite complicated. In the following an alternative readout method is described which makes use of 2 cathode planes.

THE DETECTOR

Our detector[1] has been developed for 1D X-ray diffraction measurements and fits on a standard Guinier camera. It consists of a gas-filled chamber which contains two cathode planes and one razorblade anode for gas-amplification (Fig. 1). The distance between the anode and each of the cathode planes (A&B) is about 3.2 mm. X-rays enter the detector from the right, passing a Beryllium window. The space charge created near the anode will drift to both the cathode planes, each plane receiving a certain fraction of the total charge. Using gas-fillings with a pressure above 0.5 atm., this fraction is roughly independent of both the position (depth) of absorption of an X-ray quantum and the amount of space charge created near the anode (Fig. 2).

Plane A consists of 36 strips of 4 mm width which are connected in 6 groups of 6 strips. Each group is connected to one charge amplifier.

---

*Work supported by the Netherlands Technology Foundation (STW) and the Foundation for Fundamental Research and Matter (FOM).

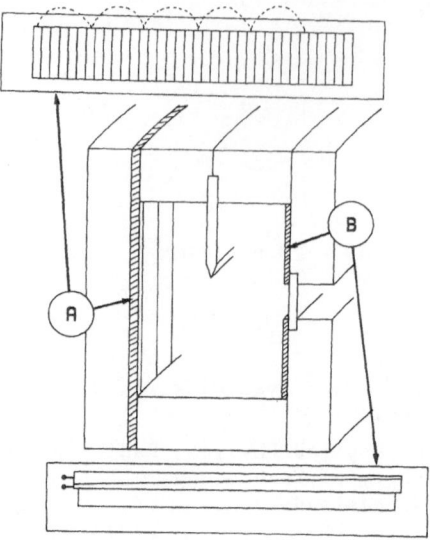

Fig. 1. Cross-section of detector with cathode planes A (strips) and B (wedges).

Plane B consists of two wedge-shaped strips which are both connected to amplifiers. The length of the strips being large compared to the width of the charge distribution induced on plane A, almost all of the charge is collected by the strips. Compared to the width of the charge distribution induced on plane B, the width of the wedges is quite small (about 4 mm) in order to have better linearity for the position reconstsruction.

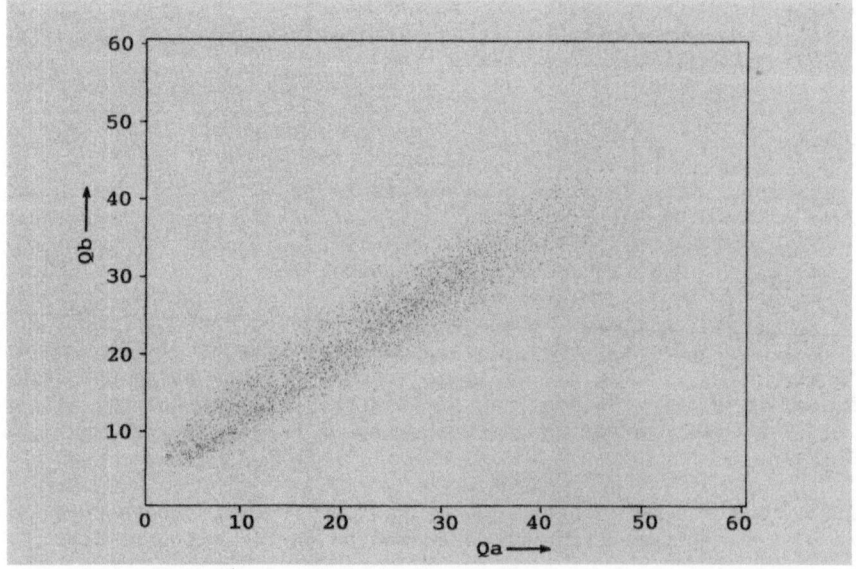

Fig. 2. Measurement of atmospheric pressure of the charge collected on plane A (Qa) as function of the charge collected on plane B (Qb).

S\W	1	2	3	4	5	6	7	. . .	30	31	32	33	34	35	36
1	1	1	1	0	7	7	7	. . .	31	31	31	31	0	0	0
2	2	2	2	2	0	8	8	. . .	32	32	32	32	32	0	0
3	3	3	3	3	3	0	9	. . .	0	33	33	33	33	33	0
4	0	4	4	4	4	4	0	. . .	28	0	34	34	34	34	34
5	0	0	5	5	5	5	5	. . .	29	29	0	35	35	35	35
6	0	0	0	6	6	6	6	. . .	30	30	30	0	36	36	36

Fig. 3. Matrix used to determine stripnumber from the coarse position W and the amplifier number S.

POSITION RECONSTRUCTION

To reconstruct the position of incidence of an X-ray quantum we have the 2 wedge signals and 6 strip signals.

At first we use the 2 wedge signals (W1 and W2) to compute a coarse position W. We determine the ratio W1/(W1+W2) and by interpolation between some known ratios at fixed points in the detector a fairly good coarse position (about 6 mm FWHM) can be obtained. Furthermore from plane A we know which amplifier (S) is giving the maximum signal. Combining coarse position (W) and amplifier number (S) using a 2D matrix (Fig. 3) we know exactly which strip out of 36 is nearest to the X-ray incidence.

To reconstruct the fine position we select the two largest strip signals. The ratio of these signals is used as input for an array which contains information about the charge distribution in the detector (Fig. 4). The output of the array gives us the fine position.

THE READOUT SYSTEM

At present we use 2 different systems to do position reconstruction: a fully hardware reconstruction and a mixture of hard- and software. The

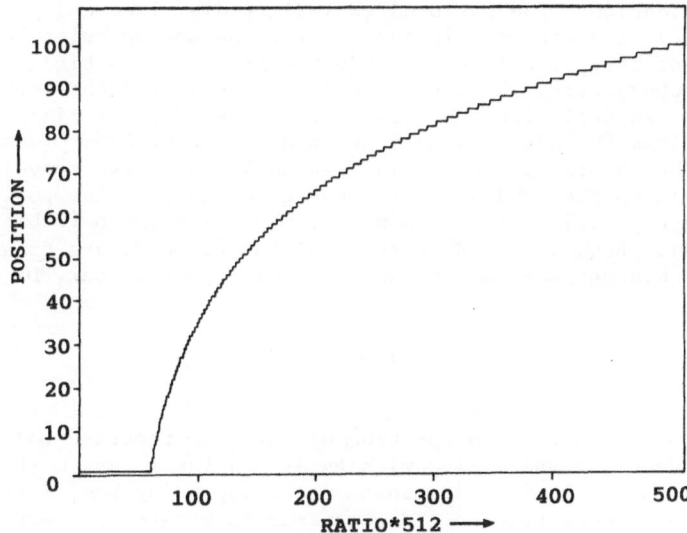

Fig. 4. Contents of array to determine the fine position from the ratio of the two largest strip signals.

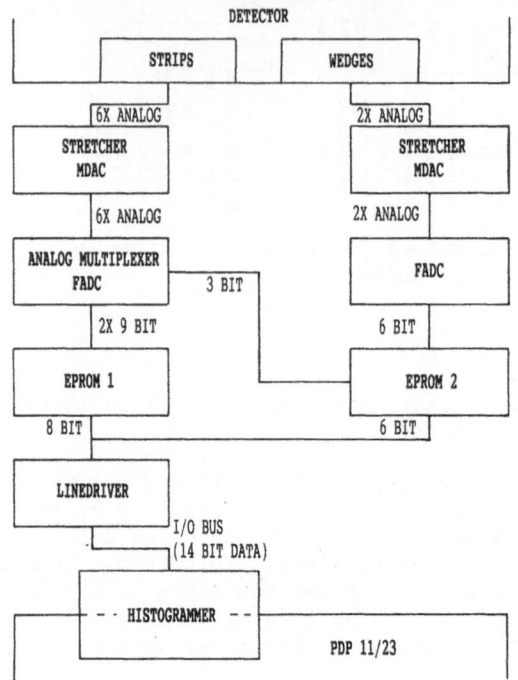

Fig. 5. Outline of the hardware position reconstruction system.

last one has already been mentioned in (ref.[1]). The hardware reconstruction, which is the fastest, will now be discussed briefly. In Fig. 5 an outline is given of our hardware reconstruction system. The detector supplies us with 6 strip signals, 2 wedge signals and an anode signal. The anode signal is used for triggering of the readout system. Using some operational amplifiers we process the wedge signals to W1 and W1+W2. All of the signals are now led through a peak stretcher and an MDAC. The last device is used as a computer controlled analog attenuator to correct the signals for differences in amplification. The ratio W1/(W1+W2) is digitized using an TRW1014 Flash ADC (6 bit) as a ratiometric converter[2]. An analog multiplexer switches pairs of stripsignals to 2 TRW1049 Flash ADC's (9 bit), also used as ratiometric converters. S (see above) is derived from the overflow bits of these ADC's. We deal with the right ratios when both overflow bits are low. The data from the ADC's is used as input for 2 EPROMS. These EPROMS (1&2) contain the information shown in Figures 4 and 5 respectively. EPROM 1 having the data from the TRW1049's as input gives us the fine position (8 bit). The coarse position is determined by EPROM 2 using S (3 bit) and data from the TRW1014. Both fine and coarse position (14 bit) are transported by an I/O bus to a histogrammer memory which forms a part of our PDP 11/23 minicomputer.

MEASUREMENTS

Fig. 6 shows a diffraction spectrum of corundum recorded with our new detector. The detector was filled with an Ar-CH4 (10 %) gas mixture at atmospheric pressure. It is a 10-minute recording using CuK_α X-rays and software position reconstruction. The maximum countrate is about 200/s and the linewidth about 300 μm FWHM, similar to previous results. With a XeCO2 gas filling 200 μm linewidth is expected. A similar spectrum has been recorded with hardware reconstruction.

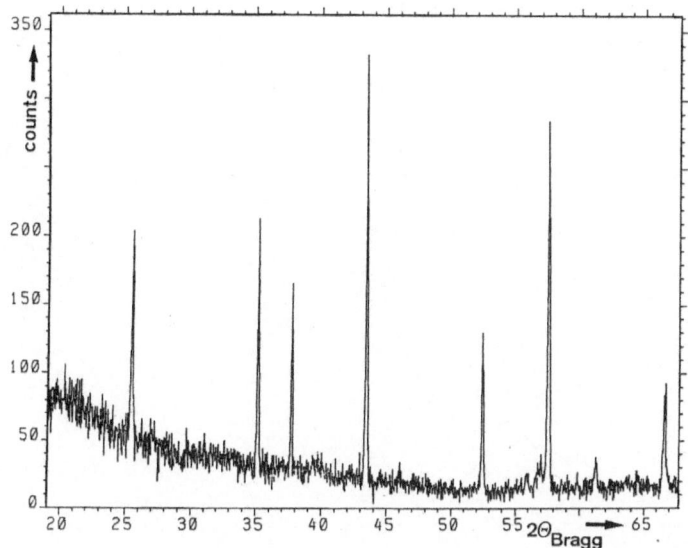

Fig. 6. Diffraction spectrum of corundum recorded with the new detector.

With this method countrates of about 20000/s could be handled. Further improvements are under way in order to achieve countrates up to 100000/s.

REFERENCES

1. J. H. Duijn, C. W. E. van Eijk, R. W. Hollander, and R. A. Marx, IEEE Trans.Nucl.Sci., Vol. NS-33, 1:388 (1986).
2. S. Kitamoto, Nucl.Instr.Meth. 198:595 (1982).
3. B. Hallgren and H. Verweij, IEEE Trans.Nucl.Sci. NS-27 (1980).

A HIGH RESOLUTION MONOCHROMATOR MODULE WITH X-RAY OPTICS FOR MULTIPURPOSE MEASUREMENT OF A^3B^5 HETEROSTRUCTURES

P. Kacerovský

Institute of Radio Engineering and Electronics
Czechoslovak Academy of Sciences
182 51 Prague 8, Czechoslovakia

A compact double-axis monochromator module with a precision single-axis goniometer is designed for multipurpose high resolution measurement of X-ray rocking curve profiles, lattice constants and topography of A^3B^5 heterostructures. The same output and input direction of the X-ray beam is used for both with and without the X-ray monochromator module. This arrangement of the X-ray optics allowed measurements either by the original Bond's method or by Bartels type four-crystal monochromator in (-++-) setting.

At the present time, characterisation of A^3B^5 heteroepitaxial layers is mainly done by reflection double-crystal method. For precision absolute lattice parameter measurements of heteroepitaxial layers, several methods have been developed, including multiple diffraction[1] and Bond's method[2] using a four-crystal monochromator as suggested by Bartels[3]. It is the aim of this work to describe the X-ray optics of a monochromator module producing a highly parallel incident beam for multipurpose measurement.

The rebuilt precision single-axis optical goniometer G 5 (made in USSR) is the basic item referring to Fig. 1 and Fig. 2. A compact apparatus with an optional double-axis monochromator module is shown in Fig. 2 and has the necessary stability, so important for triple-axis diffractometers. For the various experiment depending on the type of module used, it is possible to have following arrangements: single-axis (original Bond's method[2] without module), double-axis (diffractometry and topography) in (-+-) setting, and triple-axis (diffractometry with ability to do absolute lattice parameter measurements as suggested by Bartels[3]). The double-axis monochromator module is optional. It is fixed on a bench with a Y translation one-axis precision manual drive (the X axis is the direction of incident X-ray beam). Each axis of the module is set by Y axis positioner, a 0 rotation stage and determined by a goniometer head as a holder for the monochromator. The axis positions of the monochromator are labelled I and II, respectively in Fig. 1. The monochromators that are of the Kohra's double-reflection groove type[4] do not change the direction of the primary beam. The parallelepiped of Ge (Fig. 3), GGG or substrate type single crystals are cut as monochromators for one position (shown as I in Fig. 1) in the module. For both positions (shown as I and II in Fig. 1) in the module two Ge parallelopipeds with two symmetric (220) reflections are cut. In such a way a four-crystal monochromator arrangement as proposed by Bartels[3] is realized. Such type of Ge monochromators were provided by Martovickii[5]. The optical prism and mirrors, the monochromator sides, the slits and the surface of the sample are all used

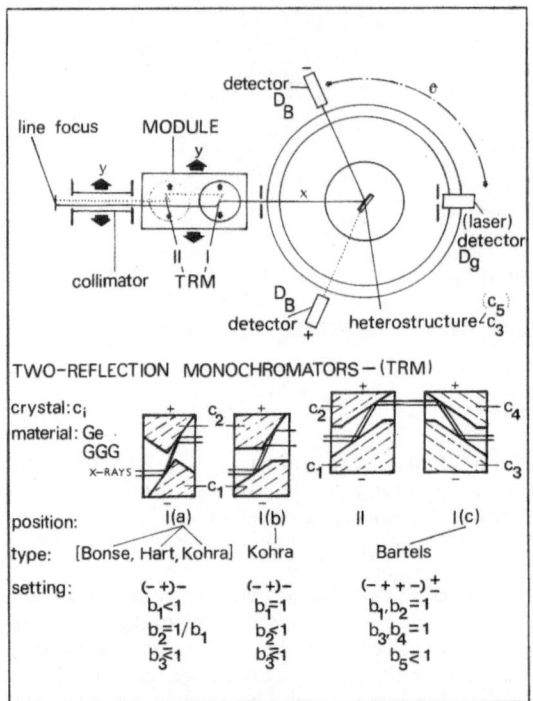

Fig. 1. Schematic diagram of the modified single-axis goniometer with the module, X-ray optics and the types of monochromators. For labelling see the text.

for a quick preliminary adjustment by laser. For this purpose the goniometer detector (labelled D_G in Fig. 1) is replaced by a small He-Ne laser. For minimum decrease of intensity of the incident X-ray beam during diffractometry of heteroepitaxial layers on GaAs and InP substrates the single-set

Fig. 2. Side view of the diffractometer.

Fig. 3. The Ge monochromator of type Ib (see Fig. 1).

module is preferred. In position I is set Ge or GGG monochromator of type Ia (Fig. 1) with two appropriate asymmetric reflections in the (-+) parallel setting. To obtain (-+-) setting (400) rocking curves there are two (400) asymmetric (-+) reflections (Fig. 1). For topography and diffractometry Ge or GGG monochromators of type Ib (Fig. 1) with two (400) symmetric-asymmetric (-+) reflections take place in the module. But this type of monochromators is not easy to make due to correction needed for refraction effect. The double-axis module with four-crystal monochromator can be set for high resolution Bond's type diffractometry of any type of heterostructure[3]. Until now, lack of a powerful X-ray generator in our laboratory prevented the use of the double-axis module. But in some cases it is possible to measure precisely, by the original Bond's method, both sides of a heterostructure to identify either the substrate peak or the thick epitaxial layer peak position in the rocking curve spectrum. In such a way, it is possible to interpret the convolution of peaks.

The potential of a single-axis goniometer with a precision absolute scale equipped with collimator and double-axis compact monochromator module as an apparatus for diffractometry and topography of A^3B^5 heterostructures is described. Even in cases of weak X-ray sources it is possible to specify the position of substrate peak on rocking curves by using Bond's method of measurement of both sides of heterostructure. The fact that the monochromators do not change the direction of the incident beam, together with the use of laser control, simplifies the procedure needed in measurements.

REFERENCES

1. B. J. Isherwood, B. R. Brown, and M. A. G. Halliwell, X-ray multiple diffraction as a tool for studying heteroepitaxial layers, J.Crystal. Growth 54-449 (1981) and Part II in J.Crystal Growth 60-33 (1982).
2. W. L. Bond, Precision lattice constant determination, Acta Cryst. 13:814 (1960).
3. W. J. Bartels, Characterisation of thin layers on perfect crystals with a multipurpose high resolution X-ray diffractometer, J.Vac.Sci.Technol. B1(2):338 (1983).
4. K. Kohra, Multiple crystal arrangements, K-1 in: "International Summer School on X-ray Dynamical Theory and Topography", Proceedings, August 18-26, Limoges (1975).
5. V. P. Martovickii and V. V. Rodin, Rastchet i izgotovleniye monokhromatorov rentgenovskikh lutchey, in: "Preprint 100", FIAN (Physical Institute, Acad.Sci. USSR), Moscow (1987).

LIST OF AUTHORS

Transmission electron microscopy
       (TEM), 333, 353
Triple point, 55
Triple-crystal diffractometer
       method (TCD), 286-287
Triplet relation, 293, 303
Twinning, 132, 345

Unoverlapped lines, 6

V, see Vanadium
Vacancy, 149-151, 358
Vanadium, 50
   standard, 149
Vapour deposition, 199-200, 205,
       253, 335
Variances of phases, 295, 297,
       300-301
Vegard's rule, 179
Vertex, 105, 296
Vickers indenter, 183

W, see Wolfram
Warren-Averbach analysis, 136
WC-Co hardmetal, 237-239
Weight fraction, 29-40
Weighting scheme, 377
Weissenberg method
Weld metal, 50
Welded steel, 50
Wide-angle X-ray scattering
       (especially stated),
       109-110, 117-121

Williamson-Hall plot, 182, 185
Wolfram, 50, 205
Wurtzitte structure, 363
Wustite (FeO), 68-70

X-ray fluorescence method, 18
X-ray diffraction
   analytical technique, 3-48, 59-88
   differential, 4
   double-crystal diffraction (XDC),
       263-264
   quantitative, 3-88
X-ray diffraction optics, 381, 389
X-ray elastic constant, 203
X-ray magnification, 381
X-ray standing wave technique, 261,
       269-271, 273-282
X-ray topography, 43, 45, 147, 389-391

Yield strength, 51
Young modulus, 179, 192, 205

Zeolite, 376
Zeolite-supported catalysts, 210
Zhdanov notation, 343
Zhdanov sequence, 349
Zinc, 85, 86, 137
Zn-Mn-S, 363
ZnO, 36, 40
ZnS, 332, 363-366
ZnS:Al, 366
Zone annealing technique, 149

ABBREVIATIONS

AA/ICP	Atomic Absorption Inductively Coupled Plasma Spectroscopy
AES	Auger Electron Spectroscopy
ADC	Analog Digital Convertor
ASTM	American Society for Testing Materials, Philadelphia
b.c.c.	Body Centered Cubic
c	Cubic Packing
CA	Chemical Analysis
CDR	Curve Diffraction Reflection Method
CEMS	Conversion Electron Mössbauer Spectroscopy
c.n.	Coordination Number
cp	Close Packed Structure
CSM	Chi-Square Minimization
CSN	Czechoslovak National Standard
CSRO	Chemical Short Range Order
CVD	Chemically Vapor Deposited Coating
1-D	One-dimensional
DGS	Directed Graph of Seminvariants
DLO	Double-Layer Octahedron
DLT	Double-Layer Tetrahedron
DRON	Difraktometer Rentgenovskii Obshchevo Naznaczeniya (X-ray diffractometer for general purposes), Technabexport, Moscow
DS	Fully Disordered Structure
DSC	(Heat Compensation) Differential Scanning Calorimetry
DTA	Differential Thermal Analysis
ED	Electron Diffraction
EDAX	Energy Dispersive Analysis of X-rays
EDXRF	Energy Dispersive X-ray Fluorescence
EGA	Evolved Gas Analysis
EGD	Evolved Gas Detection
EM	Electron Microscopy
EPMA	Electron Probe Microanalysis
EPR	Electron Paramagnetic Resonance
EPROM	Erasable Programmable Read-only Memory
f.c.c.	Face Centered Cubic
FCN	Function
FMR	Ferromagnetic Resonance
GdIG	Gadolinium Iron Garnet
GGG	Gadolinium Gallium Garnet
GPR	Graph of Phase Relations
G.P.	Guinier-Preston Zone
GS	Graph of Seminvariants
h	Hexagonal
h.c.p.	Hexagonal Close-Packing
HEED	High-Energy Electron Diffraction
HZG	Horizontal Zahlrohr Goniometer (VEB Freiberg, DDR)
IC	Incommensurate

ICDD	International Centre for Diffraction Data
IIXA	Ion Impact X-ray Analysis
IR	Infra-Red Spectroscopy
JCPDS	Joined Committee on Powder Diffraction Standards
LEED	Low Energy Electron Diffraction
LH	abbreviation for lH, 2H, ... polytypes
MDLSU	Model of Double-Layer Structural Units
MM	Multimethacrylate
MULTAN	Program for crystal structure determination using MULtisolution methods and TANgent formula
NMR	Nuclear Magnetic Resonance
OD	Order-Disorder Structure
OTP	Oblique Texture Pattern
PDF	Power Diffraction File
PIXE	Proton Induced X-ray Emission
PLM	Polarised Light Microscopy
PP	Polypropylene
PSD	Position Sensitive Detector
RDF	Radial Distribution Function
RGPR	Reduced Graph of Phase Relations
RIM	Reference Intensity Method
RIR	Reference Intensity Ratio
SANS	Small-Angle Neutron Scattering
SAXS	Small-Angle X-ray Scattering
SEM	Scanning Electron Microscopy
SIMS	Secondary Ion Mass Spectroscopy
SRO	Short-Range Ordering
SRY	Secondary Radiation Yield
STEM	Scanning Transmission Electron Diffraction
TCD	Triple-Crystal X-ray Diffractometer
TCNQ	Tetracyanoquinodimethanide
TDS	Thermal Diffuse Scattering
TG	Thermogravimetry
TMM	Thermomechanical Measurement
TMT	Thermomechanical Treatment
TP	Texture Pattern
TPA	Total Pattern Analysis
TSRO	Topological Short Range Order
TTF	Tetrathiofulvalene
TURM	Generator of Transformatoren- und Röntgenwerk Dresden, type M
VEB	Volkseigener Betrieb (national enterprise)
WAXS	Wide-Angle X-ray Scattering
XDC	X-ray Double Crystal Diffraction
XRD	X-ray Diffraction
XRF	X-ray Fluorescence
XRPD	X-ray Powder Diffraction
XSW (XRSW)	X-ray Standing Wave
YAG	Yttrium Aluminium Garnet
YIG	Yttrium Iron Garnet